Robot Soccer

Edited by

Vladan Papić

Robot Soccer

Edited by Vladan Papić

Republished by InTech

Janeza Trdine 9, 51000 Rijeka, Croatia

Technical Editor Goran Bajac

Additional hard copies can be obtained from orders@intechopen.com

Robot Soccer,
Edited by Vladan Papić
p. cm.
ISBN 978-953-307-036-0

For EU product safety concerns:
IN TECH d.o.o., Prolaz Marije Krucifikse Kozulić 3, 51000 Rijeka, Croatia,
info@intechopen.com or visit our website at intechopen.com.

Meet the editor

Born on 6th August, 1968. in Split, Croatia. B. S. in Electrotechnics, 1993, University of Split Title of diploma work: Use of PC in phase plane analysis of nelinear dynamic systems M. S. in Electrotechnics, 1996, University of Split Title of master thesis: Recognition of characteristic phases of human gait using Neural Networks Ph.D. in Electrotechnics, 2002, University of Split Title of doctoral thesis: Expert system for evaluation of human gait kinematics based on shape recognition Since 2002. assistant professor on The Faculty of natural science, mathematics and education. Classes: Electronics basics, Laboratory in electronics, Computers in technical systems. Head of Polytechnics department (PMF) 2005. - 2008. Vice-Dean for science (PMF) 2008. Professor on FESB since 2009. Full professor since 2010. Classes: Systems theory, Computer graphics, Databases. 1993. - 1997. working as a computer software developer in INFO90 and SEM-kompjuteri. Since 1998. - 2002. working as young researcher on the projectBiomechanics of human gait, control and rehabilitation

Preface

Idea of using soccer game for promoting science and technology of artificial intelligence and robotics has been presented in early 90' of the last century. Researchers in many different scientific fields all over the world recognized this idea as an inspiring challenge. Robot soccer research is interdisciplinary, complex, demanding but most of all - fun and motivational. Obtained knowledge and results of research can easily be transferred and applied to numerous applications and projects dealing with relating fields such as robotics, electronics, mechanical engineering, artificial intelligence, etc. As a consequence, we are witnesses of rapid advancement in this field with numerous robot soccer competitions and vast number of teams and team members. The best illustration is numbers from the RoboCup 2009 world championship held in Graz, Austria which gathered around 2300 participants in over 400 teams from 44 nations. Attendance numbers at various robot soccer events shows that interest in robot soccer goes beyond the academic and R&D community.

Several experts have been invited to present state of the art in this growing area. It was impossible to cover all the aspects of the research in detail but through the chapters of this book, various topics were elaborated. Among them are hardware architecture and controllers, software design, sensor and information fusion, reasoning and control, development of more robust and intelligent robot soccer strategies, AI-based paradigms, robot communication and simulations as well as some other issues such as educational aspect. Some strict partition of chapter in this book hasn't been done because areas of research are overlapping and interweaving. However, it can be said that beginning chapters are more system - oriented with wider scope of presented research while later chapters are generally dealing with some more particular aspects of robot soccer.

I would like to thank all authors for their contribution and to all those people who helped in finalisation of this project. Finally, I hope that readers will find this book interesting and informative.

Vladan Papić
University of Split

Contents

1

The Real-time and Embedded Soccer Robot Control System

Ce Li[1], Takahiro Watanabe[1], Zhenyu Wu[2], Hang Li[2] and Yijie Huangfu[2]
Waseda University[1], Japan[1]
Dalian University of Technology[2], China[2]

1. Introduction

Robot Soccer becomes more popular robot competition over the last decade. It is the passion of the robot fans. There are some international soccer robot organizations who divide the competitions into several leagues, each of these leagues focus on the different technologies.
In this chapter, the rules and history of RoboCup Small Size League games will be introduced shortly. That can make the audience understand the current design style smoothly. Comparing the small robot with our human being, we can easily find that the mechanism looks like one's body, the circuit looks like one's nerve, the control logic looks like one's cerebellum, the vision system looks like one's eyes and the off-field computer which is used for decisions looks like one's cerebrum. After all, the RoboCup motto is: "By the year 2050, develop a team of fully autonomous humanoid robots that can play and win against the human world champion soccer team" (Official RoboCup Org., 2007).
Nowadays, with the development of LSI, the applications of FPGA make the circuit design more simple and convenient, especially for the soccer robot which always needs to be programmed in the field. A soft-core CPU which can be embedded in FPGA can fill a gap in the FPGA control logic and can also make the design more flexible. In this chapter, the circuit design configuration of our soccer robot which is developed based on FPGA is introduced, including real-time control system, the function of each module, the program flow, the performance and so on.
After we got a stable control system based on single CPU in the FPGA, we start to make an attempt to embed multiple CPUs in the control system. It gets an obvious advantage of high performance that two CPUs can work at the same time. Although one CPU can meet the request of global vision, multiple CPUs could pave the way for self vision systems or more complicated control logic.

2. Background

2.1 RoboCup (Official RoboCup Org., 2007)

RoboCup is an annual international competition aimed at promoting the research and development of artificial intelligence and robotic systems. The competition focuses on the development of robotics in the areas of:

- Multi-Agent robots planning and coordination
- Pattern Recognition and real time control
- Sensing Technology
- Vision Systems (both global and local cameras)
- Mechanical design and construction

The RoboCup World Championship consists of different levels:

- Soccer Simulation League
- Small Size Robot League
- Middle Size Robot League
- Standard Platform League
- Humanoid League

RoboCup is a competition domain designed to advance robotics and AI research through a friendly competition. Small Size robot soccer focuses on the problems of intelligent multi-agent cooperation and controlling in a highly dynamic environment with a hybrid centralized or distributed system.

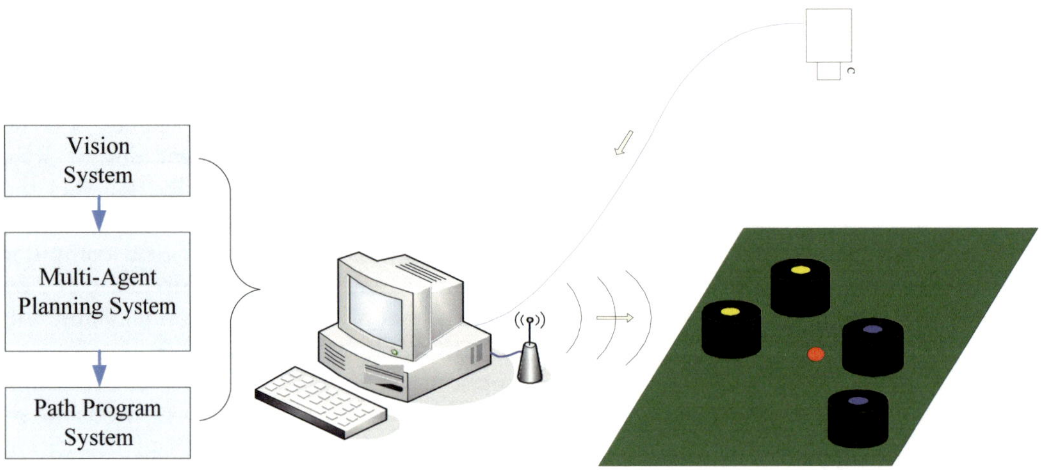

Fig. 1. Overview of the entire robot system

A Small Size robot soccer game takes place between two teams of five robots each. The environment of the game shows in Figure 1. Each robot must conform to the dimensions as specified in the rules: The robot must fit within a 180mm diameter circle and must be no higher than 15cm unless they use on-board vision. The robots play soccer (an orange golf ball) on a green carpeted field that is 6050mm long by 4050mm wide (Official RoboCup Org., 2007). For the detail rules,. For the detail rules, please refer RoboCup web site. Robots come in two ways, those with local on-board vision sensors and those with global vision. Global vision robots, by far the most common variety, use an overhead camera and off-field PC to identify and drive them to move around the field by wireless communication. The overhead camera is attached to a camera bar located 4m above the playing surface. Local vision robots have the sensing on themselves. The vision information is either processed onboard the robot or transmitted back to the off-field PC for processing. An off-field PC is used to communication referee commands and position information to the robots in the case of overhead vision. Typically the off-field PC also performs most, if not all, of the processing required for coordination and control of the robots. Communications is wireless and Wireless communication typically uses dedicated commercial transmitter/receiver units. Building a successful team requires clever design, implementation and integration of many hardware and software sub-components that makes small size robot soccer a very interesting and challenging domain for research and education.

2.2 System Overview

The robot system is a layered set containing subsystems which perform different tasks. Figure 1 shows the flow diagram that how the system is laid out. An overview of the system is given bellow by following the flow of the information from the camera to the robots actuators (motors).

The overhead digital camera captures global images of the field by 60 fps. The vision system (software installed in off-field PC) processes these images to identify and locate the robots and the ball. The environment and state information of the field are sent to the Multi-Agent Planning System (MAPS). MAPS is the highest level planner of the robot system, arranges the task of the whole team, actually, arranges each individual robot's action and its action location. Some actions include KICK and DEFEND (Ball D., 2001).

After each robot has an action, the Path Program system calculates the path for the robot to achieve its action, and optimizes path for each robot.

In the robots, there is a motion system which can accelerate and decelerate the robot to the desire speed and distance by creating force limited trajectories. The motion system ensures wheel slip to a minimum.

Fig. 2. The 3D assembly drawing of the small size robot

2.3 Mechanical Design

The mechanical design is consisted of an omni-directional drive system, a powerful crossbow kicker, a scoop shot kicker and a dribbler. It should be a compact and robust design. All the robots are of the same mechanical design, a mass of 4.2 kilograms each. The robots are constructed using aluminum that gives them a strong but light frame. The robots have a low centre of mass, achieved by placing the solenoid, the majority of the batteries and the motors on the chassis. It can reduce weight transfer between wheels as the robot accelerates based on the low centre of mass. Consistent weight transfer leads to less slip of the wheels across the playing surface. The whole assembly drawing is shown in Figure 2.

Omni-Directional Drive

For maximum agility, the robots have omni-directional drive. The robot omni-directional drive is implemented by using four motors each with one omni-directional wheel. The angle of the front wheels is 120 degree, because we should add ball control mechanism and kicker between the front wheels. It is 90 degrees between the back wheels. During the year 2007, we use the DC motors. But now, we start to use brushless motors to save more space and improve the performance. Figure 3 shows the 3D assembly drawing of the small size robot chassis with brushless motors.

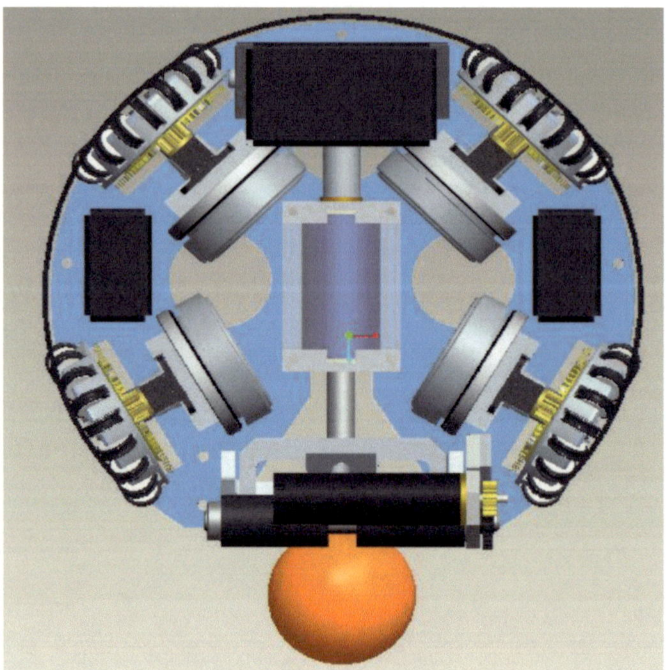

Fig. 3. The 3D assembly drawing of the small size robot chassis with brushless motors

Kicker

The robots feature a powerful kicking mechanism that is able to project the golf ball at 6.2m/s. The kicking mechanism for the Robots is a crossbow and uses a solenoid to generate the energy. The kicking mechanism, while mechanically simple, uses only one solenoid to retract the crossbow. The plate that strikes the golf ball is the same mass as the golf ball. This gives maximum efficiency of energy transfer, this experience is mentioned in reference (Ball D., 2001). There is also another solenoid and related mechanism for scoop shot.

Ball Control Mechanism

The ball control device of Robots is a rotating rubber cylinder that applies backspin to the golf ball when touching. Here we use a 6 Volt 2224 MiniMotor to drive the shaft. A 10:1 gearbox is used between the motor and the shaft. One feature of the Robots ball control mechanism is that the cylinder is separated into 2 parts. When the ball is located in this dribbler, it has the benefit as the crossbow kicker could give a more accurate and powerful kick to a ball located in the centre of the ball control mechanism.

2.4 Electrical Control System

Comparing a robot to a human, the electrical control system of the robot would be equivalent to a nervous system of human being. In humans, actions are commanded through the nervous system, and sensed information is returned through the same system. There is no difference in the robots. Sending commands and receiving sensed information are the responsibilities of a critical human organ, the brain. The micro control system in the soccer robot is equivalent to a brain.

Fig. 4. The Soccer Robot with Electronic Control system.

Acting as a metaphorical brain, the micro control system must process received information and generate the appropriate response. The off-field artificial intelligence (AI) computer does most of the required brainwork to make the robots play a recognizable game of soccer, but the on board brain translates the AI's decisions into robotic actions and does the required thought-processing which is needed to maintain these actions. Encoded commands are received from the AI computer via a wireless module.

By decoding these commands, the microcontroller system determines whether to kick, dribble or move. Onboard sensor feedback indicates if the robot should carry out a kick command. Adequate microcontrollers are necessary for quick and reliable processing of these inputs and outputs.

Microcontrollers are microprocessors with a variety of features and functionality built into one chip, allowing for their use as a single solution for control applications. The operation of a microcontroller revolves around the core Central Processing Unit (CPU), which runs programs from internal memory to carry out a task. Such a task may be as simple as performing mathematical calculations, as is done by an ordinary CPU of a personal computer. On the other hand, the task may be more complex, involving one or many of the microcontroller's hardware features including: communications ports, input/output (I/O) ports, analog-to-digital converters (A/D), timers/counters, and specialized pulse width modulation (PWM) outputs. With access to hardware ports, the CPU can interface with external devices to control the robot, gather input from sensors, and communicate with off-field PC by wireless. The control of these ports, handled by the program running on the CPU, allows for a great deal of flexibility. Inputs and outputs can be timed to occur in specific sequences, or even based on the occurrence of another input or output. A major drawback of microcontrollers is that, there are usually constraints to design and implement a system in a reasonable size for integration in compact systems, because so much processing power can be provided due to the need to fit the CPU and all of the hardware features onto the same chip.

3. Previous Control System

3.1 Control Part

The first real-time and embedded control system of our robots was designed in competitions of year 2006 by us (Zhenyu W. et al., 2007). It was remarkably reliable, and had run without big failure in 2007 China RoboCup Competitions as well as periods of testing and numerous demonstrations.

The robots' main CPU board uses TI TMS320F2812, a 32 bit DSP processor, as a CPU. With the high performance static CMOS technology, it works at a frequency of 150 MHz. This processor executes the low level motor control loop. The major benefit of using this processor is its Event Managers. The two Event Managers have 16 channels Pulse Width Modulation (PWM) generation pins that can be configured independently for a variety of tasks. Except for the kicker mechanism, the motors have encoders for fast and immediate local feedback.

A DSP and FPGA based digital control system has already been developed for the control system where FPGA gathers the data and DSP computes with it in 2006. Figure 5 describes the frame of the previous control system. This hardware architecture takes the advantage of the higher computation load of DSP and the rapid process.

In this Figure, there are two broken line frames, one is DSP board, and the other one is FPGA board. On the DSP board, DSP processor communicates with wireless module by serial interface. If the wireless module gets data from the off-field PC, it generates an interrupt to the DSP processor. When DSP is interrupted, it resets the PID parameter, then, starts a new PID control loop. After a new speed is calculated, DSP processor drivers the motors by PWM signal.

On the FPGA board, there are FPGA, RAM, flash memory, kicker control circuit, LED, etc. FPGA is used to connect with these peripherals and has the features below:

- save and fetch data in RAM and flash
- decode the signals from the 512 lines motor speed encoders
- display the system states by LED
- control the kicker
- gather the information of the acceleration

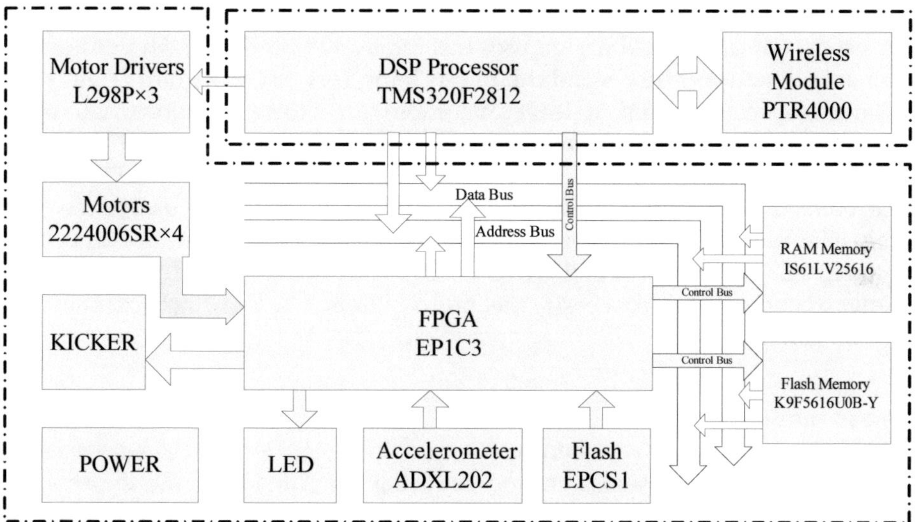

Fig. 5. The DSP and FPGA based digital control system (Zhenyu W. et al., 2007)

The innovative idea of the previous control system is that DSP controls each peripheral by writing and reading data of the registers at specific addresses in FPGA. The rest of tasks are done by FPGA except wireless communication and motor PWM control.

3.1.1 DSP

The TMS320F2812 devices, member of the TMS320C28x DSP generation, is highly integrated, high-performance solutions for demanding control applications. The C28x DSP generation is the newest member of the TMS320C2000 DSP platform. Additionally, the C28x is a very efficient C/C++ engine, hence enabling users to develop not only their system control software in a high-level language, but also enables math algorithms to be developed using C/C++. The C28x is as efficient in DSP math tasks as it is in system control tasks that typically are handled by microcontroller devices. The 32 x 32 bit MAC capabilities of the C28x and its 64-bit processing capabilities, enable the C28x to efficiently handle higher numerical resolution problems that would otherwise demand a more expensive floating-point processor solution. Add to this the fast interrupt response with automatic context save of critical registers, resulting in a device that is capable of servicing many asynchronous events with minimal latency. Special store conditional operations further improve performance (TI Corp., 2004).

3.1.2 FPGA

The Field Programmable Gate Array (FPGA) plays an important role in the sub-system on the robots. In the motion control system, the FPGA provides the interface between the motor encoder and the motion control DSP. It takes the two motor encoder signals to decode the speed of each motor. It can also be used to control the peripherals by setting the registers at the special address. During the design period of the previous control system, we compared a number of FPGAs in the MAX 7000 family and Cyclone family to choose the one that satisfied our requirements.

One major factor we considered is in-system programmability since the FPGAs are able to be programmed on board. The capacity of MAX 7000S family chip is so little, and when we select a suitable type, the price is very high. While most of the FPGAs in the EP1C3 family meet our needs, our final board design uses this family FPGA. EP1C3 devices are in-system programmable via an industry standard 10-pin Joint Test Action Group (JTAG) interface. With a large capacity of LEs, it leaves us room for future additions. For our actual implementation and functional testing, we chose the EP1C3T114C6 type FPGA. The number of available LEs determines how complicated our circuit can be. The EP1C3T114C6 has 2,910 LEs. As a point of reference, our final design utilizes 27% of all the usable resource. This covers miscellaneous logic to support enable functionality. The maximum number user I/O pins for the EP1C3T114C6 is 144 and we used 129 in our final design. These numbers are all on the order of nanoseconds and easily met our requirements. Our final circuit could run at speed of 45.7MHz.

3.2 Wireless module

Fast and compatible wireless communication module is required for the heavy task such as increasing the speed of transmission and reducing the latency of the entire system. In addition, the system must exhibit a high level of interference rejection and low error rate.

In order that a robot works successfully, it must be able to receive up-to-date game data rapidly and consistently. To meet these requirements, the following properties are necessary for a communications system:

- High transmission speed
- Low latency
- Interference rejection
- Low error rate

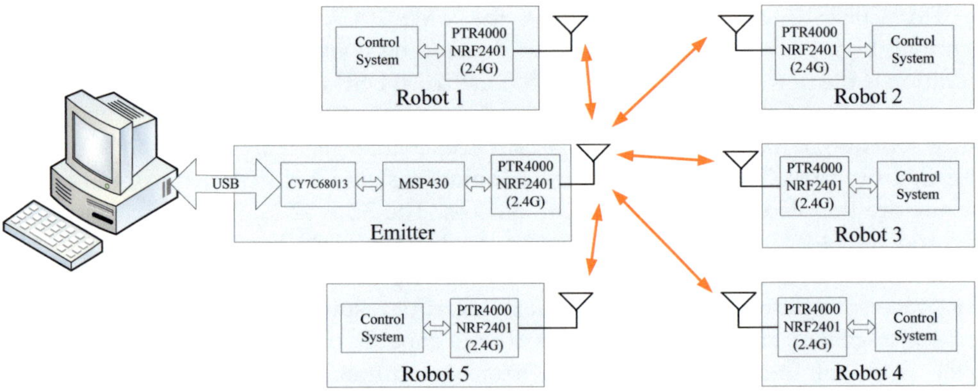

Fig. 6. Full Duplex System

Figure 6 shows the communicated method between robots and the off-field PC. The off-field PC sends motion instructions to the robot by a USB emitter, and also can get the states and information of them by the wireless communication.
For fast and steady communication, we selected PTR4000 wireless module. The central chip of PTR4000 is nRF2401 which is a single-chip radio transceiver for the world wide 2.4-2.5 GHz ISM band. The transceiver consists of a fully integrated frequency synthesizer, a power amplifier, a crystal oscillator and a modulator. Output power and frequency channels are easily programmable by use of the 3-wire serial bus. Current consumption is very low, only 10.5mA at an output power of -5dBm and 18mA in receive mode.

3.3 Motor control

The reactive control loop is initiated every millisecond by the Event managers (EVA and EVB). Once robot receives its desired velocities, it can calculate the wheel velocities and then sends and then sends them the PWM signals to the motors. Figure 7 illustrates the control model that is implemented in the reactive control loop with motors. In determining the output of the system the control loop calculates the proportional and integral errors of the wheel velocities. The velocity proportional errors for each of the wheels are calculated.

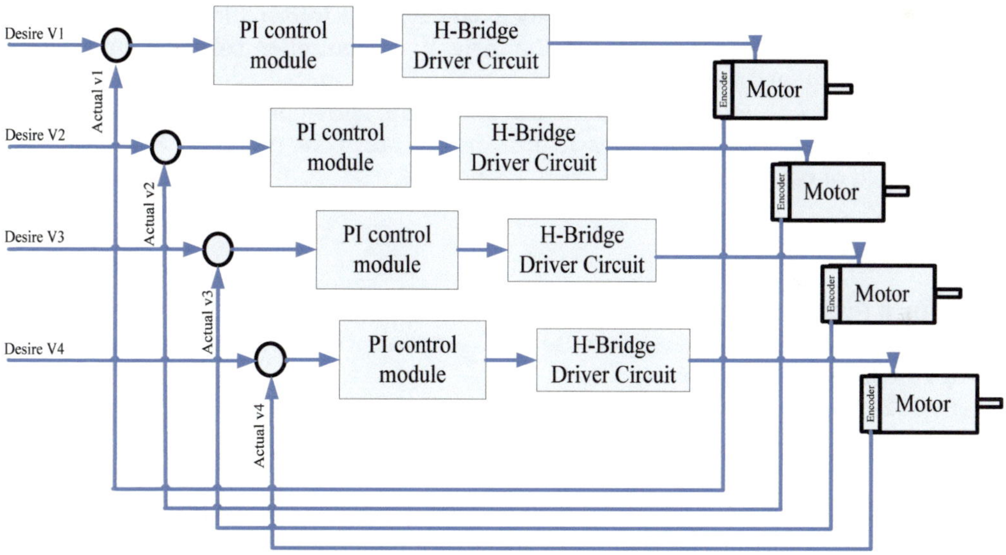

Fig. 7. The control module implemented in our robots in 2006

3.4 Evaluation

In the year 2006, the control system of the robot was a modular design that is fully capable of performing its required tasks. But the design is mainly constrained by some related factors:

- Long design period
- hard modification

There are many chips and many supply voltages, that will cause wire congestion and large area. Many interfaces between the two boards, it is easy to make mistakes. So, we explore a new method to simplify the design.

4. New System

4.1 System on a Chip

System on a Chip or System On Chip (SoC or SOC) is an idea of integrating all components of a computer or other electronic system into a single integrated circuit (chip). It may contain digital signals, analog signals, mixed-signal, and radio-frequency function on one chip. A typical application is in the area of embedded systems (Wikipedia, 2009).
With the planned improvements and designs for new control system, the previous system was no longer sufficient, thus we have to select and design a new system. This process of designing the onboard brain for the new robots involved many steps. It was important to understand the workings of the previous system in order to make educated decisions about improving upon the old design. In addition, it was important to be in contact with the other team members who were concurrently designing and developing electrical components that directly interact with the DSP. It is necessary that the interface between those components and different processors should be the easily to change. This information was the basis upon which we selected the candidate for the new microcontroller selection and also the succeeding evaluation process.
For the new control system, we decide to use SoC as the kernel of the control system. When considering the new processor, Nios II processor, a soft core processor based on FPGA, enters the field of our vision.

4.2 Nios II Processor

4.2.1 Nios II Processor

The Nios II processor is a general-purpose RISC processor core, provided by Altera Corp. Its features are (Altera Corp. 2006):

- Full 32-bit instruction set, data path, and address space
- 32 general-purpose registers
- 32 external interrupt sources
- Single-instruction 32 × 32 multiply and divide producing a 32-bit result
- Dedicated instructions for computing 64-bit and 128-bit products of multiplication
- Floating-point instructions for single-precision floating-point operations
- Single-instruction barrel shifter
- Access to a variety of on-chip peripherals, and interfaces to off-chip memories and peripherals
- Software development environment based on the GNU C/C++ tool chain and Eclipse IDE

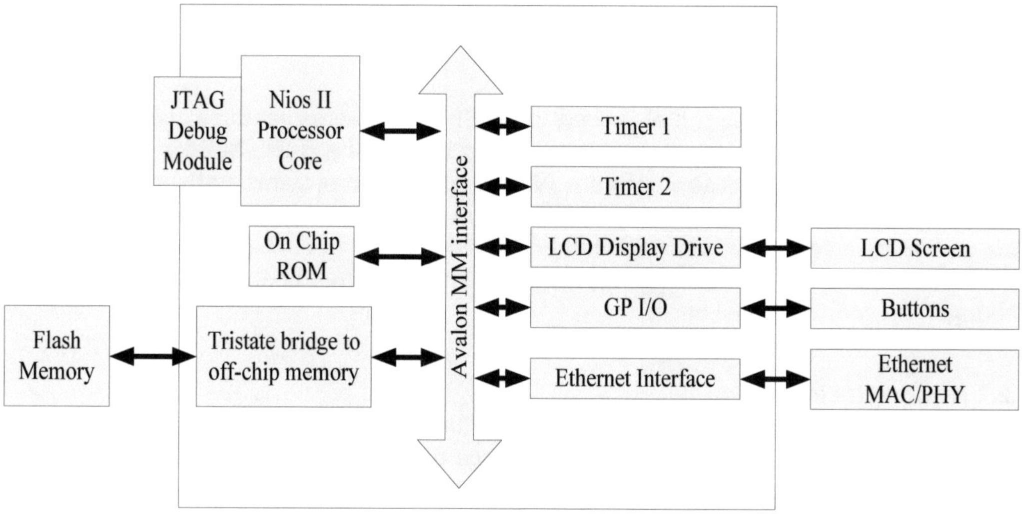

Fig. 8. Example of a Nios II Processor System (Altera Corp., 2006)

A Nios II processor system is the system that we can generate it with one or more Nios II processors, on-chip ROM, RAM, GPIO, Timer and so on. It can add or delete the peripherals and regenerate the system in minutes. Figure 8 is just an example of this system.

If the prototype system adequately meets design requirements using an Altera-provided reference design, the reference design can be copied and used as-is in the final hardware platform. Otherwise, we can customize the Nios II processor system until it meets cost or performance requirements.

4.2.2 Advantage of Nios II Processor

This section introduces Nios II concepts deeply relating to our design. For more details, refer (Altera Corp., 2006).

Configurable Soft-Core Processor

The Nios II processor is one of configurable soft-core processors provided by Altera Corp., as opposed to a fixed, off-the-shelf microcontroller. "Configurable" means that features can be added or removed on a system-by-system basis to meet performance or price goals. "Soft-core" means the CPU core is offered in "soft" design form (i.e., not fixed in silicon), and can be targeted to any FPGA. The users can configure the Nios II processor and peripherals to meet their specifications, and then program the system into an Altera FPGA, and also they can use readymade Nios II system designs. If these designs meet the system requirements, there is no need to configure the design further. In addition, software designers can use the Nios II instruction set simulator to begin writing and debugging Nios II applications before the final hardware configuration is determined.

Flexible Peripheral Set & Address Map

A flexible peripheral set is one of the most notable features of Nios II processor systems. Because of the soft-core nature of the Nios II processor, designers can easily build the Nios II processor systems with the exact peripheral set required for the target applications.

A corollary of flexible peripherals is a flexible address map. Software constructs are provided to access memory and peripherals generically, independently of address location.

Therefore, the flexible peripheral set and address map does not affect application developers.

Automated System Generation

Altera's SOPC Builder design tool is used to configure processor features and to generate a hardware design that can be programmed into an FPGA. The SOPC Builder graphical user interface (GUI) enables us to configure Nios II processor systems with any number of peripherals and memory interfaces. SOPC Builder can also import a designer's HDL design files, providing an easy mechanism to integrate custom logic into a Nios II processor system. After system generation, the design can be programmed into a board, and software can be debugged executing on the board.

4.2.3 Avalon-MM interface

The Avalon Memory-Mapped (Avalon-MM) interface specification provides with a basis for describing the address-based read/write interface found on master and slave peripherals, such as microprocessors, memory, UART, timer, etc(Altera Corp., 2006).

The Avalon-MM interface defines:

- A set of signal types
- The behavior of these signals
- The types of transfers supported by these signals

For example, the Avalon-MM interface can be used to describe a traditional peripheral interface, such as SRAM, that supports only simple, fixed-cycle read/write transfers.

4.3 Nios II Multi-Processor Systems

Multiprocessing is a generic term for the use of two or more CPUs within a single computer system. It also refers to the ability of a system to support more than one processor and/or the ability to allocate tasks between them. The CPUs are called multiprocessors. There are many variations on this basic theme, and the definition of multiprocessing can vary with context, mostly as a function of how multiprocessors are defined (multiple cores on one chip, multiple chips in one package, multiple packages in one system unit, etc.). (Wikipedia, 2009).

Multiprocessing sometimes refers to the execution of multiple concurrent software processes in a system as opposed to a single process at any one instant. However, the term multiprogramming is more appropriate to describe this concept, which is implemented mostly in software, whereas multiprocessing is more appropriate to describe the use of multiple hardware processors. A system can be both multiprocessing and multiprogramming, only one of the two, or neither of the two (Wikipedia, 2009).

Multiprocessor systems possess the benefit of increased performance, but nearly always at the price of significantly increased system complexity. For this reason, the using of multiprocessor systems has historically been limited to workstation and high-end PC computing using a complex method of load-sharing often referred to as symmetric multi processing (SMP). While the overhead of SMP is typically too high for most embedded systems, the idea of using multiple processors to perform different tasks and functions on different processors in embedded applications (asymmetrical) is gaining popularity (Altera Corp., 2007).

Multiple Nios II processors are able to efficiently share system resources using the multimaster friendly slave-side arbitration capabilities of the Avalon bus fabric. Many processors can be controlled to a system as by SOPC Builder.

To aid in the prevention of multiple processors interfering with each other, a hardware mutex core is included in the Nios II Embedded Design Suite (EDS). The hardware mutex core allows different processors to claim ownership of a shared resource for a period of time. Software debug on multiprocessor systems is performed using the Nios II IDE.

4.4 Structure of a new system

After we had selected the Nios II multiprocessor, we constructed the structure of the control system. First, we enumerated the old function of the control system, for example, motor control, speed sampling, wireless communication, kicker, LED display, flash data access and so on. The new system hardware architecture of year 2007 is shown in Figure 9.

There are many methods to separate the tasks and peripheral equipments of the control system for Multi-Processing (MP). Here we select one method which consists of two parts: a motor control part and a peripheral control part. The kernel of each part is a Nios II processor. One is used for the PID control of the motors. So that, motors have the real-time control that makes them respond quickly. The other one implements other functions of the control system, for example, wireless communication, states displaying, kicker controlling and accelerate sampling.

In the control part, using the H-bridge circuit, processor 1 controls the motors with the PID method. Each motor has a decoder, which can provide rotor position or speed information.

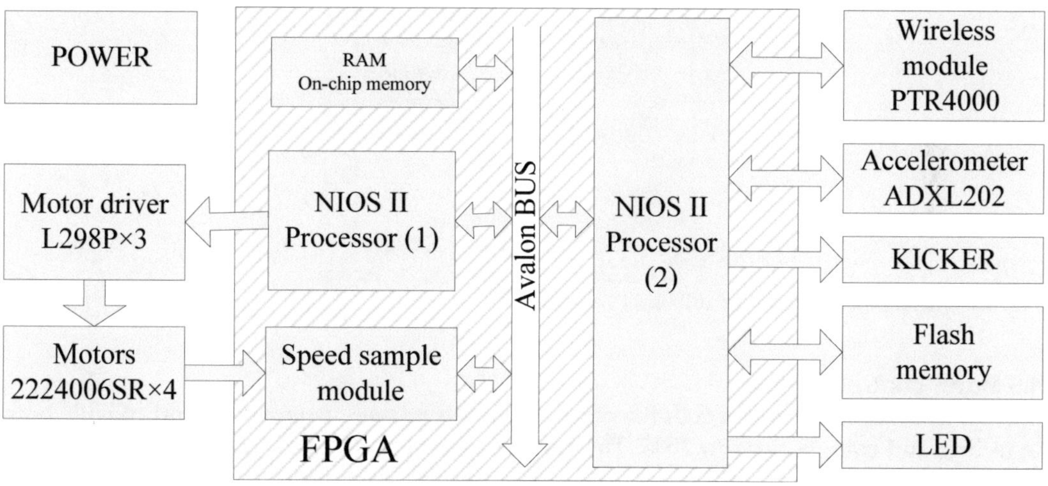

Fig. 9. The Hardware architecture of a new system

Processor 1 can read this information from speed sample module via Avalon bus. Then it compares these values with the desired value in the RAM, and outputs control signals to the motor device.

Processor 2 communicates with the off-field PC by wireless module, samples the acceleration by ADXL202, and controls the kicker and LED. It gathers the information from

PC and writes the data into the internal RAM. Then processor 1 fetches the data which is the desired value set by each control period.

The resolving of multiprocessor

Multiprocessor environments can use the mutex core with Avalon interface to coordinate accesses to a shared resource. The mutex core provides a protocol to ensure mutually exclusive ownership of a shared resource.

The mutex core provides a hardware-based atomic test-and-set operation, allowing software in a multiprocessor environment to determine which processor owns the mutex. The mutex core can be used in conjunction with shared memory to implement additional interprocessor coordination features, such as mailboxes and software mutexes.

4.5 Wireless module

The hardware of the wireless module is not changed for this new system. But because the kernel of the control system has been changed to Nios II processors, we should rewrite the wireless control model by verilog in FPGA. The interface which should accord with the avalon bus is shown in Figure 10. As mentioned in the section 3.2, the communication method is full duplex, so we should code the wireless module with the function of transmitting and receiving.

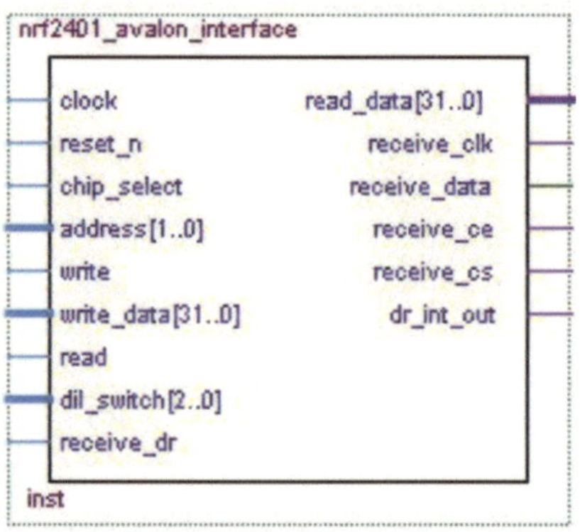

Fig. 10. The block diagram of wireless module.

4.6 Motor control module

The control loop implemented in the robot team had proven itself robust and reliable both in testing and competitions in 2006. The control system was coded by C of DSP processor. Essentially, the motor control module provides the following interface between the processor and the motors. The microcontroller sends two signals to the motor control module (a PWM and direction). These two signals get transformed to voltages applied to the DC motor terminals in our previous robot. In the reverse direction, the motor encoder sends two signals to the motor control module (channel A and B).

The motor control module is newly coded in Verilog because we select Nios II as our processor. The advantages of Verilog include a more optimized design. Also since the code is similar to C, it is easier to maintain. Most significantly, the code is portable between different FPGA families.

The motor control module consists two parts, one is speed sampling part and the other one is PWM part. Speed sampling part, as its name, is used for sampling the motor's real-time speed. Figure 11 shows the speed sampling principle circuit. The output of this circuit is the motor's digital speed, the high bit of which is the direction of the motor.

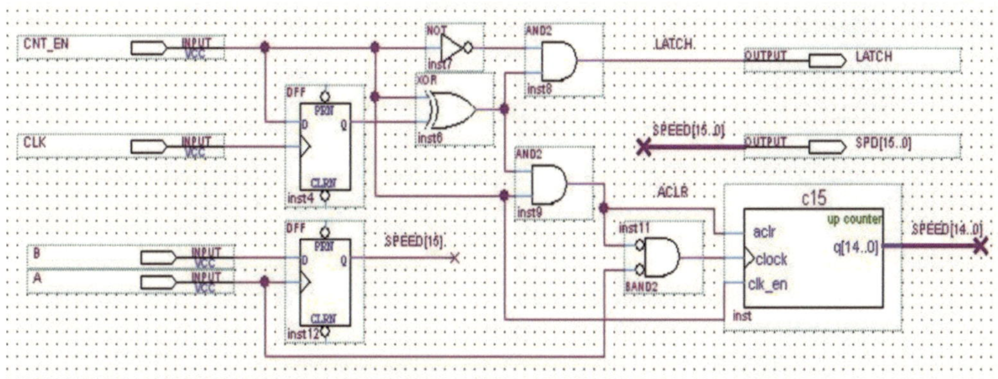

Fig. 11. The principle of speed sampling circuit

When testing the motor control module, we do the simulation with the input signals shown in Figure 12 which is the speed sampling module's simulation result. Output latch is used for latching the SPD signals (speed data) to the data bus, when Nios II processor needs to do the PI control.

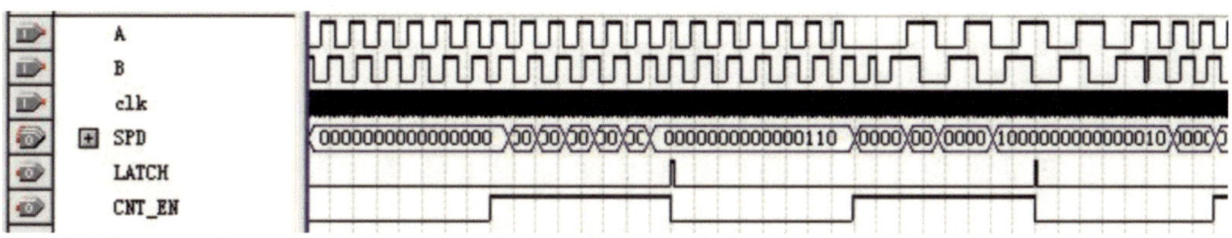

Fig. 12. The simulation result of speed sampling module

The other part of the motor control module is PWM part. The waveform shows in Figure 13. The PMU control design should be packed with Avalon MM interface. With different input number of duty_cycle, the waveform of pwm_out is changed easily. This combination of PWM part and speed sampling part is just for one motor, in our robot control system, 4 modules are mounted.

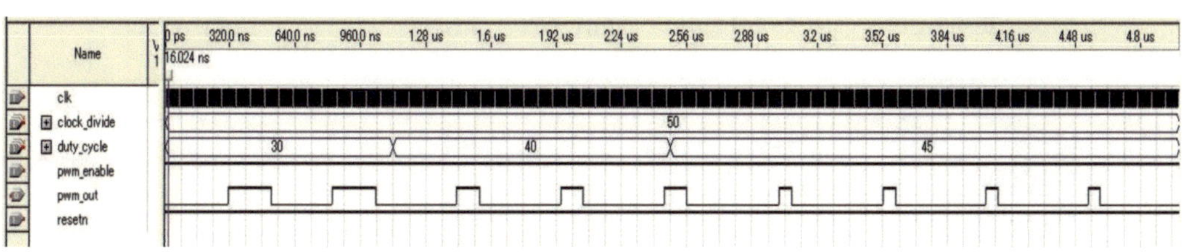

Fig. 13. The simulation result of PWM module.

5. Experiment

Before we start to do the experiment, we should know what we could do, and what the result that we want to get is. We pay more attention to the system design, and test whether the system can work with the module of motor control and wireless.

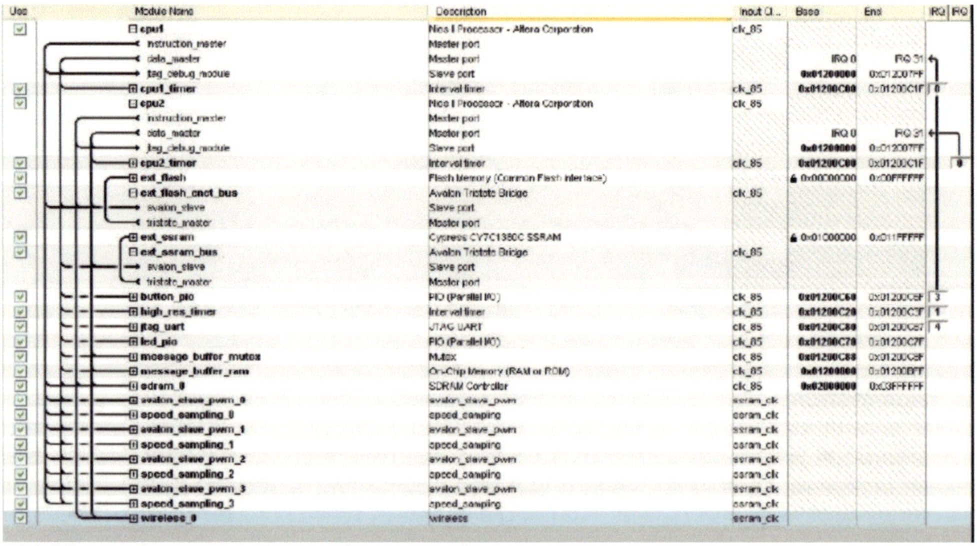

Fig. 14. The Architecture of our robot control system

The FPGA chip which contains two CPUs is EP2C8. EP2C8 is one chip of Altera Cyclone II family. With the advanced architectural features of Cyclone II FPGAs, the enhanced performance of Nios II embedded multiple processors becomes clearly. For the detailed steps of multiple processors generating, please refer (Altera Corp., 2008).

We generate the Nios II MP by SOPC Builder. In this system, as it shows, we add one timer for each Nios II CPU. One interval timer is also added. For the whole system, there are some other contents which would be used in the control system, such as LED_pio, JTAG_UART, message_buffer_mutex. The architecture of our MP robot control system is shown in Figure 14.

After the modules generation, the analysis and synthesis results of each module are shown as Table 1.

Module name	Total Logic elements	Total registers	Total PLLs
Nios CPUs related	5837	2985	1
Speed related module	426	265	0
Wireless module	519	431	0
others	1216	417	0
Total	7884	3998	1

Table 1. the analysis and synthesis results of the robot control system.

6. Conclusion

To build a reliable, robust robot control system which would improve upon the previous design, we approach the problem with a vastly different perspective which contains the MP. In some cases, it improves upon the design tremendously. But if we can not find a best arrangement of the task and cooperation for two processors, that is, it will not turn out to be a splendiferous control system for the robot.

During the design, there are many troubles we have met. The most important one is how to use the previous excellent circuit. If we achieve this, we could change our system as fast as possible.

The electrical design is slim this year. There are still areas which can be improved even more, but by and large, we are proud of the brevity and simplicity of the real-time and embedded soccer robot control system. We have read references for its feasibility and reliability, and started to embed an OS for the MP of Nios II processors.

7. Reference

Robert M. (2001). Control System Design for the RoboRoos, pp. 41~46.

Wikipedia. (2009). http://en.wikipedia.org/wiki/Multiprocessing

Wikipedia. (2009). http://en.wikipedia.org/wiki/System_on_a_Chip

Altera Corp. (2007). Creating Multiprocessor Nios II Systems Tutorial, pp. 5-11.

Altera Corp. (2006). Nios II Processor Reference Handbook, pp. 17-19.

Official Robocup Org. (2007). http://small-size.informatik.uni-bremen.de/

Official Robocup Org. (2007). http://www.robocup.org/

Ball D. (2001). Intelligence System for the 2001 RoboRoos Team. Brisbane: Univ. of Queensland

Dingle P. et al. (2004). 2002 Cornell robocup documentation. New York: Cornell Univ.

Zhenyu W.; Ce L. & Lin F. (2006). A Multi Micro-Motor Control System Based on DSP and FPGA. Small & Special Electrical Machines, vol. 35, No.1, Jan. 2007. pp 30-32, 1004-7018

TI Corp. (2004). TMS320R2811/2 Digital Signal Processors Data Manual, pp 4-7

2

CAMBADA soccer team: from robot architecture to multiagent coordination*

António J. R. Neves, José Luís Azevedo, Bernardo Cunha, Nuno Lau, João Silva, Frederico Santos, Gustavo Corrente, Daniel A. Martins, Nuno Figueiredo, Artur Pereira, Luís Almeida, Luís Seabra Lopes, Armando J. Pinho, João Rodrigues and Paulo Pedreiras
Transverse Activity on Intelligent Robotics, IEETA / DETI
University of Aveiro, Portugal

1. Introduction

Robotic soccer is nowadays a popular research domain in the area of multi-robot systems. RoboCup is an international joint project to promote research in artificial intelligence, robotics and related fields. RoboCup chose soccer as the main problem aiming at innovations to be applied for socially relevant problems. It includes several competition leagues, each one with a specific emphasis, some only at software level, others at both hardware and software, with single or multiple agents, cooperative and competitive.

In the context of RoboCup, the Middle Size League (MSL) is one of the most challenging. In this league, each team is composed of up to 5 robots with a maximum size of $50cm \times 50cm$, $80cm$ height and a maximum weight of 40Kg, playing in a field of $18m \times 12m$. The rules of the game are similar to the official FIFA rules, with minor changes required to adapt them for the playing robots

CAMBADA, *Cooperative Autonomous Mobile roBots with Advanced Distributed Architecture*, is the MSL Soccer team from the University of Aveiro. The project started in 2003, coordinated by the Transverse Activity on Intelligent Robotics group of the Institute of Electronic and Telematic Engineering of Aveiro (IEETA). This project involves people working on several areas for building the mechanical structure of the robot, its hardware architecture and controllers (Almeida et al., 2002; Azevedo et al., 2007) and the software development in areas such as image analysis and processing (Caleiro et al., 2007; Cunha et al., 2007; Martins et al., 2008; Neves et al., 2007; 2008), sensor and information fusion (Silva et al., 2008; 2009), reasoning and control (Lau et al., 2008), cooperative sensing approach based on a Real-Time Database (Almeida et al., 2004), communications among robots (Santos et al., 2009; 2007) and the development of an efficient basestation.

The main contribution of this chapter is to present the new advances in the areas described above involving the development of an MSL team of soccer robots, taking the example of the CAMBADA team that won the RoboCup 2008 and attained the third place in the last edition of the MSL tournament at RoboCup 2009. CAMBADA also won the last three editions

*This work was partially supported by project ACORD, Adaptive Coordination of Robotic Teams, FCT/PTDC/EIA/70695/2006.

of the Portuguese Robotics Open 2007-2009, which confirms the efficiency of the proposed architecture.
This chapter is organized as follows. In Section 2 it is presented the layered and modular architecture of the robot's hardware. Section 3 describes the vision system of the robots, starting in the calibration of the several parameters and presenting efficient algorithms for the detection of the colored objects and algorithms for the detection of arbitrary FIFA balls, a current challenge in the MSL. In Section 4 it is presented the process of building the representation of the environment and the algorithms for the integration of the several sources of information received by the robot. Section 5 presents the architecture used in CAMBADA robots to share information between them using a real-time database. Section 6 presents the methodology developed for the communication between robots, using an adaptive TDMA transmission control. In Section 7 it is presented the robots coordination model based on notions like *strategic positioning*, *role* and *formation*. Section 8 presents the Base Station application, responsible for the control of the agents, interpreting and sending high level instructions and monitoring information of the robots. Finally, in Section 9 we draw some conclusions.

2. Hardware architecture

The CAMBADA robots (Fig. 1) were designed and completely built in-house. The baseline for robot construction is a cylindrical envelope, with 485 mm in diameter. The mechanical structure of the players is layered and modular. Each layer can easily be replaced by an equivalent one. The components in the lower layer, namely motors, wheels, batteries and an electromagnetic kicker, are attached to an aluminum plate placed 8 cm above the floor. The second layer contains the control electronics. The third layer contains a laptop computer, at 22.5 cm from the floor, an omni-directional vision system, a frontal camera and an electronic compass, all close to the maximum height of 80 cm. The players are capable of holonomic motion, based on three omni-directional roller wheels.

Fig. 1. Robots used by the CAMBADA MSL robotic soccer team.

The general architecture of the CAMBADA robots has been described in (Almeida et al., 2004; Silva et al., 2005). Basically, the robots follow a biomorphic paradigm, each being centered on a main processing unit (a laptop), the *brain*, which is responsible for the higher-level behavior coordination, i.e. the coordination layer. This main processing unit handles external communication with the other robots and has high bandwidth sensors, typically vision, directly attached to it. Finally, this unit receives low bandwidth sensing information and sends actuating commands to control the robot attitude by means of a distributed low-level sensing/actuating system (Fig. 2), the *nervous system*.

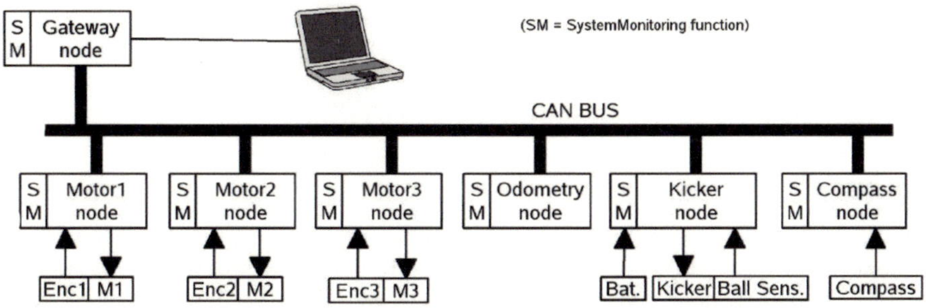

Fig. 2. Hardware architecture with functional mapping.

The low-level sensing/actuating system follows the fine-grain distributed model where most of the elementary functions, e.g. basic reactive behaviors and closed-loop control of complex actuators, are encapsulated in small microcontroller-based nodes interconnected by means of a network. For this purpose, Controller Area Network (CAN), a real-time fieldbus typical in distributed embedded systems, has been chosen. This network is complemented with a higher-level transmission control protocol to enhance its real-time performance, composability and fault-tolerance, namely the FTT-CAN protocol (Flexible Time-Triggered communication over CAN) (Almeida et al., 2002). This protocol keeps all the information of periodic flows within a master node, implemented on another basic module, which works like a maestro triggering tasks and message transmissions.
The low-level sensing/actuation system executes four main functions as described in Fig. 3, namely Motion, Odometry, Kick and System monitoring. The former provides holonomic motion using 3 DC motors. The Odometry function combines the encoder readings from the 3 motors and provides a coherent robot displacement information that is then sent to the coordination layer. The Kick function includes the control of an electromagnetic kicker and of a ball handler to dribble the ball.

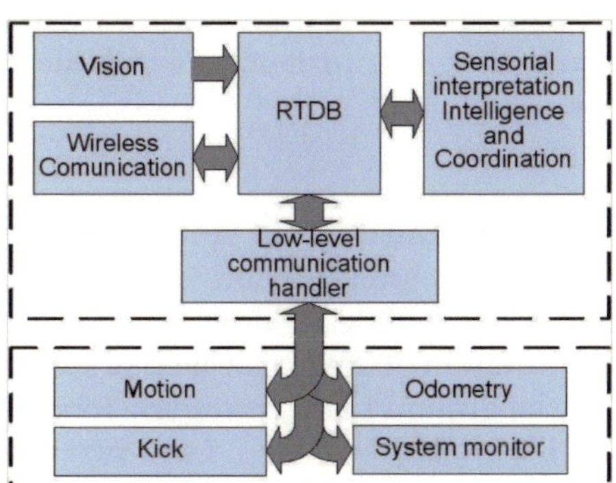

Fig. 3. Layered software architecture of CAMBADA players.

The system monitor function monitors the robot batteries as well as the state of all nodes in the low-level layer. Finally, the low-level control layer connects to the coordination layer through

a gateway, which filters interactions within both layers, passing through the information that is relevant across the layers, only. Such filtering reduces the overhead of handling unnecessary receptions at each layer as well as the network bandwidth usage at the low-level side, thus further reducing mutual interference across the layers.
A detailed description regarding the implementation of this architecture, namely the mapping between the functional architecture onto hardware and the information flows and their synchronization are presented in (Azevedo et al., 2007).

3. Vision system

The vision system of the CAMBADA robots is based on an hybrid system, formed by an omnidirectional and a perspective sub-system, that together can analyze the environment around the robots, both at close and long distances (Neves et al., 2008). The main modules of the vision system are presented in Fig. 4.

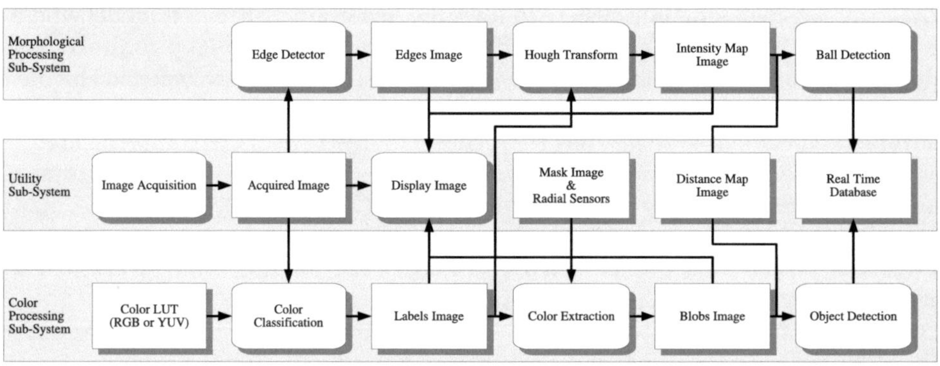

Fig. 4. The software architecture of the vision system developed for the CAMBADA robotic soccer team.

The information regarding close objects, like white lines of the field, other robots and the ball, are acquired through the omnidirectional system, whereas the perspective system is used to locate other robots and the ball at long distances, which are difficult to detect using the omnidirectional vision system.

3.1 Inverse distance map

The use of a catadioptric omni-directional vision system based on a regular video camera pointed at a hyperbolic mirror is a common solution for the main sensorial element found in a significant number of autonomous mobile robot applications. For most practical applications, this setup requires the translation of the planar field of view, at the camera sensor plane, into real world coordinates at the ground plane, using the robot as the center of this system. In order to simplify this non-linear transformation, most practical solutions adopted in real robots choose to create a mechanical geometric setup that ensures a symmetrical solution for the problem by means of single viewpoint (SVP) approach. This, on the other hand, calls for a precise alignment of the four major points comprising the vision setup: the mirror focus, the mirror apex, the lens focus and the center of the image sensor. Furthermore, it also demands

the sensor plane to be both parallel to the ground field and normal to the mirror axis of revolution, and the mirror foci to be coincident with the effective viewpoint and the camera pinhole respectively. Although tempting, this approach requires a precision mechanical setup.
We developed a general solution to calculate the robot centered distances map on non-SVP catadioptric setups, exploring a back-propagation ray-tracing approach and the mathematical properties of the mirror surface. This solution effectively compensates for the misalignment that may result either from a simple mechanical setup or from the use of low cost video cameras. Therefore, precise mechanical alignment and high quality cameras are no longer prerequisites to obtain useful distance maps. The method can also extract most of the required parameters from the acquired image itself, allowing it to be used for self-calibration purposes. In order to allow further trimming of these parameters, two simple image feedback tools have been developed.
The first one creates a reverse mapping of the acquired image into the real world distance map. A fill-in algorithm is used to integrate image data in areas outside pixel mapping on the ground plane. This produces a plane vision from above, allowing visual check of line parallelism and circular asymmetries (Fig. 5). The second generates a visual grid with 0.5m distances between both lines and columns, which is superimposed on the original image. This provides an immediate visual clue for the need of possible further distance correction (Fig. 6).

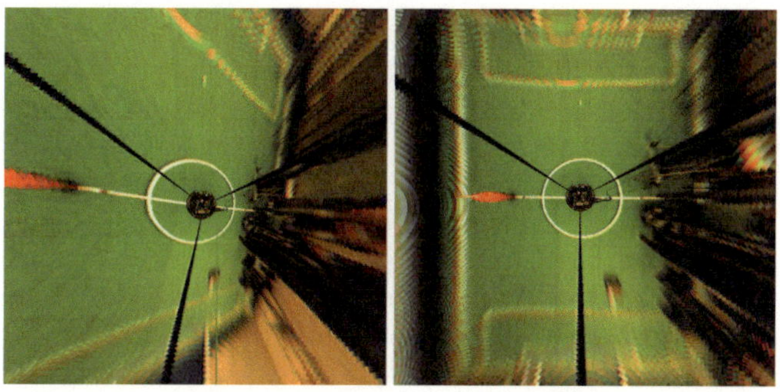

Fig. 5. Acquired image after reverse-mapping into the distance map. On the left, the map was obtained with all misalignment parameters set to zero. On the right, after automatic correction.

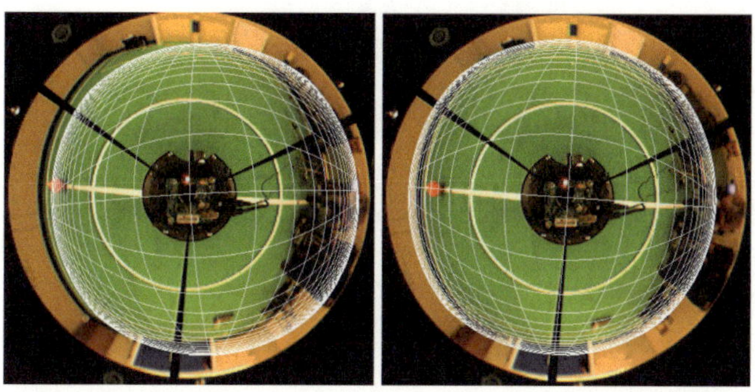

Fig. 6. A 0.5m grid, superimposed on the original image. On the left, with all correction parameters set to zero. On the right, the same grid after geometrical parameter extraction.

With this tool it is also possible to determine some other important parameters, namely the mirror center and the area of the image that will be processed by the object detection algorithms (Fig. 7). A more detailed description of the algorithms can be found in (Cunha et al., 2007).

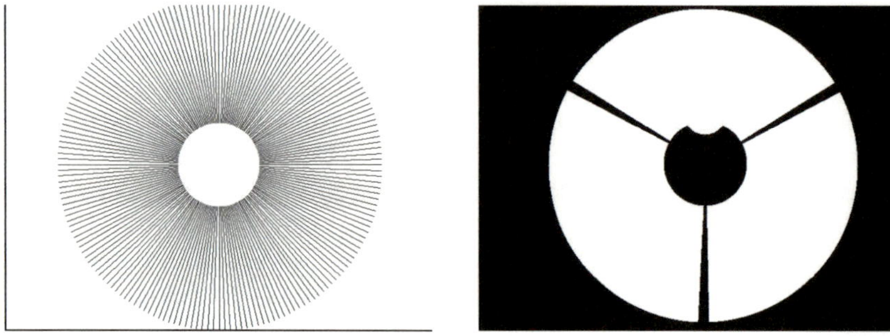

Fig. 7. On the left, the position of the radial search lines used in the omnidirectional vision system. On the right, an example of a robot mask used to select the pixels to be processed by the omnidirectional vision sub-system. White points represent the area that will be processed.

3.2 Autonomous configuration of the digital camera parameters

An algorithm was developed to configure the most important features of the cameras, namely exposure, white-balance, gain and brightness without human intervention (Neves et al., 2009). The self-calibration process for a single robot requires a few seconds, including the time necessary to interact with the application, which is considered fast in comparison to the several minutes needed for manual calibration by an expert user. The experimental results obtained show that the algorithm converges independently of the initial configuration of the camera. Moreover, the images acquired after the proposed calibration algorithm were analyzed using statistical measurements and these confirm that the images have the desired characteristics.

(a) (b) (c)

Fig. 8. An example of the autonomous configuration algorithm obtained starting with all the parameters of the camera set to the maximum value. In (a) the initial image acquired. In (b) the image obtained after applying the autonomous calibration procedure. In (c) a set of graphics representing the evolution of the camera parameters over time.

The proposed approach uses measurements extracted from a digital image to quantify the image quality. A number of typical measurements used in the literature can be computed from the image gray level histogram, namely, the mean (μ), the entropy (E), the absolute central moment (ACM) and the mean sample value (MSV). These measurements are used to calibrate the exposure and gain. Moreover, the proposed algorithm analyzes a white area in the image to calibrate the white-balance and a black area to calibrate the brightness.

3.3 Object detection

The vision software architecture is based on a distributed paradigm, grouping main tasks in different modules. The software can be split in three main modules, namely the *Utility Sub-System*, the *Color Processing Sub-System* and the *Morphological Processing Sub-System*, as can be seen in Fig. 4. Each one of these sub-systems labels a domain area where their processes fit, as the case of *Acquire Image* and *Display Image* in the *Utility Sub-System*. As can be seen in the *Color Processing Sub-System*, proper color classification and extraction processes were developed, along with an object detection process to extract information, through color analysis, from the acquired image.

Image analysis in the RoboCup domain is simplified, since objects are color coded. This fact is exploited by defining color classes, using a look-up-table (LUT) for fast color classification. The table consists of 16777216 entries (24 bits: 8 bits for red, 8 bits for green and 8 bits for blue), each 8 bits wide, occupying 16 MB in total. The pixel classification is carried out using its color as an index into the table. The color calibration is done in HSV (Hue, Saturation and Value) color space. In the current setup the image is acquired in RGB or YUV format and is then converted to an image of labels using the appropriate LUT.

The image processing software uses radial search lines to analyze the color information. A radial search line is a line that starts in the center of the robot with some angle and ends in the limit of the image. The center of the robot in the omnidirectional subsystem is approximately in the center of the image (an example is presented in Fig. 7), while in the perspective sub-system the center of the robot is in the bottom of the image. The regions of the image that have to be excluded from analysis (such as the robot itself, the sticks that hold the mirror and the areas outside the mirror) are ignored through the use of a previously generated image mask, as described in Section 3.1. The objects of interest (a ball, obstacles and the white lines) are detected through algorithms that, using the color information collected by the radial search lines, calculate the object position and/or their limits in an angular representation (distance and angle). The white lines are detected using an algorithm that, for each search line, finds the transition between green and white pixels. A more detailed description of the algorithms can be found in (Neves et al., 2007; 2008).

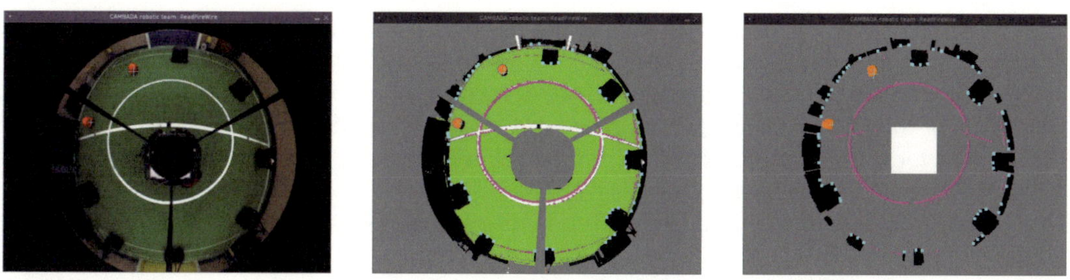

Fig. 9. On the left, the images acquired by the omnidirectional vision system. In the center, the corresponding image of labels. On the right, the color blobs detected in the images. A marks over a ball points to its center of mass. The several marks near the white lines (magenta) are the position of the white lines. Finally, the cyan marks denote the position of the obstacles.

The *Morphological Processing Sub-System* consists of a color independent ball detection algorithm, that will be described in the next section. Martins et al. (2008) presents preliminary results using this approach.
In the final of the image processing pipeline, the position of the detected objects are sent to the real-time database, described later in Section 5, after converting its position in the image into the real position in the environment, using the inverse distance map obtained with the algorithms and tools proposed in (Cunha et al., 2007) and briefly described before.

3.4 Arbitrary ball detection

The arbitrary FIFA ball recognition algorithm is based on the use of edge detection and the circular Hough transform. The search for potential ball candidates is conducted taking advantage of morphological characteristics of the ball (round shape), using a feature extraction technique known as the Hough transform. First used to identify lines in images, the Hough transform has been generalized through the years to identify positions of arbitrary shapes, most commonly circles or ellipses, by a voting procedure (Grimson and Huttenlocher, 1990; Ser and Siu, 1993; Zhang and Liu, 2000).
To feed the Hough transform process, it is necessary a binary image with the edge information of the objects. This image, *Edges Image*, is obtained using an edge detector operator. In the following, we present an explanation of this process and its implementation.
To be possible to use this image processing system in real-time, we implemented efficient data structures to process the image data (Neves et al., 2007; 2008). We used a two-thread approach to perform the most time consuming operations in parallel, namely image segmentation, edge detection and Hough transform, taking advantage of the dual core processor used by the laptop computers of our robots.
The first image processing step in the morphological detection is the edge detection. It must be as efficient and accurate as possible in order not to compromise the efficiency of the whole system. Besides being fast to calculate, the intended resulting image must be absent of noise as much as possible, with well defined contours and be tolerant to the motion blur introduced by the movement of the ball and the robots.
Some popular edge detectors were tested, namely Sobel (Zin et al., 2007; Zou et al., 2006; Zou and Dunsmuir, 1997), Laplace (Blaffert et al., 2000; Zou and Dunsmuir, 1997) and Canny (Canny, 1986). According to our experiments, the Canny edge detector was the most demanding in terms of processing time. Even so, it was fast enough for real-time operation and, because it provided the most effective contours, it was chosen.
The next step in the proposed approach is the use of the Hough transform to find points of interest containing possible circular objects. After finding these points, a validation procedure is used for choosing points containing a ball, according to our characterization. The voting procedure of the Hough transform is carried out in a parameter space. Object candidates are obtained as local maxima of a denoted *Intensity Image* (Fig. 10c)), that is constructed by the *Hough Transform* block (Fig. 4).
Due to the special features of the *Hough* circular transform, a circular object in the *Edges Image* would produce an intense peak in *Intensity Image* corresponding to the center of the object (as can be seen in Fig. 10c)). On the contrary, a non-circular object would produce areas of low intensity in the *Intensity Image*. However, as the ball moves away, its edge circle size decreases. To solve this problem, information about the distance between the robot center and the ball is used to adjust the Hough transform. We use the inverse mapping of our vision system (Cunha et al., 2007) to estimate the radius of the ball as a function of distance.

In some situations, particularly when the ball is not present in the field, false positives might be produced. To solve this problem and improve the ball information reliability, we propose a validation algorithm that discards false positives based on information from the *Intensity Image* and the *Acquired Image*. This validation algorithm is based on two tests against which each ball candidate is put through.

In the first test performed by the validation algorithm, the points with local maximum values in the *Intensity Image* are considered if they are above a distance-dependent threshold. This threshold depends on the distance of the ball candidate to the robot center, decreasing as this distance increases. This first test removes some false ball candidates, leaving a reduced group of points of interest.

Then, a test is made in the *Acquired Image* over each point of interest selected by the previous test. This test is used to eliminate false balls that usually appear in the intersection of the lines of the field and other robots (regions with several contours). To remove these false balls, we analyze a square region of the image centered in the point of interest. We discard this point of interest if the sum of all green pixels is over a certain percentage of the square area. Note that the area of this square depends on the distance of the point of interest to the robot center, decreasing as this distance increases. Choosing a square where the ball fits tightly makes this test very effective, considering that the ball fills over 90% of the square. In both tests, we use threshold values that were obtained experimentally.

Besides the color validation, it is also performed a validation of the morphology of the candidate, more precisely a circularity validation. Here, from the candidate point to the center of the ball, it is performed a search of pixels at a distance r from the center. For each edge found between the expected radius, the number of edges at that distance are determined. By the size of the square which covers the possible ball and the number of edge pixels, it is calculated the edges percentage. If the edges percentage is greater than 70, then the circularity of the candidate is verified.

Figure 10 presents an example of the of the *Morphological Processing Sub-System*. As can be observed, the balls in the *Edges Image* (Fig. 10 b)) have almost circular contours. Figure 10 c) shows the resulting image after applying the circular Hough transform. Notice that the center of the balls present a very high peak when compared to the rest of the image. The ball considered was the closest to the robot due to the fact that it has the high peak in the image.

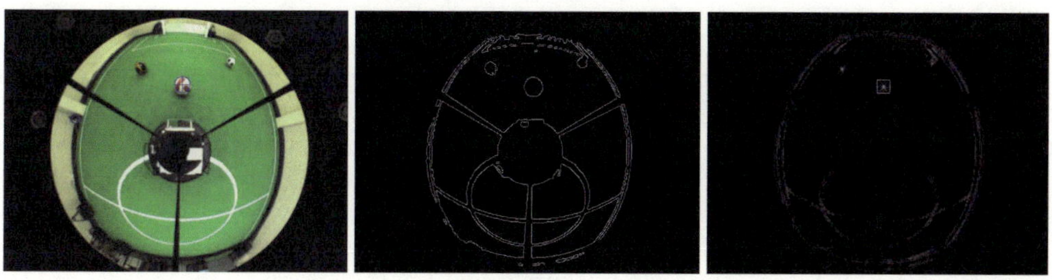

(a) (b) (c)

Fig. 10. Example of a captured image using the proposed approach. The cross over the ball points out the detected position. In b) the image a), with the Canny edge detector applied. In c), the image b) after applying the circular *Hough* transform.

4. Sensor Fusion

Having the raw information, the Integrator module is responsible for building the representation of the environment. The integration has several sources of information input, being the main input the raw information obtained by the cameras. Besides this information, the integration also uses information given by other sensors, namely an electronic compass (for localization purposes), an infra-red barrier sensor for ball engaged validation, odometry information given by the motors encoders, robot battery status, past cycles worldstate data, shared information obtained from team mate robots and coach information, both concerning game states and team formation, obtained from an external agent acting as a coach.
The first task executed by the integration is the update of the low level internal status, by updating the data structure values concerning battery and infra red ball barrier sensor. This is information that goes directly into the structure, because no treatment or filtering is needed. Afterwards, robot self-localization is made, followed by robot velocity estimation. The ball information is then treated, followed by obstacle treatment. Finally, the game state and any related issue are treated, for example, reset and update of timers, concerning setpieces.

4.1 Localization

Self-localization of the agent is an important issue for a soccer team, as strategic moves and positioning must be defined by positions on the field. In the MSL, the environment is completely known, as every agent knows exactly the layout of the game field. Given the known mapping, the agent has then to locate itself on it.
The CAMBADA team localization algorithm is based on the detected field lines, with fusion information from the odometry sensors and an electronic compass. It is based on the approach described in (Lauer et al., 2006), with some adaptations. It can be seen as an error minimization task, with a derived measure of reliability of the calculated position so that a stochastic sensor fusion process can be applied to increase the estimate accuracy (Lauer et al., 2006).
The idea is to analyze the detected line points, estimating a position, and through an error function describe the fitness of the estimate. This is done by reducing the error of the matching between the detected lines and the known field lines (Fig. 9). The error function must be defined considering the substantial amount of noise that affect the detected line points which would distort the representation estimate (Lauer et al., 2006).
Although the odometry measurement quality is much affected with time, within the reduced cycle times achieved in the application, consecutive readings produce acceptable results and thus, having the visual estimation, it is fused with the odometry values to refine the estimate. This fusion is done based on a Kalman filter for the robot position estimated by odometry and the robot position estimated by visual information. This approach allows the agent to estimate its position even if no visual information is available. However, it is not reliable to use only odometry values to estimate the position for more than a very few cycles, as slidings and frictions on the wheels produce large errors on the estimations in short time.
The visually estimated orientation can be ambiguous, i.e. each point on the soccer field has a symmetric position, relatively to the field center, and the robot detects exactly the same field lines. To disambiguate, an electronic compass is used. The orientation estimated by the robot is compared to the orientation given by the compass and if the error between them is larger than a predefined threshold, actions are taken. If the error is really large, the robot assumes a mirror position. If it is larger than the acceptance threshold, a counter is incremented. This counter forces relocation if it reaches a given threshold.

4.2 Ball integration

Within RoboCup several teams have used Kalman filters for the ball position estimation (Ferrein et al., 2006; Lauer et al., 2005; Marcelino et al., 2003; XU et al., 2006). In (Ferrein et al., 2006) and (Marcelino et al., 2003) several information fusion methods are compared for the integration of the ball position using several observers. In (Ferrein et al., 2006) the authors conclude that the Kalman reset filter shows the best performance.

The information of the ball state (position and velocity) is, perhaps, the most important, as it is the main object of the game and is the base over which most decisions are taken. Thus, its integration has to be as reliable as possible. To accomplish this, a Kalman filter implementation was created to filter the estimated ball position given by the visual information, and a linear regression was applied over filtered positions to estimate its velocity.

4.2.1 Ball position

It is assumed that the ball velocity is constant between cycles. Although that is not true, due to the short time variations between cycles, around 40 milliseconds, and given the noisy environment and measurement errors, it is a rather acceptable model for the ball movement. Thus, no friction is considered to affect the ball, and the model doesn't include any kind of control over the ball. Therefore, given the Kalman filter formulation (described in (Bishop and Welch, 2001)), the assumed state transition model is given by

$$X_k = \begin{bmatrix} 1 & \Delta T \\ 0 & 1 \end{bmatrix} X_{k-1}$$

where X_k is the state vector containing the position and velocity of the ball. Technically, there are two vectors of this kind, one for each cartesian dimension (x,y). This velocity is only internally estimated by the filter, as the robot sensors can only take measurements on the ball position. After defining the state transition model based on the ball movement assumptions described above and the observation model, the description of the measurements and process noises are important issues to attend. The measurements noise can be statistically estimated by taking measurements of a static ball position at known distances.

The standard deviation of those measurements can be used to calculate the variance and thus define the measurements noise parameter.

A relation between the distance of the ball to the robot and the measurements standard deviation can be modeled by a 2nd degree polynomial best fitting the data set in a least-squares sense. Depending on the available data, a polynomial of another degree could be used, but we should always keep in mind the computational weight of increasing complexity.

As for the process noise, this is not trivial to estimate, since there is no way to take independent measurements of the process to estimate its standard deviation. The process noise is represented by a matrix containing the covariances correspondent to the state variable vector. Empirically, one could verify that forcing a near null process noise causes the filter to practically ignore the read measures, leading the filter to emphasize the model prediction. This makes it too smooth and therefore inappropriate. On the other hand, if it is too high, the read measures are taken into too much account and the filter returns the measures themselves.

To face this situation, one have to find a compromise between stability and reaction. Given the nature of the two components of the filter state, position and speed, one may consider that their errors do not correlate.

Because we assume a uniform movement model that we know is not the true nature of the system, we know that the speed calculation of the model is not very accurate. A process

noise covariance matrix was empirically estimated, based on several tests, so that a good smoothness/reactivity relationship was kept.
Using the filter *a-priori* estimation, a system to detect great differences between the expected and read positions was implemented, allowing to detect hard deviations on the ball path.

4.2.2 Ball velocity

The calculation of the ball velocity is a feature becoming more and more important over the time. It allows that better decisions can be implemented based on the ball speed value and direction. Assuming a ball movement model with constant ball velocity between cycles and no friction considered, one could theoretically calculate the ball velocity by simple instantaneous velocity of the ball with the first order derivative of each component $\frac{\Delta D}{\Delta T}$, being ΔD the displacement on consecutive measures and ΔT the time interval between consecutive measures. However, given the noisy environment it is also predictable that this approach would be greatly affected by that noise and thus its results would not be satisfactory (as it is easily visible in Fig. 11.a).
To keep a calculation of the object velocity consistent with its displacement, an implementation of a linear regression algorithm was chosen. This approach based on linear regression (Motulsky and Christopoulos, 2003) is similar to the velocity estimation described in (Lauer et al., 2005). By keeping a buffer of the last m measures of the object position and sampling instant (in this case buffers of 9 samples were used), one can calculate a regression line to fit the positions of the object. Since the object position is composed by two coordinates *(x,y)*, we actually have two linear regression calculations, one for each dimension, although it is made in a transparent way, so the description is presented generally, as if only one dimension was considered.
When applied over the positions estimated by the Kalman filter, the linear regression velocity estimations are much more accurate than the instant velocities calculated by $\frac{\Delta D}{\Delta T}$, as visible in Fig. 11.b).

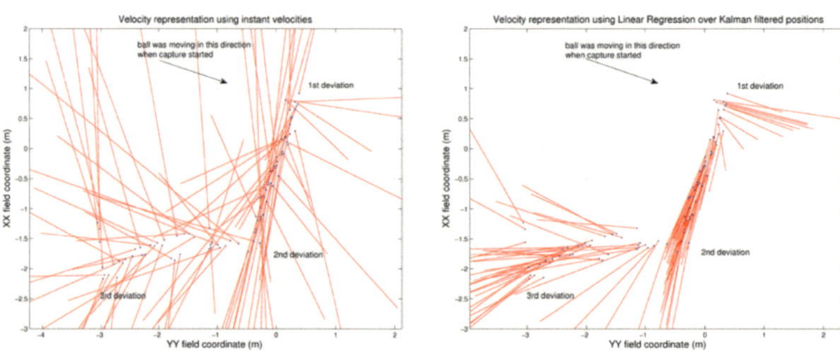

a) **b)**

Fig. 11. Velocity representation using: In **a)**: consecutive measures displacement; In **b)**: linear regression over Kalman filtered positions.

In order to try to make the regression converge more quickly on deviations of the ball path, a reset feature was implemented, which allows deletion of the older values, keeping only the n most recent ones, allowing a control of the used buffer size. This reset results from the

interaction with the Kalman filter described earlier, which triggers the velocity reset when it detects a hard deviation on the ball path.
Although in this case the Kalman filter internal functioning estimates a velocity, the obtained values were tested to confirm if the linear regression of the ball positions was still needed. Tests showed that the velocity estimated by the Kalman filter has a slower response than the linear regression estimation when deviations occur. Given this, the linear regression was used to estimate the velocity because quickness of convergence was preferred over the slightly smoother approximation of the Kalman filter in the steady state. That is because in the game environment, the ball is very dynamic, it constantly changes its direction and thus a convergence in less than half the cycles is much preferred.

4.2.3 Team ball position sharing

Due to the highly important role that the ball has in a soccer game, when a robot cannot detect it by its own visual sensors (omni or frontal camera), it may still know the position of the ball, through sharing of that knowledge by the other team mates.
The ball data structure include a field with the number of cycles it was not visible by the robot, meaning that the ball position given by the vision sensors can be the "last seen" position. When the ball is not visible for more than a given number of cycles, the robot assumes that it cannot detect the ball on its own. When that is the case, it uses the information of the ball communicated by the other running team mates to know where the ball is. This can be done through a function to get the statistics on a set of positions, mean and standard deviation, to get the mean value of the position of the ball seen by the team mates.
Another approach is to simply use the ball position of the team mate that have more confidence in the detection. Whatever the case, the robot assumes that ball position as its own. When detecting the ball on its own, there is also the need to validate that information. Currently the seen ball is only considered if it is within a given margin inside the field of play as there would be no point in trying to play with a ball outside the field. Fig. 12 illustrates the general ball integration activity diagram.

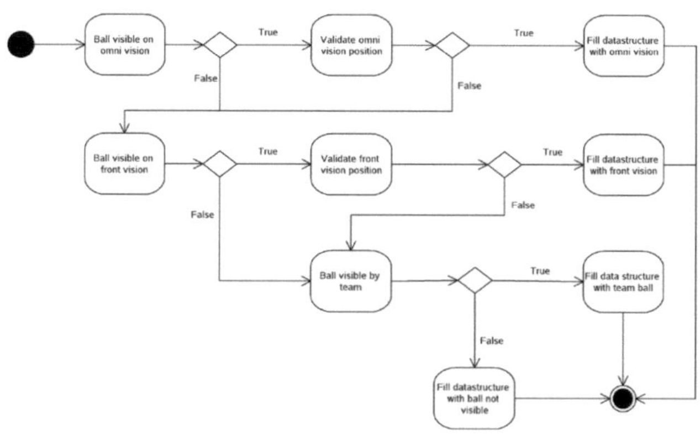

Fig. 12. Ball integration activity diagram.

4.3 Obstacle selection and identification

With the objective of refining the information of the obstacles, and have more meaningful and human readable information, the obstacles are selected and a matching is attempted, in order to try to identify them as team mates or opponents.
Due to the weak precision at long distances, a first selection of the obstacles is made by selecting only the obstacles closer than a given distance as available for identification (currently 5 meters). Also, obstacles that are smaller than 10 centimeters wide or outside the field of play margin are ignored. This is done because the MSL robots are rather big, and in game situations small obstacles are not present inside the field. Also, it would be pointless to pay attention to obstacles that are outside the field of play, since the surrounding environment is completely ignorable for the game development.
To be able to distinguish obstacles, to identify which of them are team mates and which are opponent robots, a fusion between the own visual information of the obstacles and the shared team mates positions is made. By creating a circle around the team mate positions, a matching of the estimated center of visible obstacle is made (Fig. 13), and the obstacle is identified as the corresponding team mate in case of a positive matching (Figs. 14c)). This matching consists on the existence of interception points between the team mate circle and the obstacle circle or if the obstacle center is inside the team mate circle (the obstacle circle can be smaller, and thus no interception points would exist).

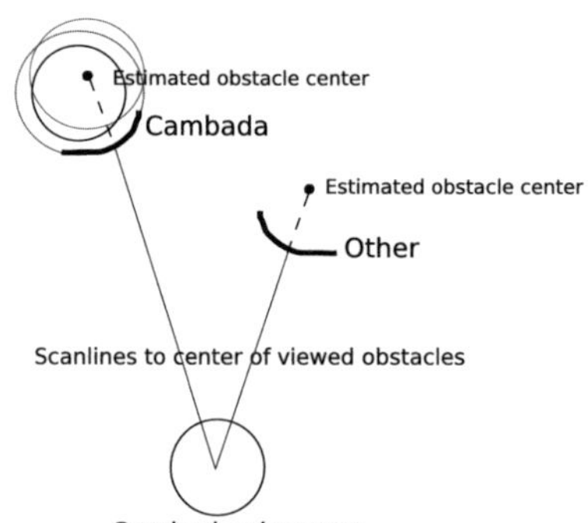

Fig. 13. When a CAMBADA robot is on, the estimated centers of the detected obstacles are compared with the known position of the team mates and tested; the left obstacle is within the CAMBADA acceptance radius, the right one is not.

Since the obstacles detected can be large blobs, the above described identification algorithm cannot be applied directly to the visually detected obstacles. If the detected obstacle fulfills the minimum size requisites already described, it is selected as candidate for being a robot obstacle. Its size is evaluated and classified as robot if it does not exceed the maximum size allowed for MSL robots (MSL Technical Committee 1997-2009, 2008) (Fig. 14a) and 14b)).
If the obstacle exceeds the maximum size of an MSL robot, a division of the obstacle is made, by analyzing its total size and verifying how many robots are in that obstacle. This is a common situation, robots clashing together and thus creating a compact black blob, originating a big obstacle. After completing the division, each obstacle is processed as described before.

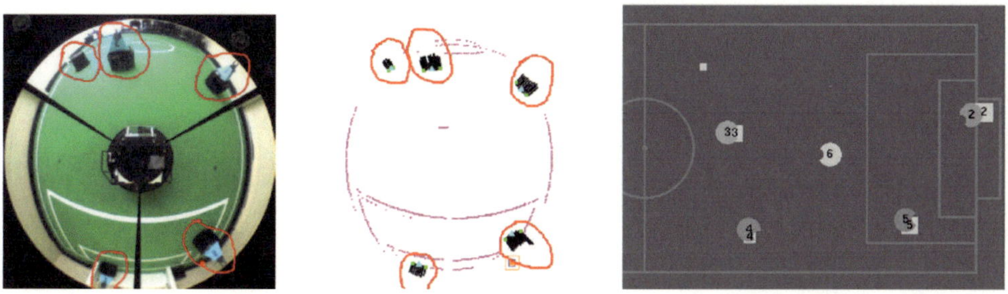

a) b) c)

Fig. 14. Illustration of single obstacles identification. In **a)**: image acquired from the robot camera (obstacles for identification are marked); In **b)**: the same image after processing; In **c)**: image of the control station. Each robot represents itself and robot 6 (the lighter gray) draws all the 5 obstacles evaluated (squares with the same gray scale as itself). All team mates were correctly identified (marked by its corresponding number over the obstacle square) and the opponent is also represented with no number.

5. Real-time database

Similarly to other teams, our team software architecture emphasizes cooperative sensing as a key capability to support the behavioral and decision-making processes in the robotic players. A common technique to achieve cooperative sensing is by means of a blackboard, which is a database where each agent publishes the information that is generated internally and that maybe requested by others. However, typical implementations of this technique seldom account for the temporal validity (coherence) of the contained information with adequate accuracy, since the timing information delivered by general-purpose operating systems such as Linux is rather coarse. This is a problem when robots move fast (e.g. above 1m/s) because their state information degrades faster, too, and temporal validity of state data becomes of the same order of magnitude, or lower, than the operating system timing accuracy.
Another problem of typical implementations is that they are based on the client-server model and thus, when a robot needs a datum, it has to communicate with the server holding the blackboard, introducing an undesirable delay. To avoid this delay, we use two features: firstly, the dissemination of the local state data is carried out using broadcasts, according to the producer-consumer cooperation model, secondly, we replicate the blackboard according to the distributed shared memory model. In this model, each node has local access to all the process state variables that it requires. Those variables that are remote have a local image that is updated automatically by an autonomous communication system (Fig. 15).
We call this replicated blackboard the Real-time Data Base (RTDB), (Almeida et al., 2004) which holds the state data of each agent together with local images of the relevant state data of the other team members. A specialized communication system triggers the required transactions at an adequate rate to guarantee the freshness of the data.
Generally, the information within the RTDB holds the absolute positions and postures of all players, as well as the position of the ball, goal areas and corners in global coordinates. This approach allows a robot to easily use the other robots sensing capabilities to complement its own. For example, if a robot temporarily loses track of the ball, it might use the position of the ball as detected by another robot.

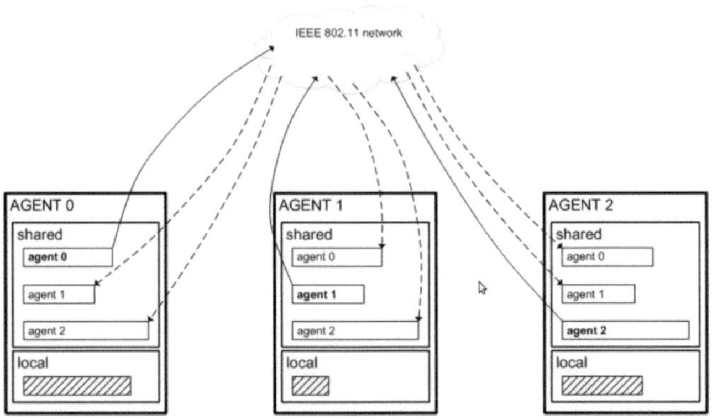

Fig. 15. Each agent broadcasts periodically its subset state data that might be required by other agents.

5.1 RTDB implementation

The RTDB is implemented over a block of shared memory. It contains two main areas: a private area for local information, only, i.e., which is not to be broadcast to other robots; and a shared area with global information. The shared area is further divided into a number of areas, one corresponding to each agent in the team. One of the areas is written by the agent itself and broadcast to the others while the remaining areas are used to store the information received from the other agents.

The allocation of shared memory is carried out by means of a specific function call, `DB_init()`, called once by every Linux process that needs access to the RTDB. The actual allocation is executed only by the first such call. Subsequent calls just return the shared memory block handler and increment a process count. Conversely, the memory space used by the RTDB is freed using the function call `DB_free()` that decreases the process count and, when zero, releases the shared memory block.

The RTDB is accessed concurrently from Linux processes that capture and process images and implement complex behaviors, and from tasks that manage the communication both with the lower-level control layer (through the CAN gateway) and with the other agents (through the wireless interface). The Linux processes access the RTDB with local non-blocking function calls, `DB_put()` and `DB_get()` that allow writing and reading records, respectively. `DB_get()` further requires the specification of the agent from which the item to be read belongs to, in order to identify the respective area in the database.

6. Communications

In the MSL, the agents communicate using an IEEE 802.11 network, sharing a single channel with the opposing team and using managed communication (through the access point), i.e., using a base station, and it is constrained to using a single channel, shared by, at least, both teams in each game. In order to improve the timeliness of the communications, our team uses a further transmission control protocol that minimizes collisions of transmissions within the team. Each robot regularly broadcasts its own data while the remaining ones receive such data and update their local structures. Beyond the robotic agents, there is also a coaching and monitoring station connected to the team that allows following the evolution of the robots status on-line and issuing high level team coordination commands.

As referred above, agents communicate using an IEEE 802.11 network, sharing a single channel with the opposing team and using managed communication (through the access point). This raises several difficulties because the access to the channel cannot be controlled and the available bandwidth is roughly divided by 2.
Therefore, the only alternative left for each team is to adapt to the current channel conditions and reduce access collisions among team members. This is achieved using an adaptive TDMA transmission control, with a predefined round period called team update period (T_{tup}) that sets the responsiveness of the global communication. Within such round, there is one single slot allocated to each team member so that all slots in the round are separated as much as possible (Fig. 16). This allows calculating the target inter-slot period T_{xwin} as $\frac{T_{tup}}{N}$, where N is the number of running agents.

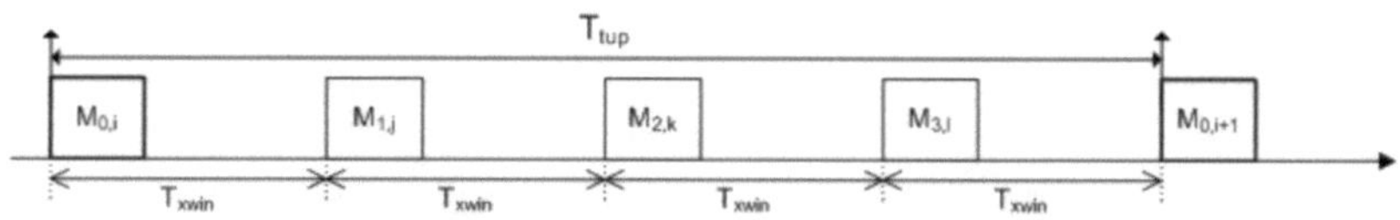

Fig. 16. TDMA round indicating the slots allocated to each robot.

The transmissions generated by each running agent are scheduled within the communication process, according to the production periods specified in the RTDB records. Currently a rate-monotonic scheduler is used. When the respective TDMA slot comes, all currently scheduled transmissions are piggybacked on one single 802.11 frame and sent to the channel. The required synchronization is based on the reception of the frames sent by the other agents during T_{tup}. With the reception instants of those frames and the target inter-slot period T_{xwin} it is possible to generate the next transmission instant. If these delays affect all TDMA frames in a round, then the whole round is delayed from then on, thus its adaptive nature. Figure 17 shows a TDMA round indicating the slots allocated to each agent and the adaptation of the round duration.

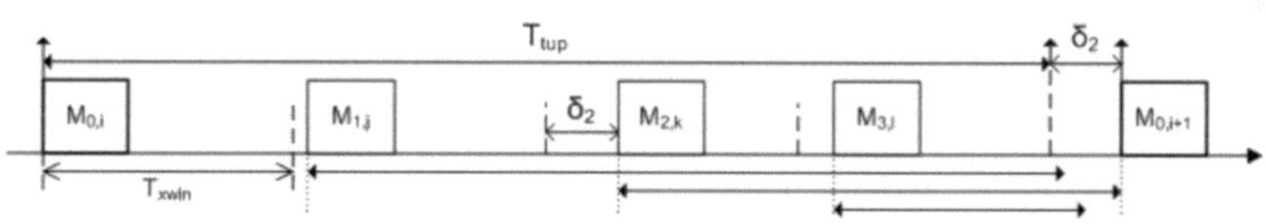

Fig. 17. An adaptive TDMA round.

When a robot transmits at time t_{now} it sets its own transmission instant $t_{next} = t_{now} + T_{tup}$, i.e. one round after. However, it continues monitoring the arrival of the frames from the other robots. When the frame from robot k arrives, the delay δ_k of the effective reception instant with respect to the expected instant is calculated. If this delay is within a validity window $[0, \Delta]$, with Δ being a global configuration parameter, the next transmission instant is delayed according to the longest such delay among the frames received in one round (Fig. 17), i.e.,

$$t_{next} = t_{now} + T_{tup} + max_k(\delta_k)$$

On the other hand, if the reception instant is outside that validity window, or the frame is not received, then (δ_k) is set to 0 and does not contribute to update t_{next}.

The practical effect of the protocol is that the transmission instant of a frame in each round may be delayed up to Δ with respect to the predefined round period T_{tup}. Therefore, the effective round period will vary between T_{tup} and $T_{tup} + \Delta$. When a robot does not receive any frame in a round within the respective validity windows, it updates t_{next} using a robot specific configuration parameter β_k in the following way

$$t_{next} = t_{now} + T_{tup} + \beta_k$$

with $0 \leq \beta_k \leq \Delta$.
This is used to prevent a possible situation in which the robots could all remain transmitting but unsynchronized, i.e. outside the validity windows of each other, and with the same period T_{tup}. By imposing different periods in this situation we force the robots to resynchronize within a limited number of rounds because the transmissions will eventually fall within the validity windows of each other.
One of the limitations of the adaptive TDMA protocol as proposed is that the number of team members was fixed, even if the agents were not active, causing the use of T_{xwin} values smaller than needed. Notice that a smaller T_{xwin} increases the probability of collisions in the team. Therefore, a self-configuration capability was added to the protocol, to cope with variable number of team members. This is the specific mechanism described in this section, which supports the dynamic insertion / removal of agents in the protocol. Currently, the T_{tup} period is still constant but it is divided by the number of running agents at each instant, maximizing the inter-slot separation between agents T_{xwin} at each moment.
However, the number of active team members is a global variable that must be consistent so that the TDMA round is divided in the same number of slots in all agents. To support the synchronous adaptation of the current number of active team members a membership vector was added to the frame transmitted by each agent in each round, containing its perception of the team status.
When a new agent arrives it starts to transmit its periodic information in an unsynchronized mode. In this mode all the agents, including the new one, continue updating its membership vector with the received frames and continue refreshing the RTDB shared areas, too. The T_{xwin} value, however, is not yet adapted and thus the new agent has no slot in the round. When all the team members reach the same membership vector, the number of active team members is updated, so as the inter-slot period T_{xwin}. The protocol enters then in the scan mode in which the agents, using their slightly different values of T_{tup}, rotate their relative phases in the round until they find their slots. From then on, all team members are again synchronized.
The removal of an absent agent uses a similar process. After a predefined number of rounds without receiving frames from a given agent, each remaining member removes it from the membership vector. The change in the vector leads to a new agreement process similar to described above.
More details about the communication process in the CAMBADA team, as well as in the MSL, can be found in (Almeida et al., 2004; Santos et al., 2009; 2007).

7. Coordination and strategy

In CAMBADA each robot is an independent agent and coordinates its actions with its teammates through communication and information exchange. The resulting behavior of the individual robot should be integrated into the global team strategy, thus resulting in cooperative actions by all the robots. This is done by the use of *roles* and *behaviors* that define each robot attitude in the field and resulting individual actions. Behaviors are the basic sensorimotor

skills of the robot, like moving to a specific position or kicking the ball, while roles select the active behavior at each time step, according to the attitude in cause.
For open play, CAMBADA uses an implicit coordination model based on notions like *strategic positioning*, *role* and *formation*. These notions and related algorithms have been introduced and/or extensively explored in the RoboCup Soccer Simulation League (Reis et al., 2001; Stone and Veloso, 1999). In order to apply such algorithms in the MSL, several changes had to be introduced.
A formation defines a movement model for the robotic players. Formations are sets of strategic positioning, where each positioning is a movement model for a specific player. The assignment of players to specific positioning is dynamic, and it is done according to some rules concerning ball and team mates positions. Each positioning is specified by three elements: Home position, which is the target position of the player when the ball is at the centre of the field, Region of the field where the player can move, and Ball attraction parameters, used to compute the target position of the player in each moment based on the current ball position All these items of information are given in a strategy configuration file. Using different home positions and attraction parameters for the positioning allows a simple definition of defensive, wing, midfielder and attack strategic movement models. Figure 18 shows an example of formation of the team for several ball positions.

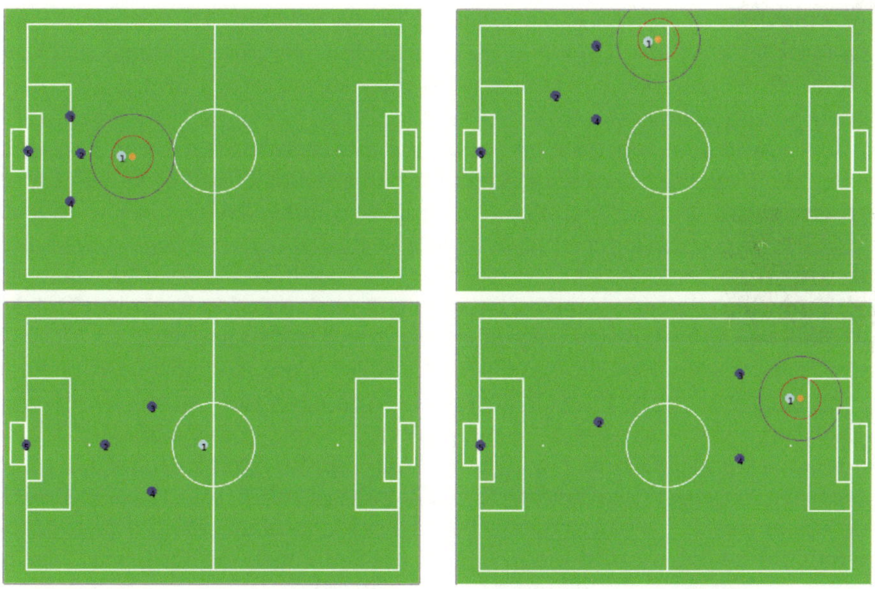

Fig. 18. CAMBADA Robots in some different game situations.

During open play, the CAMBADA agents use only three roles: `RoleGoalie`, `RoleMidfielder` and `RoleStriker`. The `RoleGoalie` is activated for the goalkeeper. `RoleMidfielder` moves according to its strategic positioning, defined as stated earlier. `RoleStriker` is an active player role, which substitutes the highest priority position of the formation, the one closer to the ball. It tries to catch the ball and score goals.

The striker activates several behaviors that try to engage the ball (`bMove`, `bMoveToAbs`), get into the opponent's side avoiding obstacles (`bDribble`) and shoot to the goal (`bKick`). The `bKick` behavior can perform 180° turns while keeping possession of the ball.
In a consistent role assignment, only one player at a time takes on the role of striker. The other teammates take on `RoleMidfielder` (Lau et al., 2008). Midfielders maintain their target positions as determined by their current positioning assignments and the current ball position. As a result, they accompany the striker as it plays along the field, without interfering. In case the ball is captured by the opponent, some of the midfielders hopefully will be in a good position to become the new striker. Occasionally, midfielders can take a more active behavior. This happens when the striker can't progress with the ball towards the opponent goal according to defined parameters. In this case, the closest midfielder to the ball also approaches the ball, acting as backup striker.
The role and position assignment in open play is based on considering different priorities for the different roles and positionings, so that the most important ones are always covered. The positioning is dynamically defined (not considering the goal keeper, which has a fixed role) by evaluating the distances of each of the robots to each of the target positions. Then the striker role is assigned to the robot that is closest to the highest priority strategic positioning, which is in turn the closest to the ball. The second position to be assigned is the defense position, then the two side positions. This algorithm results in the striker role having top priority, followed by the defensive positioning, followed by the wingers. The assignment algorithm may be performed by the coach agent in the base station, ensuring a coordinated assignment result, or locally by each robot, in which case the inconsistencies of world models may lead to unsynchronized assignments.
More explicit coordination is present on passes and setplays. Passing is a coordinated behavior involving two players, in which one kicks the ball towards the other, so that the other can continue with the ball. In the general case, the player running `RoleStriker` may decide to take on `RolePasser`, choosing the player to receive the ball. After being notified, the second player takes on the `RoleReceiver`. These roles have not been used yet for open play in international competition games, but they have been demonstrated in RoboCup 2008 MSL Free Technical Challenge and are used in a similar way in ball stoppage situations.
Another methodology implemented in CAMBADA is the use of coordinated procedures for setplays, i.e. situations when the ball is introduced in open play after a stoppage, such as kick-off, throw-in, corner kick, free kick and goal kick. Setplay procedures define a sequence of behaviors for several robots in a coordinated way. `RoleReplacer` and `RoleReceiver` are two exclusive roles used to overcome the MSL indirect rule in the case of indirect setplays against the opponent. The replacer passes the ball to the receiver which tries to score a goal, while the replacer assumes a position to defend the team mate shooting line. They position themselves as close to the shoot alignment as possible, so that a shot can be taken soon after the pass. If desired, a second receiver `RoleReceiver2` can be assigned to provide a second pass option for the replacer.
Finally, in the case of setplays against CAMBADA, `RoleBarrier` is used to protect the goal from a direct shoot. The line connecting the ball to the own goal defines the barrier positions. One player places itself on this line, as close to the ball as it is allowed. Two players place themselves near the penalty area. One player is placed near the ball (as much as allowed), 45° degrees from the mentioned line, so that it can observe the ball coming into play and report that to team mates.

8. The Base Station Application

In robotic soccer, the game is refereed by a human and his orders are communicated to the teams using an application called "Referee Box". No human interference is allowed during the games except for removing malfunctioning robots and re-entering robots in the game. The base station, a software application as described in this section, has a determinant role during the development of a robotic soccer team and also during a game. This application must control the agents interpreting and sending high level instructions, like `Start` or `Stop`, and monitor information of the robots, for example the position and velocity, allowing easily to attest the feasibility of the robots behavior.

The base station application must provide a set of tools to perform the activities mentioned above. Regarding the control activity, this application must allow high level control of the robots sending basic commands/information in particular the run and stop commands, play mode, role assignment, etc.

This application must also provide a high level monitoring of the robots internal states, namely the position in the field, velocity, battery charge, among other relevant information related with the robots and the game.

Furthermore, this application should provide an easy mechanism that can be used to easily show a specific behavior of the robot, allowing debugging.

Besides that, the base station has a fundamental role during a game, while receiving the commands from the referee box, translating them to internal game states and broadcasting the results to the robots. During a game, no human interference is allowed except for removing malfunctioning robots and re-entering robots in the game.

The role of the base station during these phases, development and games, demands the fulfillment of some requirements, being the most important the following:

Reliability / Stability: during the game, the base station is not accessible for human interaction of any kind and thus, it has to be a very robust application, all team depends on that.

Usability: the information displayed in the base station should be easy to interpret, allowing, for instance, a fast detection of a problem in a robot. It should be possible to choose different levels of details in the displayed information. Moreover, the base station has to be easy to use, allowing an intuitive management of the robots.

Adaptability: a robotic soccer team is in a permanent development stage, which may lead to significant changes within a short period of time.

Specifically to each phase the base station should provide the following features:

- **Development phase**

  Manual role assignment: acting as a cooperative team, each robot has a specific role which is, during a real game, dynamically assigned. In the development phase, it should be possible to manually assign a role to a specific robot.

  Local referee box: the base station should provide an interface widget to emulate a real referee box in order to simulate events of a real game.

  Visualization Tool: the application should provide a representation of the field and the robots in that context. Moreover, some visual information should be attached in order to improve the visual perception of the internal states of each robot.

Multi-windows solution: the application should be a multi-window environment, allowing the user to choose between different levels of information. At least, three different levels of information should be provided: a work level that presents the controls of the robots and basic status information; a visual level that presents visual information of the position of the robots and, finally a detailed level that shows all the information related to the robots.

Adaptable windows geometry: the multi-windows system should adapt to monitors with different resolutions. According to the new MSL rules, the base stations of each team must use an external monitor provided by the organizing committee.

- **Game situation**

Robust communication skills: the correct operation of the team during the game is fully dependent on the communication between the robots, the base station and the referee box. Hence, the base station should provide a robust communication layer.

Automatic processing of the game states: the base station should process the commands received from the referee box allowing the robots to change their internal game states accordingly. One specific action should be the changing of the field side at half time.

8.1 The CAMBADA base station approach

Regarding the challenges and requirements mentioned before, we will detail the used approaches in the conception of the base station application. We have divided, once again, the description in the mentioned activities in order to more precisely describe the features implemented for each one.

8.1.1 Performing control

Merging the available methods provided by the communication protocol of the team we were able to implement a widget that allows an high level control of the robots. All the actions were grouped to each robot and are accessible in a delimited space in order to improve the usability. These actions represents the enable/disable of each robot, the team color and goal color (in spite of the current rules don't specify goal colors, we decide keep it in order to facilitate the monitoring process), the role of the player, the re-location button, the start and stop that controls remotely the software in each robot and the bottom to launch remotely the agent.

Additionally, were created two other widgets, one to control all the team and one that implements a local referee box.

8.1.2 Monitoring

Regarding the monitoring activity, we developed a visualization widget that makes a representation of the field and the robots. This visualization widget shows the robots number, position and orientation and the ball that each robot view. Additionally was implemented a mechanism that allows to change the orientation of the field, in order to turn possible to monitor the robots in any position of the field, increasing the usability.

The base station has three separated windows representing three different levels of information. The main level shows the controls and relevant information about the robots state, other window only shows the visualization widget (this is the window to monitor the game, according with the new rules) and finally we implemented a window with full information of the robots, all the information available in the RTDB is shown in this window. In Fig. 19 is shown the main window.

Fig. 19. The base station `Main Window`.

In order to perform debugging in the development phase, it was implemented, in the visualization widget, a debug mechanism. Is possible to enable this mechanism writing in a specific field of the RTDB. This field is a vector with two dimensions representing a position on the game field. There are one point of debug *per* robot and if enabled in the base station this point can be shown in the game field together with the representation of the robots. This point is free to use and can represent whatever the developer wants.

Additionally, the third window, considered as the full information window, allows to perform debug to the robot state, more precisely in the transition between the roles and behaviors states.

All the project were developed using the Qt library using a modular architecture. This increased the reliability and stability allowing to test each module before the integration in the project.

9. Conclusions

This chapter presented the new advances in several areas that involves the development of an MSL team of soccer robots, namely the mechanical structure of the robot, its hardware architecture and controllers, the software development in areas such as image processing, sensor and information fusion, reasoning and control, cooperative sensing approach, communications among robots and some other auxiliary software.

The CAMBADA soccer robots have a hierarchical distributed hardware architecture with a central computer to carry out vision sensing, global coordination and deliberative functions and a low-level distributed sensing and actuation system based on a set of simple microcontroller nodes interconnected with a Controller Area Network (CAN). The advantages of distributed architectures extend from improved composability, allowing a system to be built by putting together different subsystems, to higher scalability, allowing to add functionality to the system by adding more nodes, more flexibility, allowing to reconfigure the system easily, better maintainability, due to the architecture modularity and easiness of node replacement, and higher reduction of mutual interference, thus offering a strong potential to support reactive behaviors more efficiently. Moreover, distributed architectures may also provide benefits in terms of dependability by creating error-containment regions at the nodes and opening the way for inexpensive spatial replication and fault tolerance.

The vision system of the CAMBADA robots is based on an hybrid system, formed by an omnidirectional and a perspective sub-system, that together can analyze the environment around the robots, both at close and long distances. We presented in this chapter several algorithms for the calibration of the most important parameters of the vision system and we propose efficient color-based algorithms for object detection. Moreover, we proposed a promising solution for the detection of arbitrary FIFA balls, as demonstrated by the first place obtained in the mandatory technical challenge in the Robocup 2009, where the robots had to play with an arbitrary standard FIFA ball.

Sensor and information fusion is important to maintain a more reliable description of the state of the world. The techniques proposed for information and sensor fusion proved to be effective in accomplishing their objectives. The Kalman filter allows to filter the noise on the ball position and provides an important prediction feature which allows fast detection of deviations of the ball path. The linear regression used to estimate the velocity is also effective, and combined with the deviation detection based on the Kalman filter prediction error, provides a faster way to recalculate the velocity in the new trajectory. The increasing reliability of the ball position and velocity lead to a better ball trajectory evaluation. This allowed the development of a more effective goalie action, as well as other behaviors, such as ball interception behaviors and pass reception. The results regarding obstacle identification, provide tools for an improvement of the overall team coordination and strategic play.

The robots coordination is based on a replicated database, the Real-Time Data Base (RTDB) that includes local state variables together with images of remote ones. These images are updated transparently to the application software by means of an adequate real-time management system. Moreover, the RTDB is accessible to the application using a set of non-blocking primitives, thus yielding a fast data access.

Regarding the communication between robots, is was developed a wireless communication protocol that reduces the probability of collisions among the team members. The protocol called adaptive TDMA, adapts to the current channel conditions, particularly accommodating periodic interference patterns. It was also developed an extended version of the protocol with on-line self-configuration capabilities that allow reconfiguring the slots structure of the TDMA round to the actual number of active team members, further reducing the collision probability.

In the CAMBADA MSL team, each robot is an independent agent and coordinates its actions with its teammates through communication and information exchange. The resulting behavior of the individual robot should be integrated into the global team strategy, thus resulting in cooperative actions by all the robots. This is done by the use of *roles* and *behaviors* that define each robot attitude in the field and resulting individual actions. This resulted in a coordinated behavior of the team that largely contributed to its recent successes.

The base station application has a important role during the development of a robotic soccer team and also during a game. This chapter presented the approach that was used by the CAMBADA team in the design of this important application.

The CAMBADA MSL team attained the first place in the MSL at RoboCup 2008 and attained the third place in the last edition of the MSL at RoboCup 2009. CAMBADA also won the last three editions of the Portuguese Robotics Open 2007-2009. These results confirm the effectiveness of the proposed architecture.

10. References

Almeida, L., P. Pedreiras, and J. A. Fonseca (2002). The FTT-CAN protocol: Why and how. *IEEE Transactions on Industrial Electronics 49*(6), 1189–1201.

Almeida, L., F. Santos, T. Facchinetti, P. Pedreira, V. Silva, and L. S. Lopes (2004). Coordinating distributed autonomous agents with a real-time database: The CAMBADA project. In *Proc. of the 19th International Symposium on Computer and Information Sciences, ISCIS 2004*, Volume 3280 of *Lecture Notes in Computer Science*, pp. 878–886. Springer.

Azevedo, J. L., B. Cunha, and L. Almeida (2007). Hierarchical distributed architectures for autonomous mobile robots: a case study. In *Proc. of the 12th IEEE Conference on Emerging Technologies and Factory Automation, ETFA2007*, Greece, pp. 973–980.

Bishop, G. and G. Welch (2001). An introduction to the kalman filter. In *Proc of SIGGRAPH, Course 8*, Number NC 27599-3175, Chapel Hill, NC, USA.

Blaffert, T., S. Dippel, M. Stahl, and R. Wiemker (2000). The laplace integral for a watershed segmentation. In *Proc. of the International Conference on Image Processing, 2000*, Volume 3, pp. 444–447.

Caleiro, P. M. R., A. J. R. Neves, and A. J. Pinho (2007, June). Color-spaces and color segmentation for real-time object recognition in robotic applications. *Revista do DETUA 4*(8), 940–945.

Canny, J. F. (1986, November). A computational approach to edge detection. *IEEE Transactions on Pattern Analysis and Machine Intelligence 8*(6).

Cunha, B., J. L. Azevedo, N. Lau, and L. Almeida (2007). Obtaining the inverse distance map from a non-SVP hyperbolic catadioptric robotic vision system. In *Proc. of the RoboCup 2007*, Atlanta, USA.

Ferrein, A., L. Hermanns, and G. Lakemeyer (2006). Comparing sensor fusion techniques for ball position estimation. In *RoboCup 2005: Robot Soccer World Cup IX*, Volume 4020 of *LNCS*, pp. 154–165. Springer.

Grimson, W. E. L. and D. P. Huttenlocher (1990). On the sensitivity of the hough transform for object recognition. *IEEE Trans. on Pattern Analysis and Machine Intelligence 12*, 1255–1274.

Lau, N., L. S. Lopes, and G. Corrente (2008, April). CAMBADA: information sharing and team coordination. In *Proc. of the 8th Conference on Autonomous Robot Systems and Competitions, Portuguese Robotics Open - ROBOTICA'2008*, Aveiro, Portugal, pp. 27–32.

Lauer, M., S. Lange, and M. Riedmiller (2005). Modeling moving objects in a dynamically changing robot application. In *KI 2005: Advances in Artificial Intelligence*, Volume 3698 of *LNCS*, pp. 291–303. Springer.

Lauer, M., S. Lange, and M. Riedmiller (2006). Calculating the perfect match: an efficient and accurate approach for robot self-localization. In *RoboCup 2005: Robot Soccer World Cup IX*, Lecture Notes in Computer Science, pp. 142–153. Springer.

Marcelino, P., P. Nunes, P. Lima, and M. I. Ribeiro (2003). Improving object localization through sensor fusion applied to soccer robots. In *Proc. Scientific Meeting of the Portuguese Robotics Open - ROBOTICA 2003*.

Martins, D. A., A. J. R. Neves, and A. J. Pinho (2008, october). Real-time generic ball recognition in RoboCup domain. In *Proc. of the 11th edition of the Ibero-American Conference on Artificial Intelligence, IBERAMIA 2008*, Lisbon, Portugal.

Motulsky, H. and A. Christopoulos (2003). *Fitting models to biological data using linear and nonlinear regression*. GraphPad Software Inc.

MSL Technical Committee 1997-2009 (2008). Middle size robot league rules and regulations for 2009.

Neves, A. J. R., A. J. P. B. Cunha, and I. Pinheiro (2009, June). Autonomous configuration of parameters in robotic digital cameras. In *Proc. of the 4th Iberian Conference on Pattern Recognition and Image Analysis, ibPRIA 2009*, Póvoa de Varzim, Portugal.

Neves, A. J. R., G. Corrente, and A. J. Pinho (2007). An omnidirectional vision system for soccer robots. In *Progress in Artificial Intelligence*, Volume 4874 of *Lecture Notes in Artificial Inteligence*, pp. 499–507. Springer.

Neves, A. J. R., D. A. Martins, and A. J. Pinho (2008, April). A hybrid vision system for soccer robots using radial search lines. In *Proc. of the 8th Conference on Autonomous Robot Systems and Competitions, Portuguese Robotics Open - ROBOTICA'2008*, Aveiro, Portugal, pp. 51–55.

Reis, L., N. Lau, and E. Oliveira (2001). Situation based strategic positioning for coordinating a team of homogeneous agents. In *Balancing Reactivity and Social Deliberation in Multi-Agent Systems*, Volume 2103 of *Lecture Notes in Computer Science*, pp. 175–197. Springer.

Santos, F., L. Almeida, L. S. Lopes, J. L. Azevedo, and M. B. Cunha (2009). Communicating among robots in the robocup middle-size league. In *RoboCup 2009: Robot Soccer World Cup XIII*, Lecture Notes in Artificial Intelligence. Springer (In Press).

Santos, F., G. Corrente, L. Almeida, N. Lau, and L. S. Lopes (2007, December). Self-configuration of an adaptive TDMA wireless communication protocol for teams of mobile robots. In *Proc. of the 13th Portuguese Conference on Artificial Intelligence, EPIA 2007*, Guimarães, Portugal.

Ser, P.-K. and W.-C. Siu (1993). Invariant hough transform with matching technique for the recognition of non-analytic objects. In *IEEE International Conference on Acoustics, Speech, and Signal Processing, ICASSP 1993.*, Volume 5, pp. 9–12.

Silva, J., N. Lau, J. Rodrigues, and J. L. Azevedo (2008, october). Ball sensor fusion and ball interception behaviours for a robotic soccer team. In *Proc. of the 11th edition of the Ibero-American Conference on Artificial Intelligence, IBERAMIA 2008*, Lisbon, Portugal.

Silva, J., N. Lau, J. Rodrigues, J. L. Azevedo, and A. J. R. Neves (2009). Sensor and information fusion applied to a robotic soccer team. In *RoboCup 2009: Robot Soccer World Cup XIII*, Lecture Notes in Artificial Intelligence. Springer (In Press).

Silva, V., R. Marau, L. Almeida, J. Ferreira, M. Calha, P. Pedreiras, and J. Fonseca (2005). Implementing a distributed sensing and actuation system: The CAMBADA robots case study. In *Proc. of the 10th IEEE Conference on Emerging Technologies and Factory Automation, ETFA2005*, Italy, pp. 781–788.

Stone, P. and M. Veloso (1999). Task decomposition, dynamic role assignment, and low-bandwidth communication for real-time strategic teamwork. Volume 110, pp. 241–273.

XU, Y., C. JIANG, and Y. TAN (2006). SEU-3D 2006 Soccer simulation team description. In *CD Proc. of RoboCup Symposium 2006*.

Zhang, Y.-J. and Z.-Q. Liu (2000). Curve detection using a new clustering approach in the hough space. In *IEEE International Conference on Systems, Man, and Cybernetics, 2000*, Volume 4, pp. 2746–2751.

Zin, T. T., H. Takahashi, and H. Hama (2007). Robust person detection using far infrared camera for image fusion. In *Second International Conference on Innovative Computing, Information and Control, ICICIC 2007*, pp. 310–310.

Zou, J., H. Li, B. Liu, and R. Zhang (2006). Color edge detection based on morphology. In *First International Conference on Communications and Electronics, ICCE 2006*, pp. 291–293.

Zou, Y. and W. Dunsmuir (1997). Edge detection using generalized root signals of 2-d median filtering. In *Proc. of the International Conference on Image Processing, 1997*, Volume 1, pp. 417–419.

3

Small-size Humanoid Soccer Robot Design for FIRA HuroSot

Ching-Chang Wong, Chi-Tai Cheng, Kai-Hsiang Huang, Yu-Ting Yang, Yueh-Yang Hu and Hsiang-Min Chan
Department of Electrical Engineering, Tamkang University
Tamsui, Taipei, 25137, Taiwan

1. Introduction

Robot soccer games are used to encourage the researches on the robotics and artificial intelligence. FIRA (URL: http://www.fira.net) is an international robot soccer association to advance this research and hold some international competitions and congresses. There are many different leagues, such as SimuroSot, MiroSot, RoboSot, and HuroSot, in FIRA RoboWorld Cup. Each league is established for different research purposes. In the HuroSot league, many technology issues and scientific areas must be integrated to design a humanoid robot. The research technologies of mechanism design, electronic system, biped walking control, autonomous motion, direction judgment, kicking ball need to be applied on a humanoid robot (Chemori & Loria, 2004; Esfahani & Elahinia, 2007; Guan et al., 2006; Hemami et al., 2006; Hu et al., 2008; Haung et al., 2001; Miyazaki & Arimoto, 1980; Sugihara et al., 2002; Wong et al., 2005; Zhou & Jagannathan, 1996). This chapter introduces an autonomous humanoid robot, TWNHR-IV (**T**ai**w**a**n** **H**umanoid **R**obot-IV), which is able to play sports, such as soccer, basketball, weight lifting, and marathon. The robot is designed to be a vision-based autonomous humanoid robot for HuroSot League of FIRA Cup. TWNHR-IV joined FIRA Cup in 2007 and 2008. In FIRA 2007, TWNHR-IV won the first place in robot dash, penalty kick, obstacle run, and weight lifting; the second place in basketball and marathon. In FIRA 2008, TWNHR-IV won the first place in penalty kick, obstacle run, and weight lifting, the second place in robot dash and the third place in basketball. TWNHR-IV determines the environment via its sensors and executes the suitable motion by its artificial intelligent. In order to let TWNHR-IV have the environment perceptive ability, a vision sensor (a CMOS sensor), six distance sensors (six infrared sensors), a posture sensor (an accelerometer sensor) and a direction sensor (a digital compass) are equipped on the body of TWNHR-IV to obtain the information of the environment. Two processors are used to control the robot. The first one is a DSP for the high-level control purpose. The DSP receives and processes the image data from the CMOS sensor via a serial port. It is in charge of the high level artificial intelligent, such as navigation. The second one is NIOS II (an embedded soft-core processor) for the robot locomotion control. The second processor is used as a low-level controller to control the walking and other actions. TWNHR-IV is designed as a soccer player so that it can walk,

turn, and kick the ball autonomously. A control board with a FPGA chip and a 64 Mb flash memory are mainly utilized to control the robot. Many functions are implemented on this FPGA chip so that it can receive motion commands from DSP via a serial port and process the data obtained by infrared sensors, digital compass, and accelerometer. It is able to accomplish the different challenges of HuroSot, including Penalty Kick (PK), basketball, lift-and-carry, obstacle run, robot dash, weight lifting, and marathon autonomously and effectively.
The rest of this chapter is organized as follows: In Section 2, the mechanical system design of the robot TWNHR-IV is described. In Section 3, the electronic system including a vision system, a control core, and sensor systems are described. In Section 4, some simulation and experiments results of the proposed controller are described. Finally, some conclusions are made in Section 5.

2. Mechanical System Design

Mechanical system design is the first step of design a humanoid robot. Human body mechanism basically consists of bones, joints, muscles, and tendons. It is impossible to replace all of the muscular-skeletal system by current mechanical and electrical components. Therefore, the primary goal of the humanoid robot mechanical system design is to develop a robot that can imitate equivalent human motions. The degrees of freedom (DOFs) configuration of TWNHR-IV is presented in Figure 1. TWNHR-IV has 26 DOFs. In this chapter, the rotational direction of each joint is defined by using the inertial coordinate system fixed on the ground as shown in Figure 1 (Wong et al., 2008c).

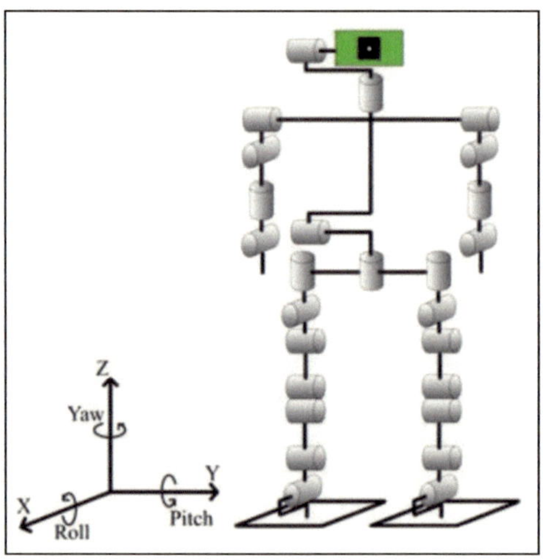

Fig. 1. DOFs configuration

A photograph and a 3D mechanical structure of the implemented robot are shown in Figure 2 The design concepts of TWNHR-IV are light weight and compact size. The height of TWNHR-IV is 43 cm and the weight is 3.4 kg with batteries. A human-machine interface is designed to manipulate the servo motors. The details of the mechanical structure of the head, arms, waist, trunk, and legs are described as follows.

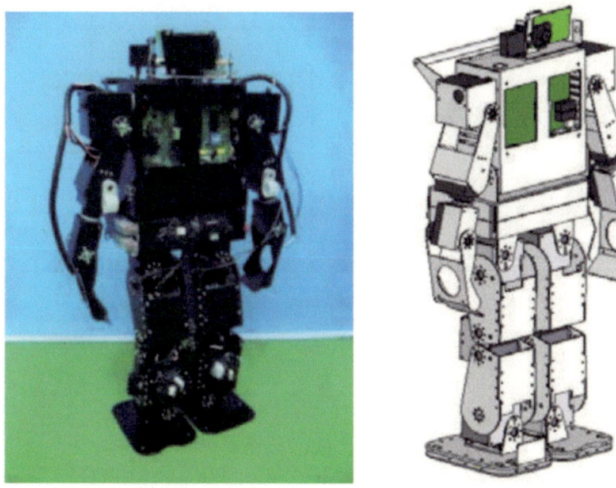

Fig. 2. Photograph and mechanical structure of TWNHR-IV

The head of TWNHR-IV has 2 DOFs. Figure 3 shows the 3D mechanism design of the head. The head is designed based on the concept that the head of the robot can accomplish the raw and pitch motion. Table 1 presents the head DOFs relation between human and TWNHR-IV.

Human Figure	TWNHR-IV	Human Figure	TWNHR-IV
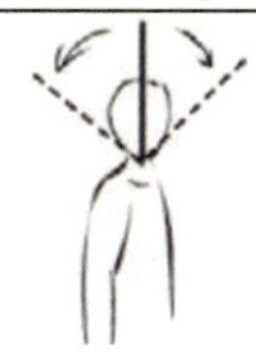			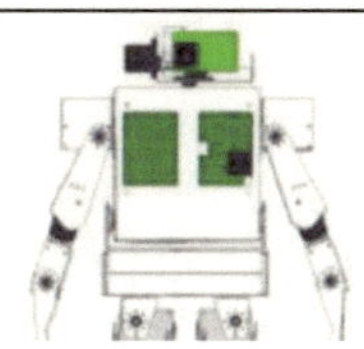

Table 1. The head DOFs relation between human and TWNHR-IV

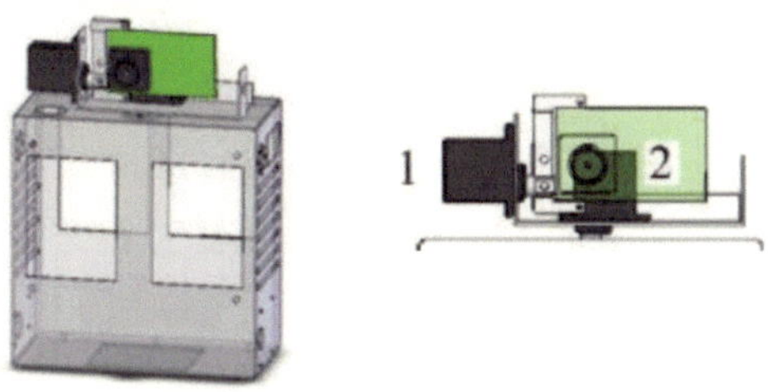

(a) Head 3D design (b) DOFs diagram

Fig. 3. Head mechanism

The head of TWNHR-IV has 2 degrees of freedom. Figure 4 shows the 3D mechanism design of the waist and trunk. The trunk is designed based on the concept that robot can walk and maintain its balance by using gyro to adjust the trunk motions to compensate for the robot's walk motion. Table 2 presents the specification of the joints for the waist and trunk.

Human Figure	TWNHR-IV	Human Figure	TWNHR-IV
	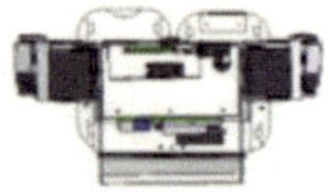		

Table 2. The waist and trunk DOFs relation between human and TWNHR-IV

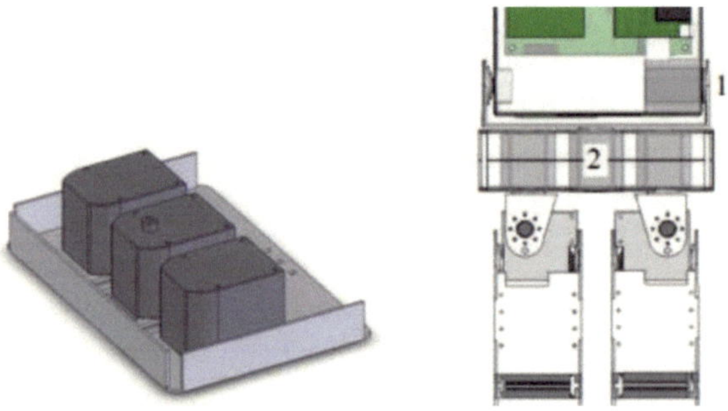

(a) Waist 3D design (b) DOFs diagram

Fig. 4. Waist and trunk mechanism

Each arm of TWNHR-IV has 4 DOFs. Figure 5 shows the 3D mechanism design of the arms. The arms are designed based on the concept of size of the general human arms. The arms of the robot can hold an object such as a ball. Table 3 presents the specification of the joints for each arm.

Human Figure	TWNHR-IV	Human Figure	TWNHR-IV
	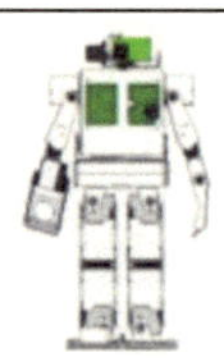		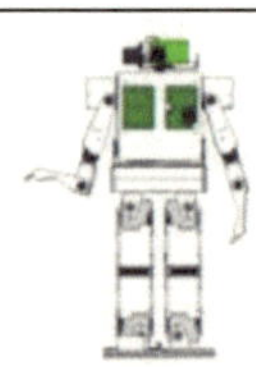

Table 3. The arms DOFs relation between human and TWNHR-IV

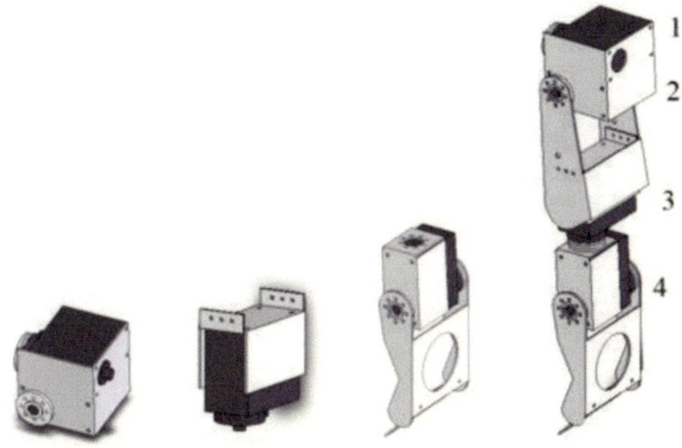

(a) Shoulder (b) Elbow (c) Wrist (d) Arm

Fig. 5. Left arm mechanism

Each leg of TWNHR-IV has 7 Degrees of freedom. Figure 6 shows the 3D mechanism design of the legs. The legs are designed based on the concept that robot can accomplish the human walking motion. Table 4 presents the specification of the joints for each leg.

Human Figure	TWNHR-IV	Human Figure	TWNHR-IV
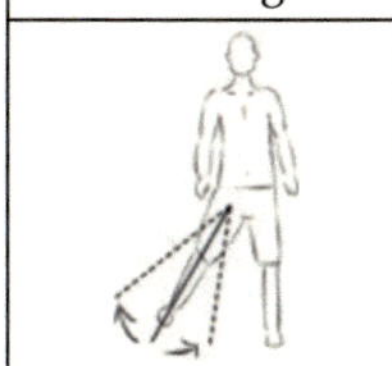		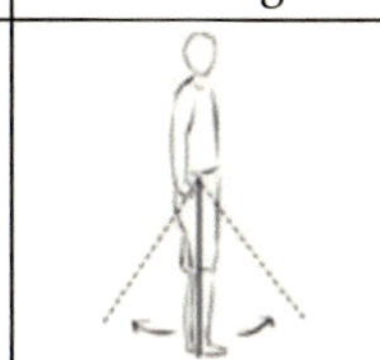	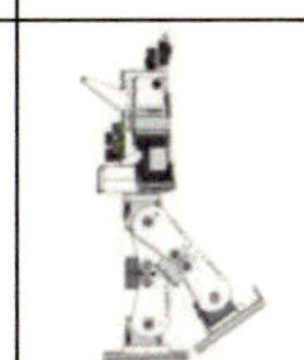
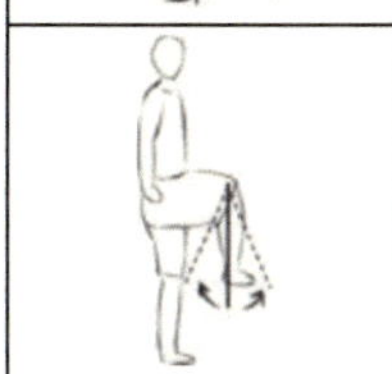		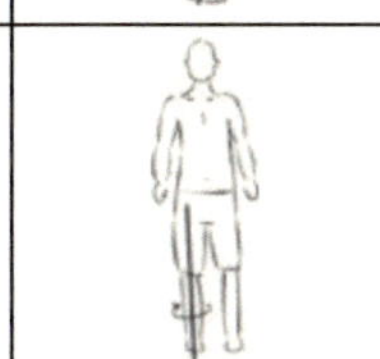	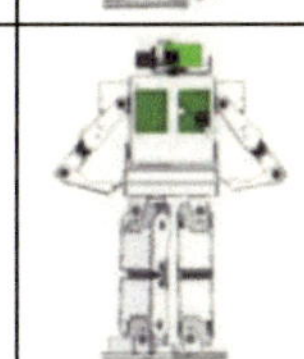
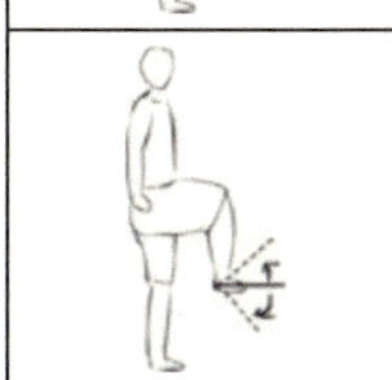	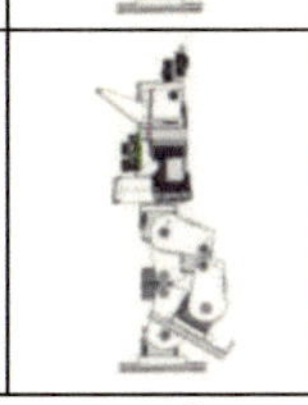		

Table 4. The legs DOFs relation between human and TWNHR-IV

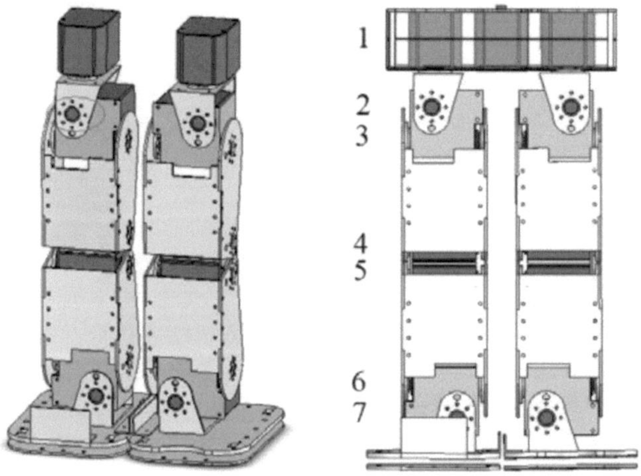

(a) 3D design (b) DOFs diagram

Fig. 6. Legs mechanism

The head of TWNHR-IV has 2 DOFs. The head is designed based on the concept that the head of the robot can accomplish the raw and pitch motion. The trunk of TWNHR-IV has 2 DOFs. The trunk is designed based on the concept that robot can walk to adjust the trunk motions to compensate for the robot's walk motion. Each arm of TWNHR-IV has 4 DOFs. The arms are designed based on the concept of size of the general human arms. The arms of the robot can hold an object such as a ball. Each leg of TWNHR-IV has 7 DOFs. The legs are designed based on the concept that robot can accomplish the human walking motion. The specification is shown in Table 5.

Specification		
Height		43 cm
Weight		3.4 kg
Degree of Freedom & Motor Configuration		
	DOFs	Torque (kg/cm)
Head	2	1.7
Thunk	2	40.8
Legs	7(x2)	37.5
Arms	4(x2)	20
Total	26	

Table 5. Mechanism specification

3. Electronic System

The electronic system diagram is show in Figure 7, where NIOS II is a 32-bit embedded soft-core processor implement on a FPGA chip of a development board. TWNHR-IV is using the NIOS II development board to control all of the servo motors and communicate with sensors. The DSP processor μ'nsp decides motions and gives the NIOS II development board order commands to do such as walk forward, turn right and left. The motions through the RS-232 download to the NIOS II development board.

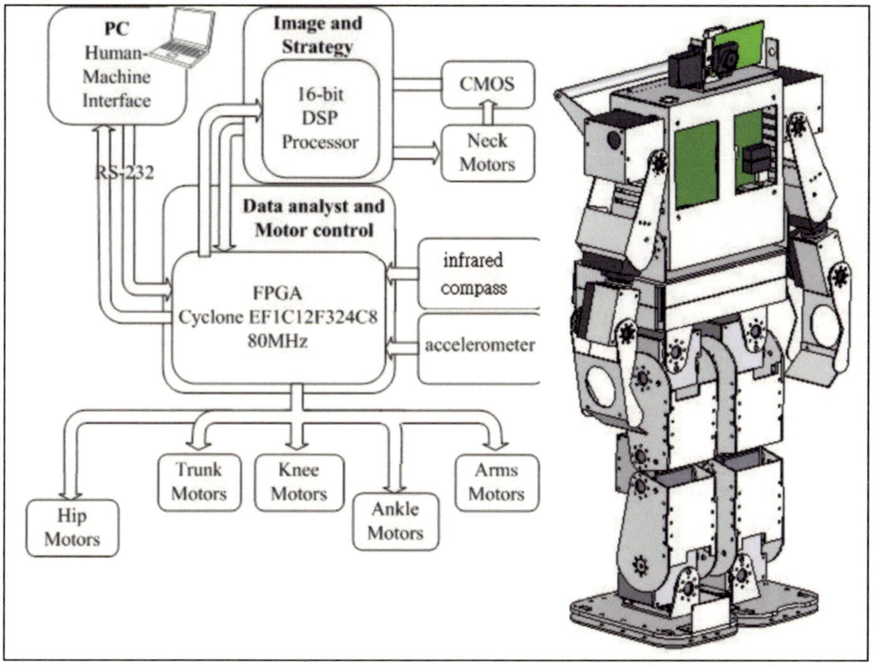

Fig. 7. System block diagram of the electronic system used for TWNHR-IV

3.1 Vision System

A 16-bits DSP processor named μ'nsp is used to receive and process the image data from the CMOS image sensor via the serial transmission. The CMOS sensor is mounted on the head of the robot so that the vision information of the field can be obtained. Two main electrical parts in the vision system of the robot are a CMOS sensor and a 16-bit DSP processor. The captured image data by the CMOS sensor is transmitted to the DSP processor via a serial port. Based on the given color and size of the object, the DSP processor can process the captured image data to determine the position of the object in this image. The noise of the environmental image can be eliminated by the DSP processor. It is shown an example of color image in Figure 8. In this image, two balls are detected. The cross marks in Figure 8 (b) denote center of each color region. Based on the extracted position information, an appropriate strategy is made and transmitted to the FPGA chip via a serial transmission.

(a) Picture from original image

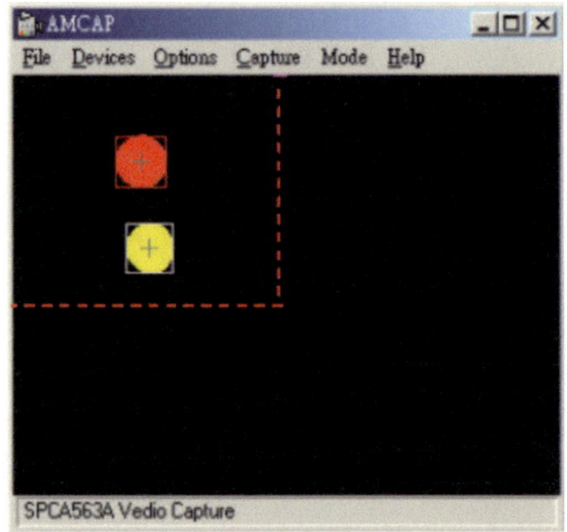

(b) Picture after processing the image

Fig. 8. Color detection from the input image.

3.2 Control Core

The NIOS II development board is used to process the data transmission and motion control. The motions of the robot are stored in the flash memory of NIOS II development board. The internal block diagram of NIOS II development is shown in Figure 9. There are several blocks in NIOS II, such as RS232 Transmission module, Receive module, Data Analysis module, Motion Execution module, Flash Access module, and Motor controller. These blocks are used to download and execute the motions, and these blocks are accomplished by the VHDL language. (Wong et al., 2008a)

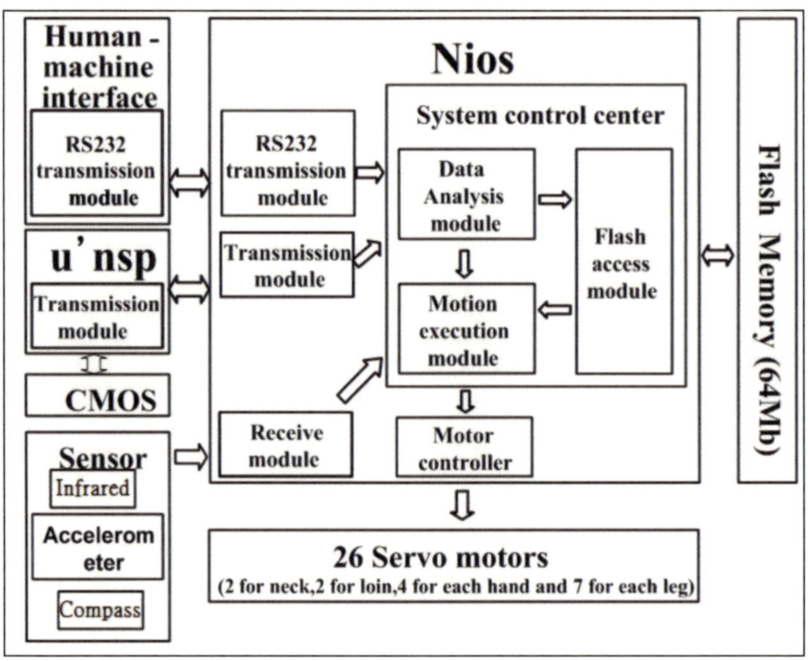

Fig. 9. The internal block diagram of NIOS II development

The motions of the robot are designed on a PC, and downloaded to the RS232 transmission module of the robot. Two different data will be sent to the RS232 transmission module, motion data and motion execution command. The Data analysis module analyzes the data from the RS232 transmission module. If the command is motion data, this data will be sent to the Flash access module and stored in the flash memory. If the command is motion execution, the Motion execution module will get the motion data from the flash memory and execute it. The diagram of flash access is shown in Figure 10.

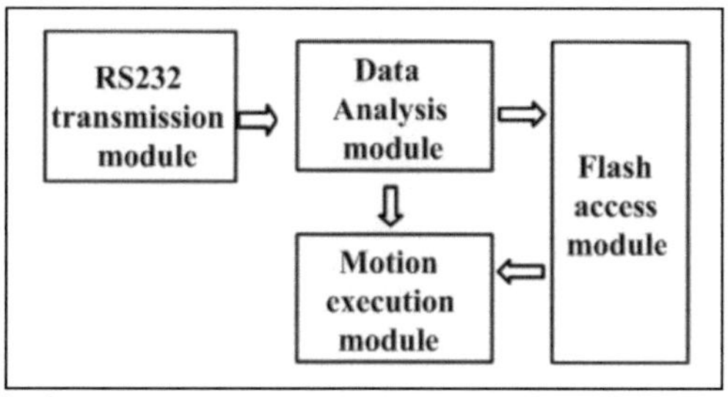

Fig. 10. The diagram of flash access

3.3 Sensors and Application

In order to let TWNHR-IV can go to an appointed place accurately, a digital compass is installed on the body of TWNHR-IV to determine the head direction of the robot. The indispensable magnetism information of the digital compass can provide the important direction information for the robot navigation. The digital compass picks the angle of the robot and the north geomagnetic pole. It can be used in the competition of robot dash, marathon, or others competition.

In order to let TWNHR-IV can avoid the obstacle, six infrared sensors are installed on the robot for detecting the distance between the obstacle and the robot. Two sensors are installed on the back of hands, two on the waist, and two on the legs. The distance information is sent to NIOS II decelopment board by RS232. The infrared is used in the competition of obstacle run, robot dash and weight lifting to avoid the obstacle on the field.

In order to measure the motion state of the robot. An accelerometer is utilized on the body of the robot for the balancing purpose. It provides important motion state information for robot balancing. The accelerometer can offer the indispensable acceleration information of the robot's motion. It is used to detect and examine the acceleration of the robot on the center of robot. This sensor can measure the speed of the earth's axis. The acceleration information is sent to the controller by a serial port. Then the robot can correct its actions by itself. The robot falling down can be detected by this accelerometer so that the robot can autonomously decide to stand up from the ground.

4. Experiments

4.1 Penalty Kick

The penalty kick is one of competitions in HuroSot League of FIRA RoboWorld Cup. In the competition, the robot needs to approach and kick a ball positioned somewhere in the ball area. The robot recognizes the ball by using the CMOS image sensor according to color. In the strategy design of penalty kick, a architecture of decision based on the finite-state

transition mechanism is proposed to solve varied situations in the competition. Figure 11 is the architectonic diagram of penalty kick strategy.
There are three states (Find ball, Track ball, and Shoot ball) in the strategy of penalty kick. According to the information of environment through the CMOS image sensor, the robot can decide what state is the next state, and decide the best behavioral motion via the relative location and distance between the ball and the robot. For example, when the robot can see the ball (Find ball state), then the next state is "Track ball state". But if the robot loses the ball in "Track ball state", the robot state will change to "Find ball state" to search the ball again. The behavioral motions of robot according to the relative location and distance between the ball and the robot are showed in Table 6 and Table 7.

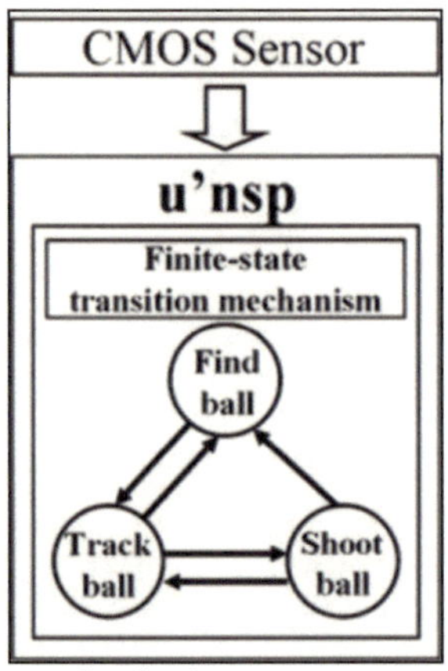

Fig. 11. Architectonic diagram of penalty kick strategy

Relative location between the ball and the robot			
Angle of head in horizontal axis			
Image frame	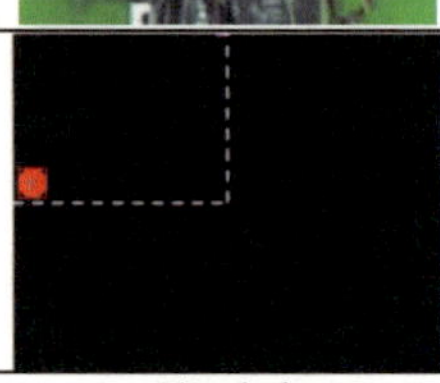		
Behavioral motion	Slip left	Go straight	Slip right

Table 6. The behavioral motion of robot according the location between object and robot

Distance between the ball and the robot			
Angle of head in vertical axis			
Image frame		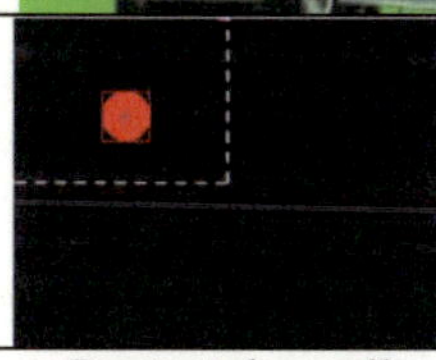	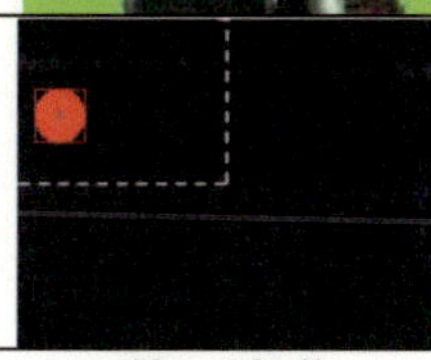
Behavioral motion	Go straigh	Go straigh small	Shoot ball

Table 7. The behavioral motion of robot according the distance between object and robot

Some pictures of TWNHR-IV playing the competition event: Penalty Kick (PK) are shown in Figure 12, where four pictures of TWNHR-IV are described: (a) Search and toward the ball, (b) Search the goal, (c) Kick the ball toward the goal, and (d) Goal. In this competition event, TWNHR-IV can use the CMOS sensor to track the ball and search the goal. The maximum effective distance of the CMOS sensor is 200 cm. When TWNHR-IV kicks the ball, the maximum shooting distance is 250 cm. (Wong et al., 2008b)

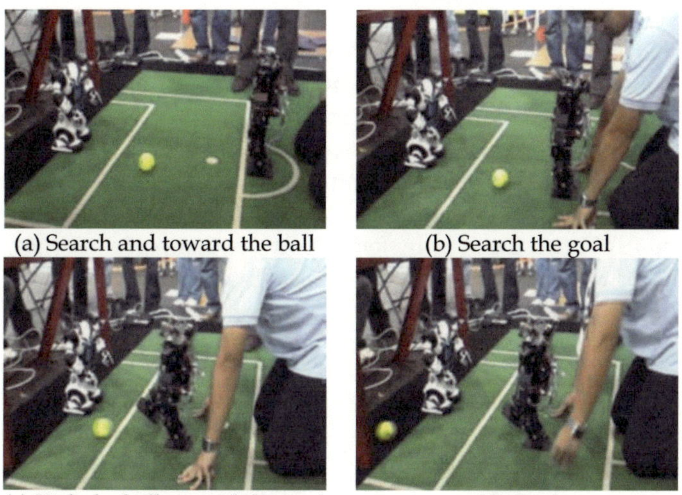

(a) Search and toward the ball (b) Search the goal

(c) Kick the ball toward the goal (d) Goal

Fig. 12. Photographs of TWNHR-IV for the penalty kick

4.2 Basketball

The basketball is one of competitions in HuroSot League of FIRA RoboWorld Cup. In the competition, the robot needs to throw the ball into a red basket. The robot stands in start point and then the robot need to move out of the start area. When the robot move out the start area, the robot could throw the ball into the basket. In the competition of basketball, the robot hold the ball and stay in the start area. When the robot move out the start area, the robot start to search the basket. The robot moves itself to face the baseket, and shoots the ball. Some pictures of TWNHR-IV playing the competition event: the basketball are shown in Figure 13, where we can see four pictures: (a) Search and hold the ball, (b) Slip to the right side, (c) Search the basket, and (d) Shoot. In this competition event, TWNHR-IV can use its arms to hold the ball and shoot the ball. The maximum shooting distance is 45cm.

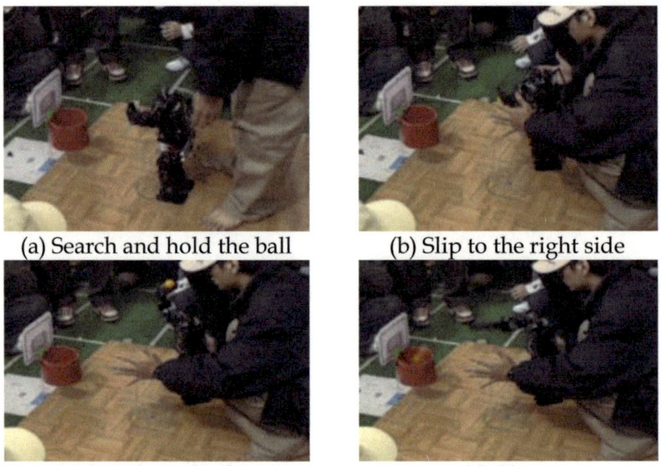

(a) Search and hold the ball (b) Slip to the right side

(c) Search the basket (d) Shoot

Fig. 13. Photographs of TWNHR-IV for the basketball

4.3 Lift-and-Carry

The lift-and-carry is one of competitions in HuroSot League of FIRA RoboWorld Cup. In the competition, the robot needs to carry some batteries and cross an uneven surface. There are four colors on the uneven surface, each color means different height of the field surface. The robot need to cross the field by passsing these color steps.

Some pictures of TWNHR-IV playing the competition event: the lift-and-carry are shown in Figure 14, where we can see four pictures: (a) Lift right leg, (b) Touch the stair by right leg, (c) Lift left leg, and (d) Touch the stair by left leg. In this competition event, TWNHR-IV can use the CMOS sensor to determine these stairs. The maximum crossing height is 2 cm, and the maximum crossing distance is 11 cm.

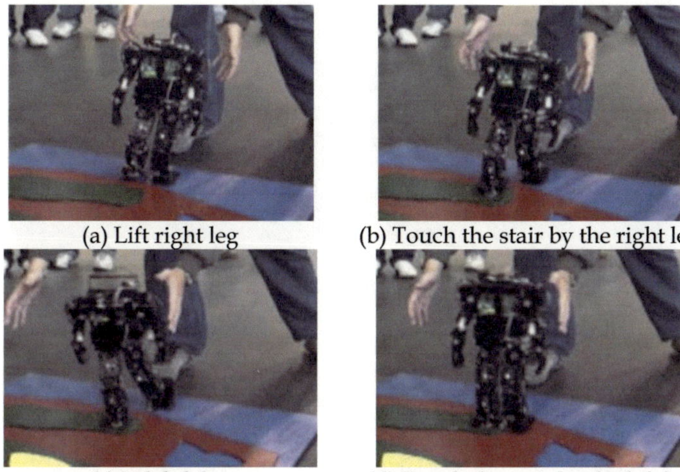

(a) Lift right leg (b) Touch the stair by the right leg

(c) Lift left leg (d) Touch the stair by the left leg

Fig. 14. Photographs of TWNHR-IV for the lift-and-carry

4.4 Obstacle Run

The obstacle run is one of competitions in HuroSot League of FIRA RoboWorld Cup. In the competition, the robot needs to avoid obstacles and arrive the goal area. In this competition, six infrared sensors are installed on the robot to detect the obstacle, as shown in Figure 15. The digital compass sensor is used to correct the head direction of the robot, when the obstacles are detected in safe range, the robot modify its head direction to the goal direction. (Wong et al., 2005; Wong et al., 2007a; Wong et al., 2007b)

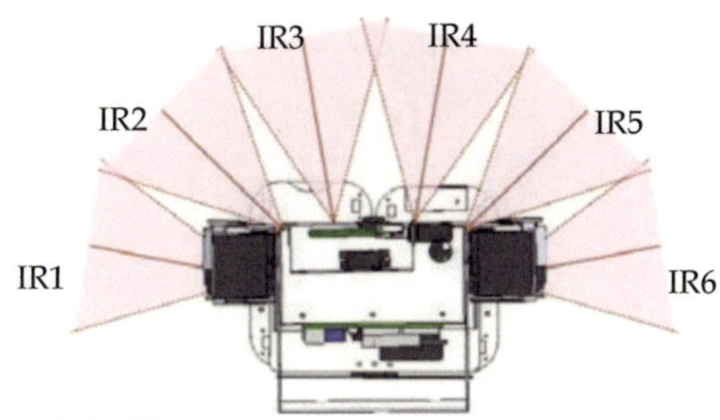

Fig. 15. Detectable range of six IR sensors

Some pictures of TWNHR-IV playing the competition event: the obstacle run are shown in Figure 16, where we can see two pictures: (a) Avoid obstacles, and (b) Move forward. In this competition event, TWNHR-IV can use the CMOS sensor and 6 infrared sensors to detect obstacles. Furthermore, TWNHR-IV can use the digital compass to determine its head direction. The maximum effective distance of the infrared sensor is 150 cm. The digital compass can determine 0 to 360 degree.

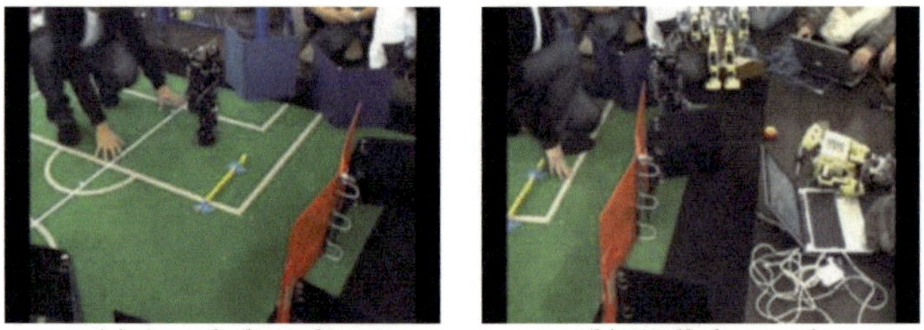

(a) Avoid obstacles (b) Walk forward

Fig. 16. Photographs of TWNHR-IV for the obstacle run

4.5 Robot Dash

The robot dash is one of competitions in HuroSot League of FIRA RoboWorld Cup. In the competition, the robot needs to go forward and backward as soon as possible. The digital compass sensor is used to correct the head direction. As shown in Figure 17, the robot direction is different to the goal direction that detected by a digital compass sensor, the correct motion is executed, that is shown in Table 8. There are three motions module when robot is walking forward. When the goal direction is on the right of the robot, the "Turn Left Forward" is executed, it is indicated that the turn left and forward are executed at the same time. When the goal direction is on the left of the robot, the "Turn Right Forward" is executed, it is indicated that turn left and forward are executed at the same time.. Normally, the "Forward" is executed. In that way, the robot does not waste time to stop and turn.

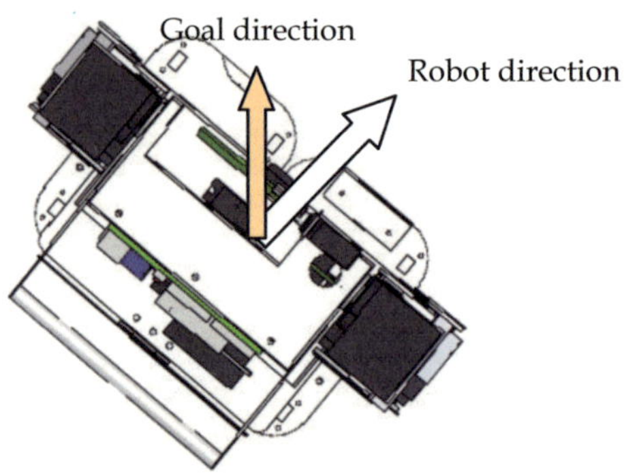

Fig. 17. Description of the relative angle of the goal direction and the robot direction

Turn Left Forward	Forward	Turn Right Forward
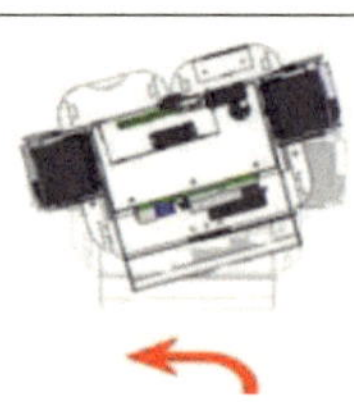	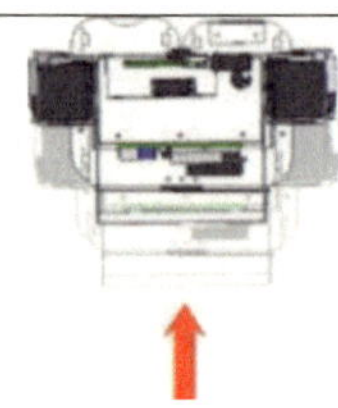	

Table 8. Three motions mode

Some pictures of TWNHR-IV playing the competition event: the robot dash are shown in Figure 18, where we can see two pictures: (a) Walk forward, and (b) Walk backward. In this competition event, TWNHR-IV can use the 2 infrared sensors to detect the wall. If the infrared sensors detect the wall, TWNHR-IV will change the motion of walk forward to the motion of walk backward. The walking velocity of TWNHR-IV is 12 cm per second.

(a) Walk forward (b) Walk backward

Fig. 18. Photographs of TWNHR-IV for the robot-dash

4.6 Weight Lifting

The weight lifting is one of competitions in HuroSot League of FIRA RoboWorld Cup. In the competition, the robot needs to go forward and lift the weight. The mechanism of TWNHR-IV is challenged in this competition, it also needs stable walking to complete.

Some pictures of TWNHR-IV playing the competition event: the weight lifting are shown in Figure 19, where we can see four pictures: (a) Hold the dumbbells, (b) Walk forward 15 cm, (c) Lift up the dumbbells, and (d) Walk forward 15 cm. In this competition event, TWNHR-IV can use infrared sensors to detect the wall. If the infrared sensors detect the wall, TWNHR-IV will lift up the discs and walk forward 15 cm. The maximum disc number lifted by TWNHR-IV is 43.

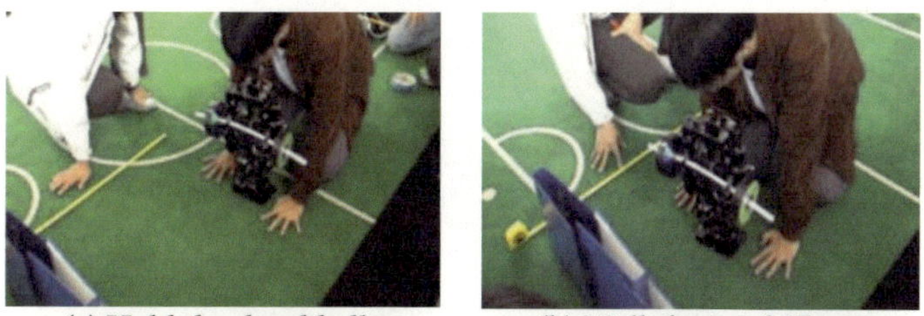

(a) Hold the dumbbells (b) Walk forward 15 cm

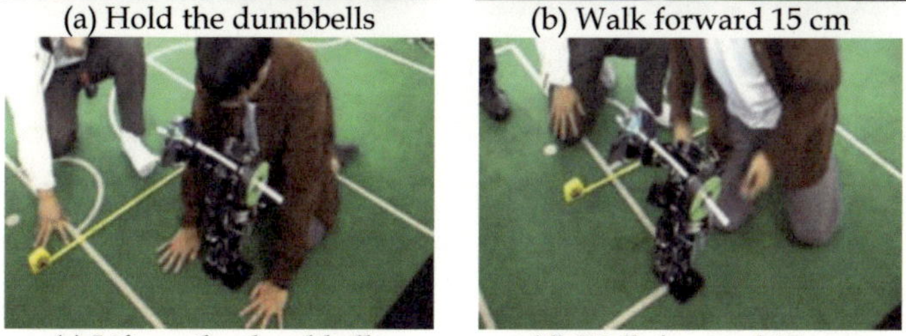

(c) Lift up the dumbbells (d) Walk forward 15 cm

Fig. 19. Photographs of TWNHR-IV for the weight lifting

4.7 Marathon

The marathon is one of competitions in HuroSot League of FIRA RoboWorld Cup. In the competition, the robot needs to track a visible line for a distance of 42.195 m (1/1000 of a human marathon distance) and finish it as quickly as possible. The robot recognizes the visible line by using the CMOS image sensor. From Figure 20, the image which we select is circumscribed by a white frame. (Wong et al., 2008d)

Every selected object is marked by a white fram. Through the CMOS image sensor, we can obtain the digital image data in the visible line. From Figure 20, P_{End} is the terminal point in the white frame and P_{Sta} is the starting point in the white frame. We transform the coordinate system from the white frame to the full image. The pixel of the full image are 120×160. According to the relation between the coordinate of the terminal point (P_{End}) and the coordinate of the starting point (P_{Sta}), we can obtain the trend of the visible line from one field to another field. The humanoid soccer robot decides the best appropriate strategy for the movement.

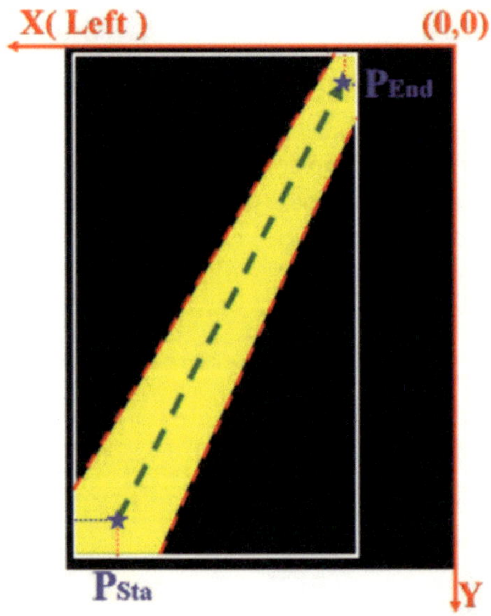

Fig. 20. Reference points of image information in the relative coordinate

Some pictures of TWNHR-IV playing the competition event: the marathon is shown in Figure 21, where we can see the TWNHR-IV follow the line. In this competition event, TWNHR-IV can use the CMOS sensor to follow the blue line and walk about 1 hour.

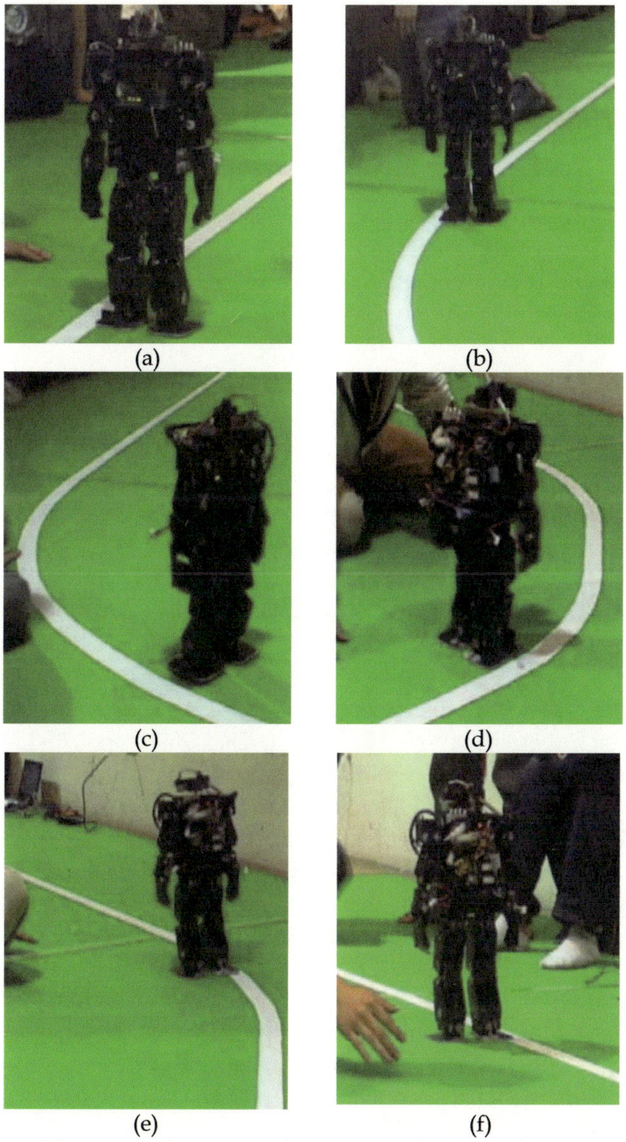

Fig. 21. A vision-based humanoid soccer robot can walk along the white line autonomously in the competition of marathon

5. Conclusions

A humanoid soccer robot named TWNHR-IV is presented. A mechanical structure is proposed to implement a humanoid robot with 26 degrees of freedom in this chapter. This robot has 2 degrees of freedom on the head, 2 degrees of freedom on the trunk, 4 degrees of freedom on each arm, and 7 degrees of freedom on each leg. A CMOS sensor, a digital compass, an accelerometer, and six infrared sensors are equipped on the body of TWNHR-IV to obtain the information of the environment. A CMOS sensor is installed on the head of the humanoid robot so that it can find the ball, track the line, and others image process. A digital compass is installed to detect the direction of the robot. An accelerometer is installed in the body of the humanoid robot so that it can detect the posture of robot. The implemented TWNHR-IV can autonomously detect objects, avoid obstacles, cross uneven surface, and walk to a destination. TWNHR-IV joined FIRA HuroSot 2007 and 2008, and won 7 times first place, 3 times second place, and one third place in different competitions. In the future, the localization ability is going to build in TWNHR-IV. Besides, fuzzy system and some evolutional algorithms such as GA and PSO will be considered to improve the speed of the motions and the movements.

6. References

Chemori, A. & Loria, A. (2004). Control of a planar underactuated biped on a complete walking cycle, *IEEE Transactions on Automatic Control,* Vol.49, Issue 5, May 2004, pp. 838-843, ISBN 0018-9286.

Esfahani, E. T. & Elahinia, M. H. (2007). Stable walking pattern for an SMA-actuated biped, *IEEE/ASME Transactions on Mechatronics,* Vol. 12, Issue 5, Oct. 2007, pp. 534-541, ISBN 1083-4435.

Guan, Y.; Neo, E.S.; Yokoi, K. & Tanie, K. (2006). Stepping over obstacle with humanoid robot, *IEEE Transaction on Robotics*, Vol. 22, Oct. 2006, pp. 958-973, ISBN 1552-3098.

Hemami, H.; Barin, K. & Pai, Y.C. (2006). Quantitative analysis of the ankle strategy under translational platform disturbance, *IEEE Transactions on Neural Systems and Rehabilitation Engineering,* Vol. 14, Issue 4, Dec. 2006, pp. 470-480, ISBN 1534-4320.

Hu, L.; Zhou C. & Sun Z. (2008). Estimating biped gait using spline-based probability distribution function with Q-learning, *IEEE Transactions on Industrial Electronics*, Vol. 55, Issue 3, Mar. 2008, pp. 1444-1452, ISBN 0278-0046.

Huang, Q.; Li, K. & Nakamura, Y. (2001). Humanoid walk control with feedforward dynamic pattern and feedback sensory reflection, *IEEE International Symposium on Computational intelligence in Robotics and Automation*, pp. 29-34, ISBN 0-7803-7203-4, Aug. 2001.

Miyazaki, F. & Arimoto S. (1980). A control theoretic study on dynamical biped locomotion, *ASME J. Dyna. Syst. Meas. Contr.*, Vol.102, pp.233-239, 1980.

Pauk J. H. & Chung H. (1999). ZMP compensation by on-line trajectory generation for biped robots, *IEEE International Conference on Systems, Man, and Cybernetics,* Vol.4, pp. 960-965, ISB 0-7803-5731-0, Oct. 1999.

Sugihara, T.; Nakamura, Y. & Inoue, H. (2002). Real time humanoid motion generation through ZMP manipulation based on inverted pendulum control, *IEEE International Conference on Robotics and Automation,* Vol.2, pp. 1404-1409, ISBN 0-7803-7272-7, 2002.

Wong, C.C.; Cheng, C.T.; Wang, H.Y.; Li, S.A.; Huang, K.H.; Wan, S.C. Yang, Y.T.; Hsu, C.L.; Wang, Y.T.; Jhou, S.D.; Chan, H.M.; Huang, J.C.; Hu, Y.Y. (2005). Description of TKU-PaPaGo team for humanoid league of RoboCup 2005, *RoboCup International Symposium*, 2005.

Wong, C.C.; Cheng, C.T.; Huang, K.X.; Wu, H.C.; Hsu, C.L.; Yang, Y.T.; Wan, S.C.; Chen, L.C. & Hu, Y.Y. (2007a). Design and implementation of humanoid robot for obstacle avoidance, *FIRA Robot World Congress*, San Francisco, USA, 2007.

Wong, C.C.; Cheng, C.T.; Huang, K.H.; Yang, Y.T. & Chan, H.M. (2007b). Humanoid robot design and implementation for obstacle run competition of FIRA Cup, *2007 CACS International Automatic Control Conference*, pp. 876-881, Nov. 2007.

Wong, C.C.; Cheng, C.T. & Chan, H.M. (2008a). TWNHR-IV: Humanoid soccer robot, *5th International Conference on Computational Intelligence, Robotics and Autonomous Systems* (CIRAS 2008), pp. 173-177, Jun. 2008.

Wong, C.C. ; Cheng, C.T. ; Huang, K.H. ; Yang, Y.T.; Hu, Y.Y.; Chan, H.M. & Chen, H.C. (2008b). Humanoid soccer robot: TWNHR-IV, *Journal of Harbin Institute of Technology*, Vol.15, Sup. 2, Jul. 2008, pp. 27-30, ISSN 1005-9113.

Wong, C.C.; Cheng, C.T.; Huang, K.H.; Yang, Y.T.; Chan, H.M.; Hu, Y.Y. & Chen, H.C. (2008c). Mechanical design of small-size humanoid robot: TWNHR-IV, *Journal of Harbin Institute of Technology*, Vol.15, Sup. 2, Jul. 2008, pp. 31-34, ISSN 1005-9113.

Wong, C.C.; Huang, K.H.; Yang, Y.T.; Chan, H.M.; Hu, Y.Y.; Chen, H.C.; Hung, C.H. & Lo, Y.W. (2008d). Vision-based humanoid soccer robot design for marathon, *2008 CACS International Automatic Control Conference*, pp. 27-29, ISBN 978-986-84845-0-4, Nov. 2008.

Zhou, C. & Jagannathan, K. (1996). Adaptive network based fuzzy of a dynamic biped walking control robot, *IEEE Int. Conf. on Robotics and Automation*, pp. 109-116, ISBN 0-8186-7728-7, Nov. 1996.

4

Humanoid soccer player design

Francisco Martín, Carlos Agüero, José María Cañas and Eduardo Perdices
Rey Juan Carlos University
Spain

1. Introduction

The focus of robotic research continues to shift from industrial environments, in which robots must perform a repetitive task in a very controlled environment, to mobile service robots operating in a wide variety of environments, often in human-habited ones. There are robots in museums (Thrun et al, 1999), domestic robots that clean our houses, robots that present news, play music or even are our pets. These new applications for robots make arise a lot of problems which must be solved in order to increase their autonomy. These problems are, but are not limited to, navigation, localisation, behavior generation and human-machine interaction. These problems are focuses on the autonomous robots research.

In many cases, research is motivated by accomplishment of a difficult task. In Artificial Intelligence research, for example, a milestone was to win to the chess world champion. This milestone was achieved when *deep blue* won to Kasparov in 1997. In robotics there are several competitions which present a problem and must be solved by robots. For example, Grand Challenge propose a robotic vehicle to cross hundred of kilometers autonomously. This competition has also a urban version named Urban Challenge.

Fig. 1. Standard Platform League at RoboCup.

Our work is related to RoboCup. This is an international initiative to promote research on the field of Robotics and Artificial Intelligence. This initiative proposes a very complex problem, a soccer match, in which several techniques related to these field can be tested, evaluated and compared. The long term goal of the RoboCup project is, by 2050, develop a team of fully autonomous humanoid robots that can win against the human world champion team in soccer.

This work is focused on the *Standard Platform League*. In this league, all the teams use the same robot and changes in hardware are not allowed. This is the key factor that makes that the efforts concentrate on the software aspects rather than in the hardware. This is why this league is known as The *Software League*. Until 2007, the chosen robot to play in this league was Aibo robot. But since 2008 there is a new platform called Nao (figure 1). Nao is a biped humanoid robot, this is the main difference with respect Aibo that is a quadruped robot. This fact has had a big impact in the way the robot moves and its stability while moving. Also, the sizes of both robots is not the same. Aibo is 15 cm tall while Nao is about 55 cm tall. That causes the big difference on the way of perception. In addition to it, both robots use a single camera to perceive. In Aibo the perception was 2D because the camera was very near the floor. Robot Nao perceives in 3D because the camera is at a higher position and that enables the robot to calculate the position of the elements that are located on the floor with one single camera.

Many problems have to be solved before having a fully featured soccer player. First of all, the robot has to get information from the environment, mainly using the camera. It must detect the ball, goals, lines and the other robots. Having this information, the robot has to self-localise and decide the next action: move, kick, search another object, etc. The robot must perform all these tasks very fast in order to be reactive enough to be competitive in a soccer match. It makes no sense within this environment to have a good localisation method if that takes several seconds to compute the robot position or to decide the next movement in few seconds based on the old percerpetion. The estimated sense-think-act process must take less than 200 millisecond to be truly eficient. This is a tough requirement for any behavior architecture that wishes to be applied to solve the problem.

With this work we are proposing a behavior based architecture that meets with the requirements needed to develop a soccer player. Every behavior is obtained from a combination of reusable components that execute iteratively. Every component has a specific function and it is able to activate, deactivate o modulate other components. This approach will meet the vivacity, reactivy and robustness needed in this environment. In this chapter we will show how we have developed a soccer player behavior using this architecture and all the experiments carried out to verify these properties.

This paper is organised as follows: First, we will present in section 2 all relevant previous works which are also focused in robot behavior generation and following behaviors. In section 3, we will present the Nao and the programming framework provided to develop the robot applications. This framework is the ground of our software. In section 4, the behavior based architecture and their properties will be described. Next, in section 5, we will describe how we have developed a robotic soccer player using this architecture. In section 6, we will introduce the experiment carried out to test the proposed approach and also the robotic soccer player. Finally, section 7 will be the conclusion.

2. Related work

In this section, we will describe the previous works which try to solve the robot behavior generation and the following behaviors. First of all, the classic approaches to generate robot behaviors will be described. These approaches have been already successfully tested in wheeled robots. After that, we will present other approaches related to the RoboCup domain. To end up, we will describe a following behavior that uses an approach closely related to the one used in this work.

There are many approaches that try to solve the behavior generation problem. One of the first successful works on mobile robotics is Xavier (Simmons et al, 1997). The architecture used in these works is made out of four layers: obstacle avoidance, navigation, path planning and task planning. The behavior arises from the combination of these separate layers, with an specific task and priority each. The main difference with regard to our work is this separation. In our work, there are no layers with any specific task, but the tasks are broken into components in different layers.

Another approach is (Stoytchev & Arkin, 2000), where a hybrid architecture, which behavior is divided into three components, was proposed: deliberative planning, reactive control and motivation drives. Deliberative planning made the navigation tasks. Reactive control provided with the necessary sensorimotor control integration for response reactively to the events in its surroundings. The deliberative planning component had a reactive behavior that arises from a combination of schema-based motor control agents responding to the external stimulus. Motivation drives were responsible of monitoring the robot behavior. This work has in common with ours the idea of behavior decomposition into smaller behavioral units. This behavior unit was explained in detail in (Arkin, 2008).

In (Calvo et al, 2005) a follow person behavior was developed by using an architecture called JDE (Cañas & Matellán, 2007). This reactive behavior arises from the activation/deactivation of components called schemes. This approach has several similarities with the one presented in this work.

In the RoboCup domain, a hierarchical behavior-based architecture was presented in (Lenser et al, 2002). This architecture was divided in several levels. The upper levels set goals that the bottom level had to achieve using information generated by a set of virtual sensors, which were an abstraction of the actual sensors.

Saffiotti (Saffiotti & Zbigniew, 2003) presented another approach in this domain: the *ThinkingCap* architecture. This architecture was based in a fuzzy approach, extended in (Gómez & Martínez, 1997). The perceptual and global modelling components manage information in a fuzzy way and they were used for generating the next actions. This architecture was tested in the four legged league RoboCup domain and it was extended in (Herrero & Martínez, 2008) to the Standar Platform League, where the behaviors were developed using a LUA interpreter. This work is important to the work presented in this paper because this was the previous architecture used in our RoboCup team.

Many researches have been done over the Standar Platform League. The B-Human Team (Röfer et al, 2008) divides their architecture in four levels: perception, object modelling, behavior control and motion control. The execution starts in the upper level which perceives the environment and finishes at the low level which sends motion commands to actuators. The behavior level was composed by several basic behavior implemented as finite state machines. Only one basic behavior could be activated at same time. These finite state machine was written in XABSL language (Loetzsch et al, 2006), that was interpreted at runtime and let change and reload the behavior during the robot operation. A different approach was presented by Cerberus Team (Akin et al, 2008), where the behavior generation is done using a four layer planner model, that operates in discrete time steps, but exhibits continuous behaviors. The topmost layer provides a unified interface to the planner object. The second layer stores the different roles that a robot can play. The third layer provides behaviors called "Actions", used by the roles. Finally, the fourth layer contains basic skills, built upon the actions of the third layer.
The behavior generation decomposition in layers is widely used to solve the soccer player problem. In (Chown et al 2008) a layered architecture is also used, but including coordination among the robots. They developed a decentralized dynamic role switching system that obtains the desired behavior using different layers: strategies (the topmost layer), formations, roles and sub-roles. The first two layers are related to the coordination and the other two layers are related to the local actions that the robot must take.

3. Nao and NaoQi framework

The behavior based architecture proposed in this work has been tested using the Nao robot. The applications that run in this robot must be implemented in software. The hardware cannot be improved and all the work must be focused in improving the software. The robot manufacturer provides an easy way to access to the hardware and also to several high level functions, useful to implement the applications.

Our soccer robot application uses some of the functionality provided by this underlying software. This software is called NaoQi[1] and provides a framework to develop applications in C++ and Python.

NaoQi is a distributed object framework which allows to several distributed binaries be executed, all of them containing several software modules which communicate among them. Robot functionality is encapsulated in software modules, so we can communicate to specific modules in order to access sensors and actuators.

1 http://www.aldebaran-robotics.com/

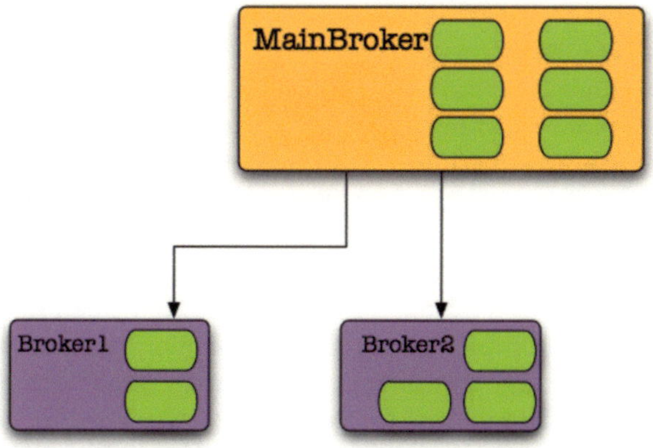

Fig. 2. Brokers tree.

Every binary, also called *broker*, runs independently and is attached to an address and port. Every broker is able to run both in the robot (cross compiled) and the computer. Then we are able to develop a complete application composed by several brokers, some running in a computer and some in the robot, that communicate among them. This is useful because high cost processing tasks can be done in a high performance computer instead of in the robot, which is computationally limited.

The broker's functionality is performed by modules. Each broker may have one or more modules. Actually, brokers only provide some services to the modules in order to accomplish their tasks. Brokers deliver call messages among the modules, subscription to data and so on. They also provide a way to solve module names in order to avoid specifying the address and port of the module.

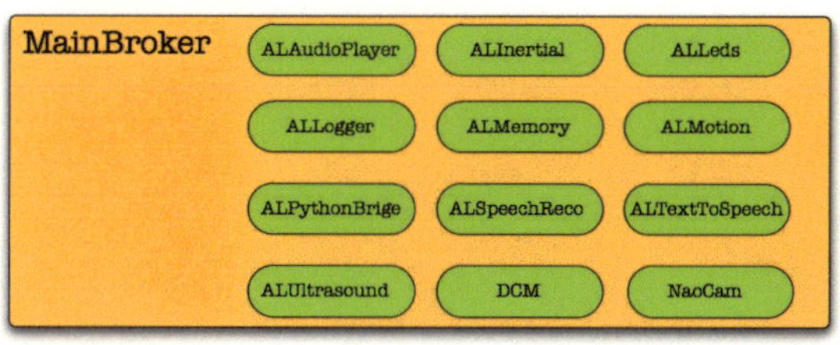

Fig. 3. Modules within MainBroker.

A set of brokers are hierarchically structured as a tree, as we can see in figure 2. The most important broker is the *MainBroker*. This broker contains modules to access to robot sensors and actuators and other modules provide some interesting functionality (figure 3). We will describe some of the modules intensively used in this work:

- The main information source our application is the camera. The images are fetched by *ALVideoDevice* module. This module uses the Video4Linux driver and makes the images available for any module that create a proxy to it, as we can observe in figure 4. This proxy can be obtained locally or remotelly. If locally, only a reference

to the data image is obtained, but if remotelly all the image data must be encapsulated in a SOAP message and sent over the network.
To access the image, we can use the normal mode or the direct raw mode. Figure 5 will help to explain the difference. Video4Linux driver maintains in kernel space a buffer where it stores the information taken from the camera. It is a round robin buffer with a limited capacity. NaoQi unmaps one image information from Kernel space to driver space and locks it. The difference in the modes comes here. In normal mode, the image transformations (resolution and color space) are applied, storing the result and unlocking the image information. This result will be accessed, locally or remotelly, by the module interested in this data. In direct raw mode, the locked image information is available (only locally and in native color space, YUV422) to be accessed by the module interested in this data. This module should manually unlock the data before the driver in kernel space wants to use this buffer position (around 25 ms). Fetching time varying depending on the desired color space, resolution and access mode, as we can see in figure 6.

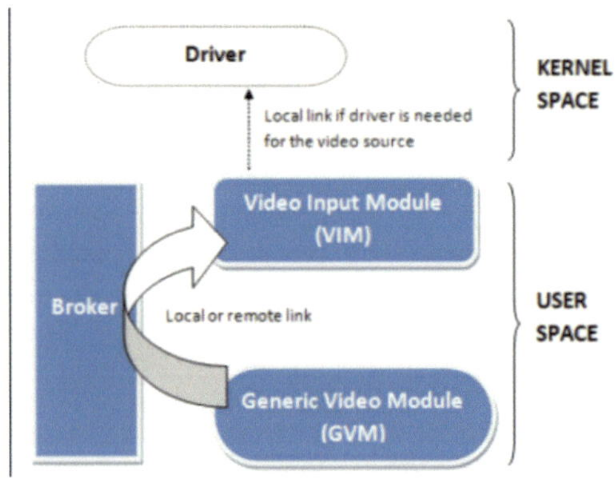

Fig. 4. NaoQi vision architecture overview.

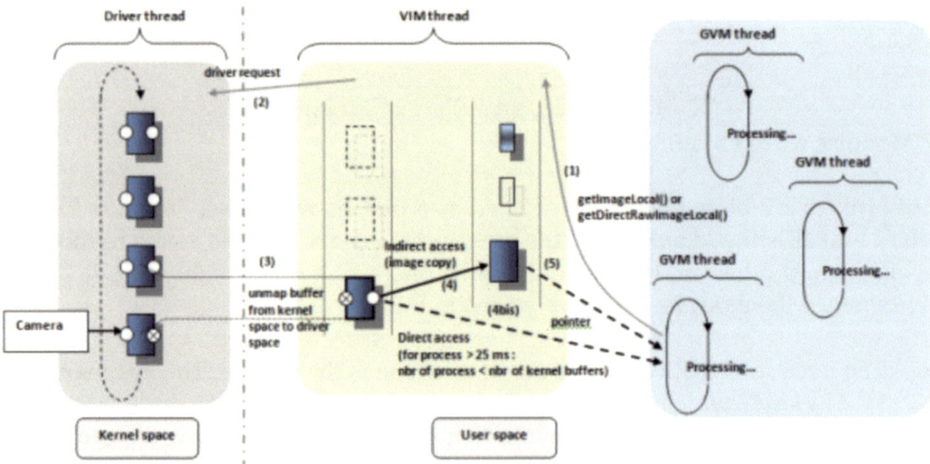

Fig. 5. Access to an image in the NaoQi framework.

Fetching time	RGB	HSV	YUV	YUV422 Interlaced	YUV422 Interlaced RAW
KQQVGA (160x120)	26617 usec	16489 usec	10524 usec	4042 usec	1150 usec
KQVGA (320x240)	78697 usec	39055 usec	12425 usec	8674 usec	1465 usec
KVGA (640x480)	558900 usec	162840 usec	52697 usec	31597 usec	1591 usec

Fig. 6. Access time to the image depending on resolution and space color. Last column is *direct raw* mode.

- In order to move the robot, NaoQi provides the *ALMotion* module. This module is responsible for the actuators of the robot. This module's API let us move a single joint, a set of joints or the entire body. The movements can be very simple (p.e. set a joint angle with a selected speed) or very complex (walk a selected distance). We use these high level movement calls to make the robot walk, turn o walk sideways. As a simple example, the `walkStraight` function is:

```
void walkStraight (float distance, int pNumSamplesPerStep)
```

  This function makes the robot walk straight a `distance`. If a module, in any broker, wants to make the robot walk, it has to create a proxy to the *ALMotion* module. Then, it can use this proxy to call any function of the *ALMotion* module.

  The movement generation to make the robot walk is a critical task that NaoQi performs. The operations to obtain each joint position are critical. If these real time operations miss the deadlines, the robot may lost the stability and fall down.

- NaoQi provides a thread-safe module for information sharing among modules, called *ALMemory*. By its API, modules write data in this module, which are read by any module. NaoQi also provides a way to subscribe and unsubscribe to any data in *ALMemory* when it changes or periodically, selecting a class method as a callback to manage the reception. Besides this, *ALMemory* also contains all the information related to the sensors and actuators in the system, and other information. This module can be used as a *blackboard* where any data produced by any module is published, and any module that needs a data reads from *ALMemory* in order to obtain it.

As we said before, each module has an API with the functionality that it provides. Brokers also provide useful information about their modules and their APIs via *web services*. If you use a browser to connect to any broker, it shows all the modules it contains, and the API of each one.

When a programmer develops an application composed by several modules, she decides to implement it as a dynamic library or as a binary (broker). In the dynamic library (like a plug-in) way, the modules that it contains can be loaded by the *MainBroker* as its own modules. Using this mechanism the execution speeds up, from point of the view of communication among modules. As the main disadvantage, if any of the modules crashes, then *MainBroker* also crashes, and the robot falls to the floor. To develop an application as a separate broker makes the execution safer. If the module crashes, only this module is affected.

The use of NaoQi framework is not mandatory, but it is recommended. NaoQi offers high and medium level APIs which provide all the methods needed to use all the robot's functionality. The movement methods provided by NaoQi send low level commands to a microcontroller allocated in the robot's chest. This microcontroller is called DCM and is in charge of controlling the robot's actuators. Some developers prefer (and the development framework allows it) not to use NaoQi methods and use directly low level DCM functionality instead. This is much laborious, but it takes absolute control of robot and allows to develop an own walking engine, for example.

Nao robot is a fully programmable humanoid robot. It is equipped with a x86 AMD Geode 500 Mhz CPU, 1 GB flash memory, 256 MB SDRAM, two speakers, two cameras (non stereo), Wi-fi connectivity and Ethernet port. It has 25 degrees of freedom. The operating system is Linux 2.6 with some real time patches. The robot is eqquiped with a microcontroller ARM 7 allocated in its chest to controll the robot's motors and sensors, called DCM.

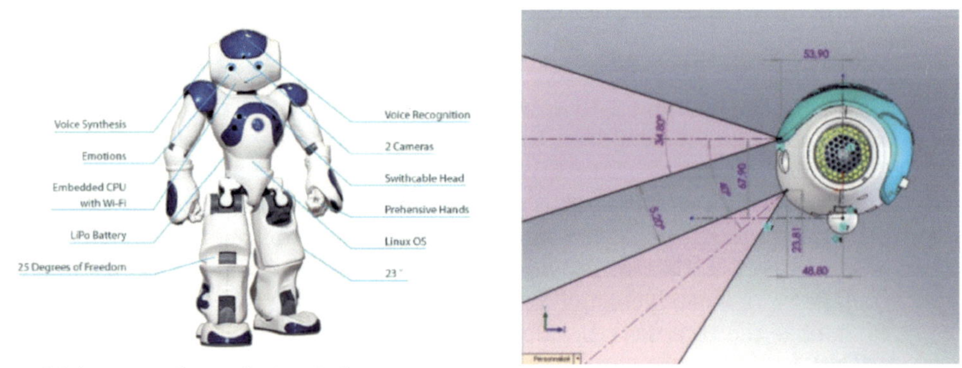

Fig. 7. Aldebaran Robotics' Nao Robot.

These features impose some restrictions to our behavior based architecture design. The microprocessor is not very powerful and the memory is very limited. These restrictions must be taken into account to run complex localization or sophisticathed image processing algorithms. Moreover, the processing time and memory must be shared with the OS itself (an GNU/Linux embedded distribution) and all the software that is running in the robot, including the services that let us access to sensors and motors, which we mentioned before. Only the OS and all this software consume about 67% of the total memory available and 25% of the processing time.

The robot hardware design also imposes some restrictions. The main restriction is related to the two cameras in the robot. These cameras are not stereo, as we can observe in the right side of the figure 7. Actually, the bottom camera was included in the last version of the robot after RoboCup 2008, when the robot designer took into account that it was difficult track the ball with the upper camera (the only present at that time) when the distance to the ball was less than one meter. Because of this non stereo camera characteristic, we can't estimate elements position in 3D using two images of the element, but supposing some other characteristics as the heigh position, the element size, etc.

Besides of that, the two cameras can't be used at same time. We are restricted to use only one camera at the time, and the switching time is not negligible (about 70 ms). All these restrictions have to taken into account when designing our software.

The software developed on top of NaoQi can be tested both in real robot and simulator. We use Webots (figure 8) (MSR is also available) to test the software as the first step before testing it in the real robot. This let us to speed up the development and to take care of the real robot, whose hardware is fragile.

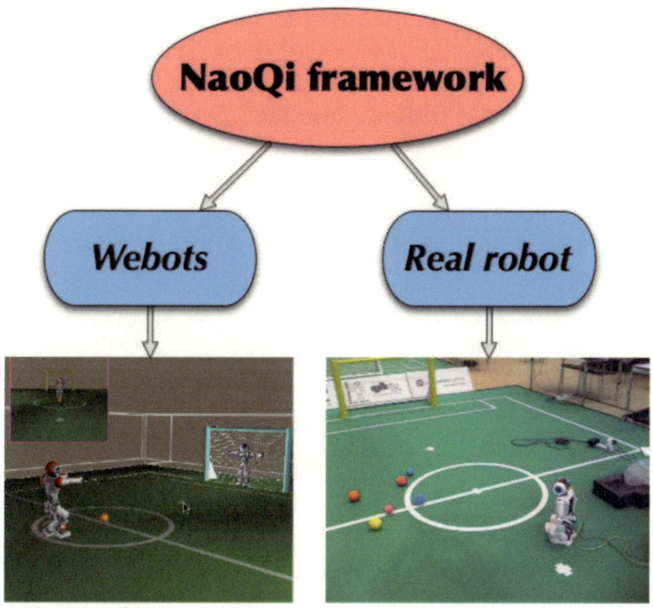

Fig. 8. Simulated and real robot.

4. Behavior based architecture for robot applications

The framework we presented in the last section provides useful functionality to develop a software architecture that makes a robot perform any task. We can decompose the functionality in modules that communicate among them. This framework also hides almost all the complexity of movement generation and makes easy to access sensors (ultrasound, camera, bumpers…) and actuators (motors, color lights, speaker…).

It is possible to develop basic behaviors using only this framework, but it is not enough for our needs. We need an architecture that let us to activate and deactivate components, which is more related to the cognitive organization of a behavior based system. This is the first step to have a wide variety of simple applications available. It's hard to develop complex applications using NaoQi only.

In this section we will describe the design concepts of the robot architecture we propose in this chapter. We will address aspects such as how we interact with NaoQi software layer, which of its functionality we use and which not, what are the elements of our architecture, how they are organized and timing related aspects.

The main element in the proposed architecture is the *component*. This is the basic unit of functionality. In any time, each component can be active or inactive. This property is set using the start/stop interface, as we can observe in figure 6. When it is active, it is running and performing a task. When inactive, it is stopped and it does not consume computation resources. A component also accepts modulations to its actuation and provides information of the task it is performing.

For example, lets suppose a component whose function is perceive the distance to an object using the ultrasound sensors situated in the robot chest. The only task of this component is to detect, using the sensor information, if a obstacle is in front of the robot, on its left, on its right or there is not obstacle in a distance less than *D* mm. If we would like to use this functionality, we have to activate this component using its start/stop interface (figure 9). We may modulate the *D* distance and ask whenever we want what is this component output (front, left, right or none). When this is information is no longer needed, we may deactivate this component to stop calculating the obstacle position, saving valuable resources.

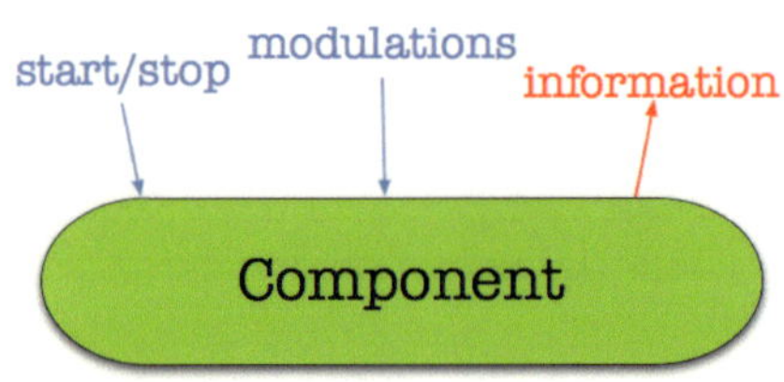

Fig. 9. Component inputs and outputs.

A component, when active, can activate another components to achieve its goal, and these components can also activate another ones. This is a key idea in our architecture. This let to decompose funtionality in several components that work together. An application is a set of components which some of them are activated and another ones are deactivated. The subset of the components that are activated and the activation relations are called *activation tree*. In figure 10 there is an example af an *activation tree*. When component A, the root component, is activated, it activates component B and E. Component B activates C and D. Component A needs all these components actived to achieve its goal. This estructure may change when a component is modulated and decides to stop a component and activate another more adequate one. In this example, component A does not need to know that B has activated C and D. The way component B performs its task is up to it. Component A is only interested in the component B and E execution results.

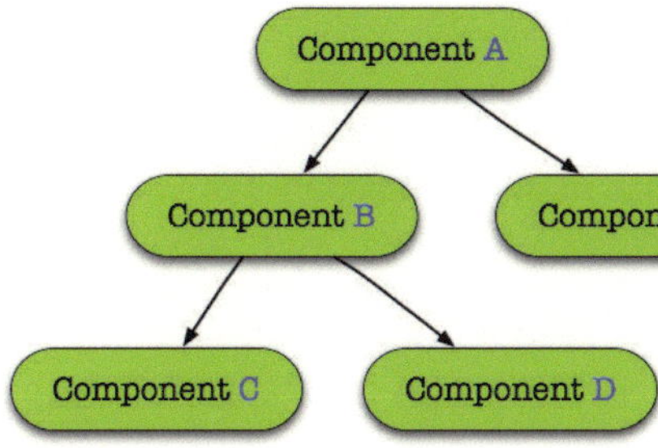

Fig. 10. *Activation tree* composed by several components.

Two differents components are able to activate the same child component, as we can observe in figure 11. This property lets two components to get the same information from a component. Any of them may modulate it, and the changes affect to the result obtained in both component.

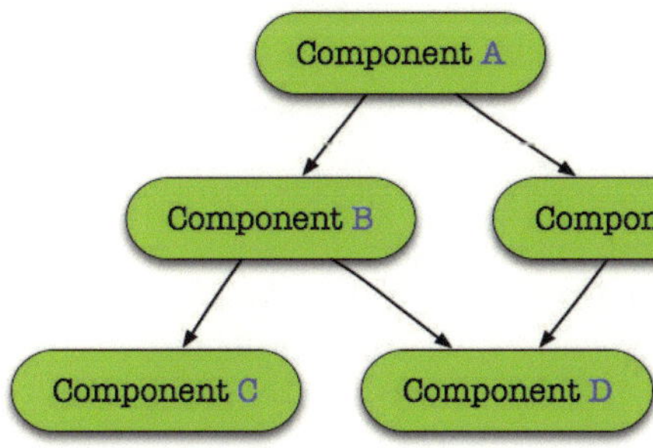

Fig. 11. *Activation tree* where B and D activates D component.

The *activation tree* is no fixed during the robot operation. Actually, it changes dinamically depending on many factors: main task, environment element position, interaction with robots or humans, changes in the environment, error or falls... The robot must adapt to the changes in these factors by modulating the lower level components or activating and deactivating components, changing in this way the static view of the tree.

The main idea of our approach is to decompose the robot functionality in these components, which cooperate among them to make arise more complex behaviors. As we said before, component can be active or inactive. When it is active, a `step()` function is called iteratively to perform the component task.

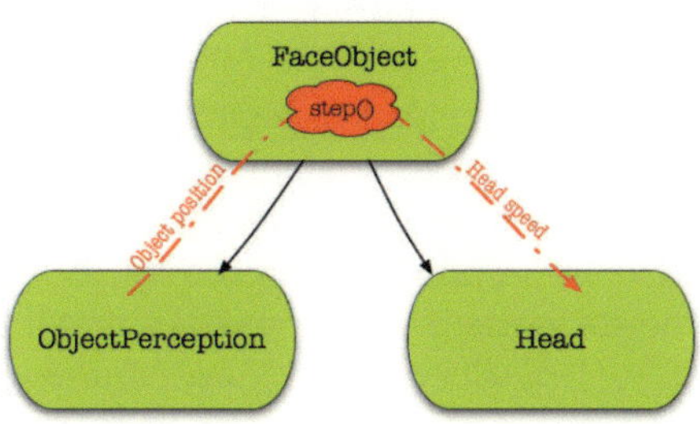

Fig. 12. Activation tree with two low level components and a high level components that modulates them.

As an example, in figure 12 we show an activation tree composed by 3 components. `ObjectPerception` is a low level component that determines the position of an interesting object in the image taken by the robot's camera. `Head` is a low level component that moves the head. These components functionality is used by a higher level component called `FaceObject`. This component activates both low level components, that execute iteratively. Each time `FaceObject` component performs its `step()` function, it asks to `FaceObject` for the object position and modulates `Head` movement to obtain the global behavior: facing the object.

Components can be very simple or very complex. For example, the `ObjectPerception` component of the example is a perceptive iterative component. It does't modulate or activate another component. It only extract information from an image. The `ObjectPerception` component is a iterative controller, that activate and modulate another components. Another components may activate and deactivate components dinamically dependining on some stimulus. They are implemented as *finite state machine*. In each state there is set of active components, and this set is eventually different to the one in other state. Transitions among states reflect the need to adapt to the new conditions the robot must face to.

Using this guideline, we have implemented our architecture in a single NaoQi module. The components are implemented as *Singleton* C++ classes and they communicate among them by method calls. It speeds up the communications with respect the SOAP message passing approach.

When NaoQi module is created, it starts a thread which continuosly call to `step()` method of the root component (the higher level component) in the *activation tree*. Each `step()` method of every component at level *n* has the same structure:

1. Calls to `step()` method of components in *n-1* level in its branch that it wants to be active to get information.
2. Performs some processing to achieve its goal. This could include calls to components methods in level *n-1* to obtain information and calls to lower level components methods in level *n-1* to modulate their actuation.
3. Calls to `step()` methods of component in *n-1* level in its branch that it wants to be active to modulate them.

Each module runs iteratively at a configured frequency. It has not sense that all the components execute at the same frequency. Some informations are needed to be refreshed very fast, and some decisions are not needed to be taken such fast. Some components may need to be configured at the maximun frame rate, but another modules may not need such high rate. When a `step()` method is called, it checks if the elapsed time since last execution is equal or higher to the established according to its frequency. In that case, it executes 1, 2 and 3 parts of the structure the have just described. If the elapsed time is lower, it only executes 1 and 3 parts. Tipically, higher level components are set up with lower frequency than lower level ones, as we can observe in figure 13.

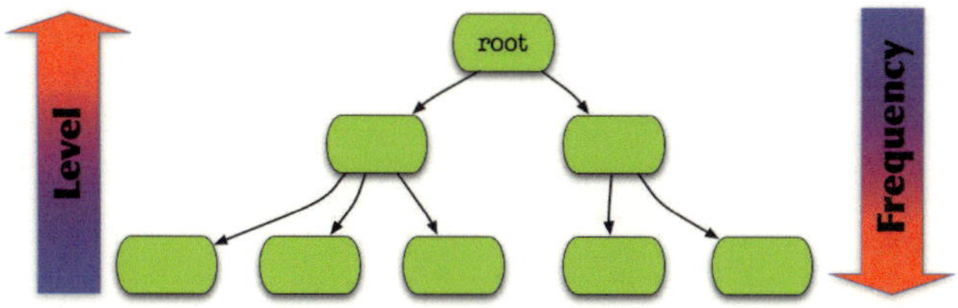

Fig. 13. Activation tree with a root component in the higher level. As higher is the level, lower is the frequency.

Using this approach, we can modulate every module frequency, and be aware of situations where the system has a high load. If a module does not meet with its (soft) deadline, it only makes the next component to executed a little bit late, but its execution is not discarted (*graceful degradation*).

In next section we will describe some of the components developed using this approach for our soccer player application, clarifying some aspects not fully described .

5. Soccer player design

The concepts presented in last section summarizes the key ideas of this architecture design. We have presented the *component* element, how these components can be activated in a activation tree and how they execute. This architecture is focused to develop robot applications using a behavioral approach. In this section we will present how, using this architecture, we solve the problem previously introduced in the section 1: play soccer.
A soccer player implementation is defined by the set of activation trees and how the components modulate another ones. These components are related to perception and actuations and are part of the basis of this architecture. High level components make use of these lower level components to achieve higher level components. So, the changes between soccer player implementations depends on these higher level components. We will review in next sections how particular components to make a robot play soccer are designed and implemented.

5.1 Soccer player perception

At RoboCup competition, the environment is designed to be perceived using vision and all the elements have a particular color and shape. Nao is equipped with two (non-stereo) cameras because they are the richest sensors available in robotics. This particular robot has also ultrasound sensors to detect obstacles in front of it, but a image processing could also detect the obstacle and, additionally, recognize whether it is a robot (and what teams it belong) or another element. This is why we have based the robot perception in vision.

The perception is carried out by the `Perception` component. This component obtains the image from one of the two cameras, process it and makes this information available to any component interested on it using the API it implements. Furthermore, it may calculate the 3D position of some elements in the environment. Finally, we have developed a novel approach to detect the goals, calculating at same time an estimation of the robot pose in 3D.

The relevant elements in the environment are the ball, the green carpet, the blue net, the yellow net, the lines and the other robots. The ilumination is not controlled, but it is supposed to be addequate and stable. The element detection is made attending to its color, shape, dimensions and position with respect the detected border of the carpet (to detect if it is in the field).

We want to use *direct raw* mode because the fetching time varying depending on the desired color space, resolution and access mode, as we can see in figure 14.

Fig. 14. Relevant elements in the environment.

Perception component can be modulated by other components that uses it to set different aspects related to the perception:

- Camera selection. Only bottom or upper camera is active at same time.
- Set the stimulus of interest.

We have designed this module to detect only one stimulus at the same time. There are four types of stimulus: ball in the image, goals in the image, ball in ground coordinates and goal in robot coordinates. This is usefull to avoid unncecessary processing when any of the elements are not usefull.

5.1.1 Ball and goal in image

These stimulus detection is performed in the `step()` method of this component. Once the image obtained is filtered attending only to the color of the element we want to detect. To speed up this process we use a lookup table. In the next step, the resulting pixels on the filtering step are grouped in blobs that indicate connected pixels with the same color. In the last step, we apply some conditions to each blob. We test the size, the density, the center mass position with respect the horizont, etc. The horizon is the line that indicates the upper border of the green carpet. Ball is always under horizon, and nets have a maximun and minimum distance to it. All this process for ball and net is shown in figure 15.

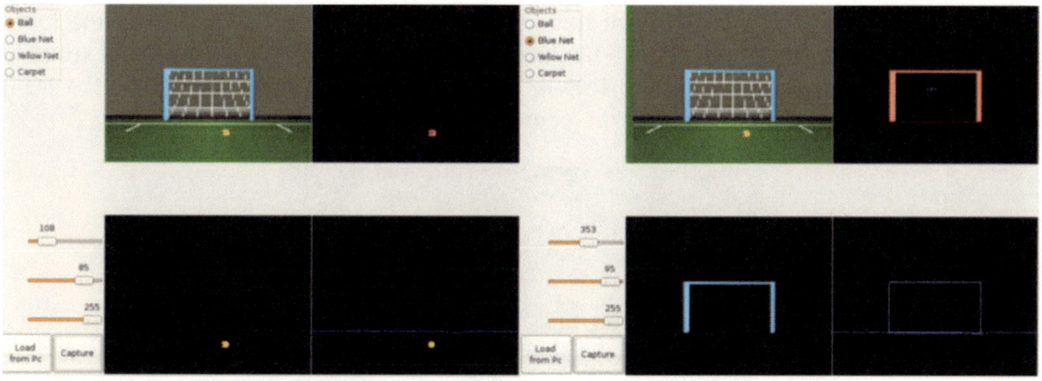

Fig. 15. Element detection process.

The element detected is coded as a tuple {[-1,1],[-1,1]}, indicating the normalized position {X,Y} of the object in the image. When `step()` method finishes, any component can ask for this information using method such us `getBallX()`,`getBlueNetY()`, etc.

Fig. 16. Tuple containing the ball position.

5.1.2 Ball in ground coordinates

In last subsection we describe how the element information is calculated. This information is 2D and is related to the image space. Sometimes it is not enough to achieve some task. For example, if the robot wants to be aligned in order to kick the ball, it is desired to have the ball position available in the robot space reference, as we can see in figure 17.

Obtain the element position in 3D is not an easy task, and it is more difficult in the case of an humanoid robot that walks and perceive an element with a single camera. We have placed the robot axes in the floor, centered under the chest, as we can see in figure 17. The 3D position $\{O_X, O_Y, O_Z=0\}$ of the observed element O (red lines in figure 11) is with respect the robot axes (blue lines in figure 17).

To calculate the 3D position, we start from the 2D position of the center of the detected element related to the image space in one camera. Using the *pinhole* model, we can calculate the a 3D point situated in the line that joints the center of the camera and the element position in the camera space.

Once obtained this point we represent this point and the center of the camera in the robot space axes. We use NaoQi functions to help to obtain the transformation from robot space to camera space. Using *Denavit and Hartenberg* method (Denavit, 1955), we obtain the *(4x4)* matrix that correspond to that transform (composed by rotations and translations).

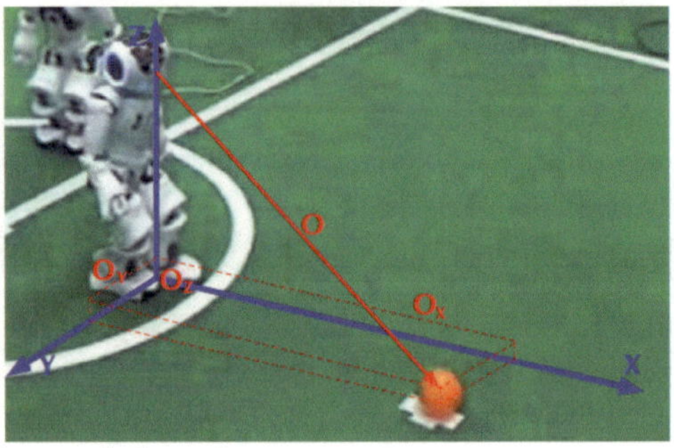

Fig. 17. Element 3D position and the robot axes.

Each time this component is asked for the 3D position of an image element, it has to calculate this transformation matrix (each time the joint angles from foots to camera are differents) and apply to to the 2D element position in the camera frame calculated in the last `step()` iteration.

5.1.3 Goal in robot coordinates

Once the goal has been properly detected in the image, spatial information can be obtained from the that goal using geometric 3D computations. Let Pix1, Pix2, Pix3 and Pix4 be the pixels of the goal vertices in the image. The position and orientation of the goal relative to the camera can be inferred, that is, the 3D points P1, P2, P3 and P4 corresponding to the goal vertices. Because the absolute positions of both goals are known (AP1,AP2,AP3,AP4) that information can be reversed to compute the camera position relative to the goal, and so, the absolute location of the camera (and the robot) in the field. In order to perform such 3D geometric computation the robot camera must be calibrated.

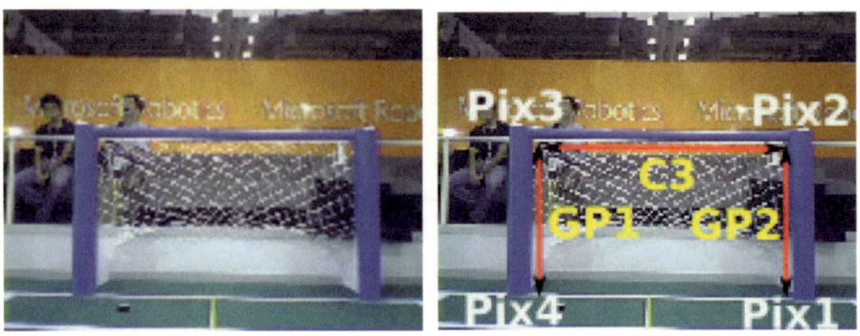

Fig. 18. Goal detection.

Two different 3D coordinates are used: the absolute field based reference system and the system tied to the robot itself, to its camera. Our algorithm deals with line segments. It works in the absolute reference system and finds the absolute camera position computing some restrictions coming from the pixels where the goal appears in the image.

There are three line segments in the goal detected in the image: two goalposts and the crossbar. Taking into consideration only one of the posts (for instance GP1 at figure 18) the way in which it appears in the image imposes some restrictions to the camera location. As we will explain later, a 3D thorus contains all the camera locations from which that goalpost is seen with that length in pixels (figure 19). It also includes the two corresponding goalpost vertices. A new 3D thorus is computed considering the second goalpost (for instance GP2 at figure 18), and a third one considering the crossbar. The real camera location belongs to the three thorus, so it can be computed as the intersection of them.

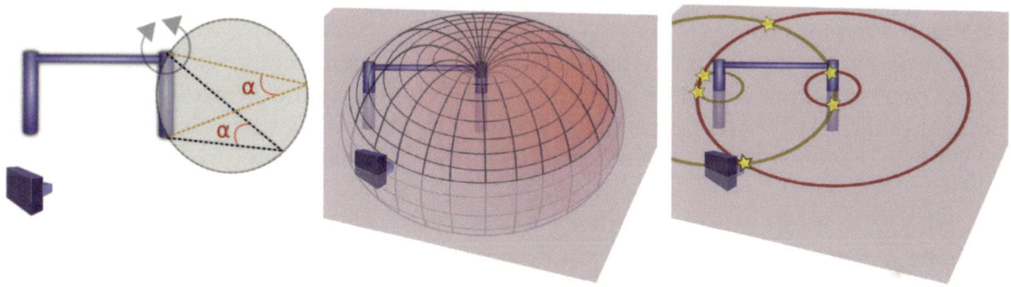

Fig. 19. Camera 3D position estimation using a 3D thorus built from the perception.

Nevertheless the analytical solution to the intersection of three 3D thorus is not simple. A numerical algorithm could be used. Instead of that, we assume that the height of the camera above the floor is known. The thorus coming from the crossbar is not needed anymore and it is replaced by a horizontal plane, at h meters above the ground. Then, the intersection between three thorus becomes the intersection between two parallel thorus and a plane. The thorus coming from the left goalpost becomes a circle in that horizontal plane, centered at the goalpost intersection with the plane. The thorus coming from the right goalpost also becomes a circle. The intersection of both circles gives the camera location. Usually, due to simmetry, two different solutions are valid. Only the position inside the field is selected.

To compute the thorus coming from one post, we take its two vertices in the image. Using projective geometry and the intrisinc parameters of the camera, a 3D projection ray can be computed that traverses the focus of the camera and the top vertex pixel. The same can be computed for the bottom vertex. The angle α between these two rays in 3D is calculated using the dot product.

Let's now consider one post at its absolute coordinates and a vertical plane that contains it. Inside that plane only the points in a given circle see the post segment with an angle α . The thorus is generated rotating such circle around the axis of the goalpost. Such thorus contains all the camera 3D locations from which that post is seen with a angle α, regardless its orientation. In other words, all the camera positions from which that post is seen with such pixel length.

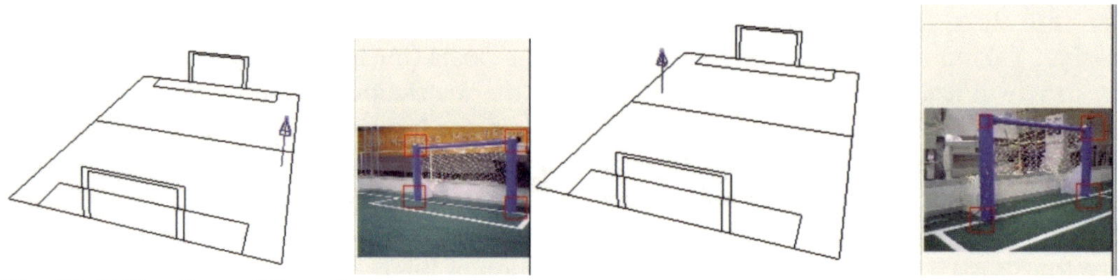

Fig. 20. Estimation of the robot position

5.2 Basic movements

Robot actuation is not trivial in a legged robot. It is even more complicated in biped robots. The movement is carried out by moving the projection of center of mass in the floor (zero moment point, ZMP) to be in the support foot. This involves the coordination of almost all the joints in the robot. In fact, it is common even use the arms to improve the balance.

It is hard to develop complete walking mechanism. This means to generate all the joint positions in every moment, which is not mathematically trivial. It also involves real time aspects because if a joint command is sent late, even few milliseconds, the result is fatal, and the robot may fall to floor. All this work is critical for any task that the robot performs, but it has not very valuable, from the scientific point of view. Sometimes there is not chance, and this work has to be done. For example, we had to calculate every joint position each 8 milliseconds to make walk the quadruped AiBo robot because there were not any library or function to make it walk. Luckily, NaoQi provides some high level functions to make the robot move. There are function to walk (straight or side), turn or move in many ways an only joint. It is not mandatory to use it, and everyone can develop his own walking mechanism, but it is difficult to improve the results that NaoQi provides.

We have chosen to use NaoQi high level functionality to move the robot. We do not use these function in every component that wants to move the robot in any way. This would incur in conflicts and it is not desiderable mix high and low level functions. For these reasons, we have developed some components to manage the robot movement, providing and standard and addequate interface for all the component that wants to perform any actuation. This interface is implemented by the `Body`, `Head` and `FixMove` components.

5.2.1 `Body` component

The `Body` component manages the robot walk. Its modulation consists in two parameters: straight velocity (v) and rotation velocity (w). Each parameters accepts values in the [-1,1]. If v is 1, the robot walks forward straight; if v is -1, the robot walks backward straight; if v is 0, robot doesn't move straight. If w is 1, the robot turn left; if w is -1, the robot turn right; if w is 0, robot doesn't turn. Unfortunately, this movements can't be combined and only one of them is active at the same time.

Actually, `Body` doesn't this work directly but it activates and modulates two lower level components: `GoStraight` and `Turn`, as we can see in figure 21. When v is different to 0, it deactivates `Turn` component if it was active, modulates and activates `GoStraight` component. When w is different to 0, it deactivates `GoStraight` and activates `Turn`.

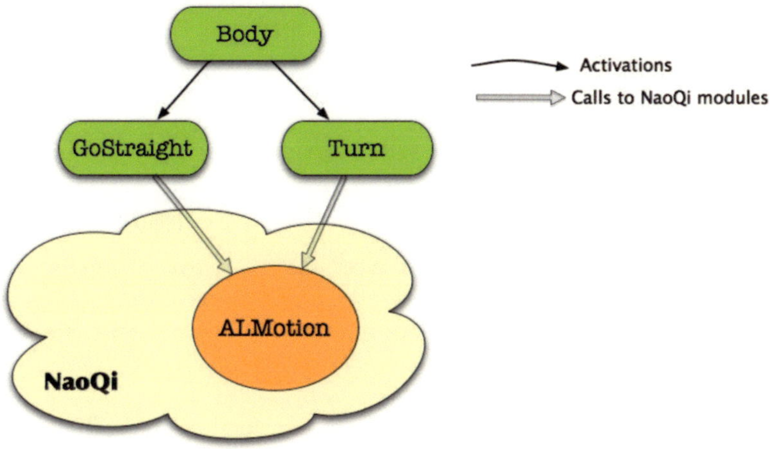

Fig. 21. `Body` component and its lower level components, which comumnicate with NaoQi to move the robot.

5.2.2 `Head` component

`Body` component makes move all the robot but the robot head. Robot head is involved in the perception and attention process and can be controlled independiently from the rest of the robot. The robot head is controlled by the `Head` component. This component, when active, can be modulated in velocity and position to control the pan and tilt movement. While the head control in position is quite simple (it sends motion commands to `ALMotion` to set the joint to the desired angle), the control in velocity is more sophisticated. We developed a *PID controller* to adjust the movement speed. The modulation parameter for this type of control, in range [-1,1] in each pan and tilt, is taken as the input of this controller. The value -1 means the maximun value in one turn sense, 1 in the other sense, and 0 means to stop the head in this axe.

5.2.3 Fixed Movement behavior

The last component involved in actuation is the `FixMove` component. Sometimes it is required to perform a fixed complex movement composed by several joint positions in determined times. For example, when we want that robot kicks the ball we have to made a coordinate movement that involves all the body joints and takes several seconds to complete. These movements are coded in several files, one for each fixed movement, that describe the joints involved in the movement, the positions and when these positions should applied. Lets look an example of this file:

```
Movement_name
name_joint_1 name _joint_2 name _joint_3 ... name _joint_n
angle_1_joint_1 angle_2_joint_1 angle_3_joint_1 ... angle_m1_joint_1
angle_1_joint_2 angle_2_joint_2 angle_3_joint_2 ... angle_m2_joint_2
angle_1_joint_3 angle_2_joint_3 angle_3_joint_3 ... angle_m3_joint_3
...
angle_1_joint_n angle_2_joint_n angle_3_joint_n ... angle_mn_joint_n
time_1_joint_1 time_2_joint_1 time_3_joint_1 ... time_m1_joint_1
time_1_joint_2 time_2_joint_2 time_3_joint_2 ... time_m2_joint_2
time_1_joint_3 time_2_joint_3 time_3_joint_3 ... time_m3_joint_3
...
time_1_joint_n time_2_joint_n time_3_joint_n ... time_mn_joint_n
```

In addition to the desired fixed movement, we can modulate two parameters that indicates a walking displacement in straight and side senses. This is useful to align the robot with the ball when kicking the ball. If this values are not zero, a walk preceed the execution of the fixed movement.

As we have just introduced, we use this component for kicking the ball and for standing up when the robot is pushed and falls to the floor.

5.3 Face Ball behavior

`FaceBall` component tries to center the ball in the image taken from the camera. To achieve this goal, when active, this component activates both `Perception` and `Head` components, as we see in figure 22.

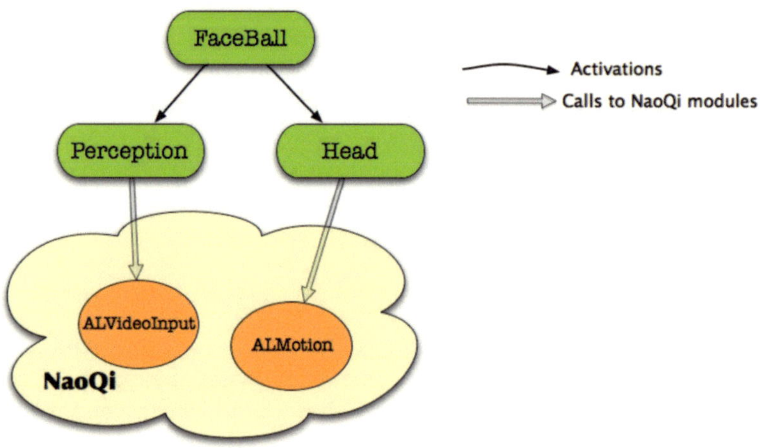

Fig. 22. `FaceBall` component.

This component activates `Perception` and `Head` while it is active. It modulates `Perception` to detect the ball. In its `step()` function, it simply takes the output of the perception component and uses this value as the input of the `Head` component. These values are in the [-1,1] range. When the ball is centered, the X and Y value of the ball are 0, so the head is stopped. If the ball is in the extreme right, the X value, 1, will be the modulation of the pan velocity, turning the head to the left. Here is the code of the `step()` funtion of `FaceBall`.

```
void
FaceBall::step(void)
{
        perception->step();

        if (isTime2Run())
        {
                head->setPan( perception->getBallX());
                head->setTilt( perception->getBallY() );
        }

        head->step();
}
```

5.4 Follow Ball behavior

The main function of `FollowBall` component is going to the ball when it is detected by the robot. This component activates FaceBall and Body components. The modulation of the Body component is the position of the robot head, that is tracking the ball. Simplifying, the code of the step function of this component is something like this:

```
void
FollowBall::step(void)
{
        faceball->step();

        if (isTime2Run())
        {
                float panAngle = toDegrees(headYaw);
                float tiltAngle = toDegrees(headPitch);

                if(panAngle > 35) body->setVel(0, panAngle/fabs(panAngle));
                else body->setVel(1, 0);

        }

        body->step();
}
```

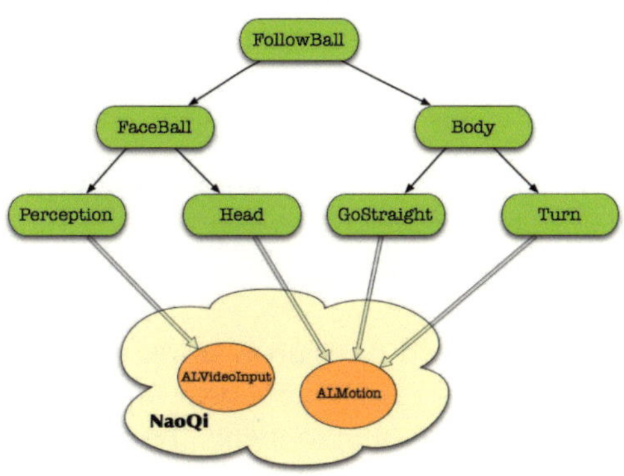

Fig. 23. `FollowBall` component.

5.5 Search Ball behavior

The main function of `SearchBall` component is search the ball when it is not detected by the robot. This component introduces the concept of finite state machine in a component.

When this component is active, it can be in two states: *HeadSearch* or *BodySearch*. It starts from *HeadSearch* state and it only moves the head to search the ball. When it has scanned all the space in front of the robot it transitates to *BodySearch* state and the robot starts to turn while it is scanning with the head. In any state, `SearchBall` component modulates `Perception` component in order to periodically change the active camera.

Depending on the state, the activation tree is different, as we can see in figure 24. At start, the active state is *HeadSearch*. In this state only `Head` component is active. During this state, `Head` component is modulated directly from `SearchBall` component. It starts moving the head up and it continues moving the head to scan the space in front of the robot. When this scan is completed, it transitates to *BodySearch* state. In this state, `Body` component is also activated in order to make turn the robot in one direction.

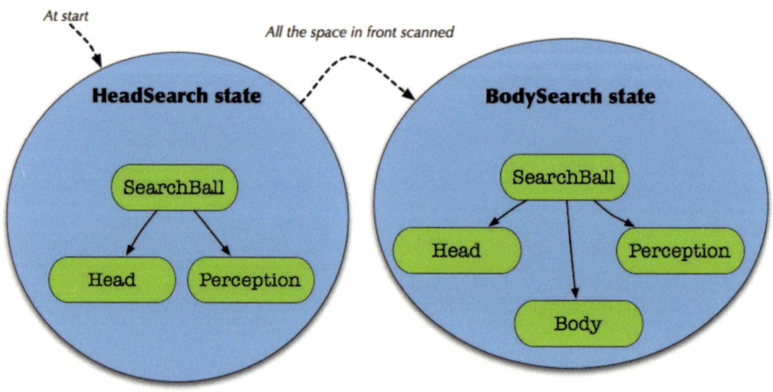

Fig. 24. The two states that this component can be and the activation tree in each state.

This component does not use the `Perception` component to get information about the ball presence. This component only manages the robot movement and the camera selection. Any other component has to activate `SearchBall` and `Perception` components, and stop `SearchBall` once the ball is found. Next we will see which component do this work.

5.6 Search Net behavior

This behavior is implemented by the `SearchNet` component is used to search the net where the robot must kick the ball to. It activates `Head` and `Perception` components. Its work is divided in two states: *Scanning* and *Recovering*.

When the *Scanning* state starts, the head position is stored (it is supposed to be tracking the ball) and the robot modulates `Perception` component to detect the nets instead of the ball. It has not sense continuing doing processing to detect the ball if now it is not the interesting element, saving processing resources.

While *Scanning* state, this component also modulates Head component to move the head along the horizont, searching the ball. When this component is active, the robot is stopped, and we can suppose where is the horinzont. If the scanning is complete, or the net is detected, this component tansitates to the *Recovering* state.

In the *Recovering* state the `Perception` component is modulated to detect the ball, and the head moves to the position stored when *Scanning* state started.

5.7 Field Player behavior

The `Player` component is the root component of the forward player behavior. Its functionality is decomposed in five states: *LookForBall, Approach, SeekNet, Fallen and Kick.* These five states encode all the behavior that makes the robot play soccer.

In *LookForBall* state, `Player` component activates SearchBall and Perception components, as is shown in figure 25. For clarity reasons, in this figure we don't display all the activation tree but the components that `Player` component directly activates.

When the `Perception` components indicates that the ball is detected, this component transitates to *Approach* state. It deactivates SearchBall State, and the ball is supposed to be in the active camera. In this state is activated the FollowBall component in order to make the robot walk to the ball.

It is common that the robot initially detects the ball with the upper camera, and it starts the approach to the ball using this camera. When the ball is nearer than one meter, it can't follow it with the upper camera bacause of the neck limitations and the ball is lost. It transitates to the *LookForBall* state again and starts searching the ball, changing the camera. When the ball is detected with the lower camera, it continues the approaching with the right camera.

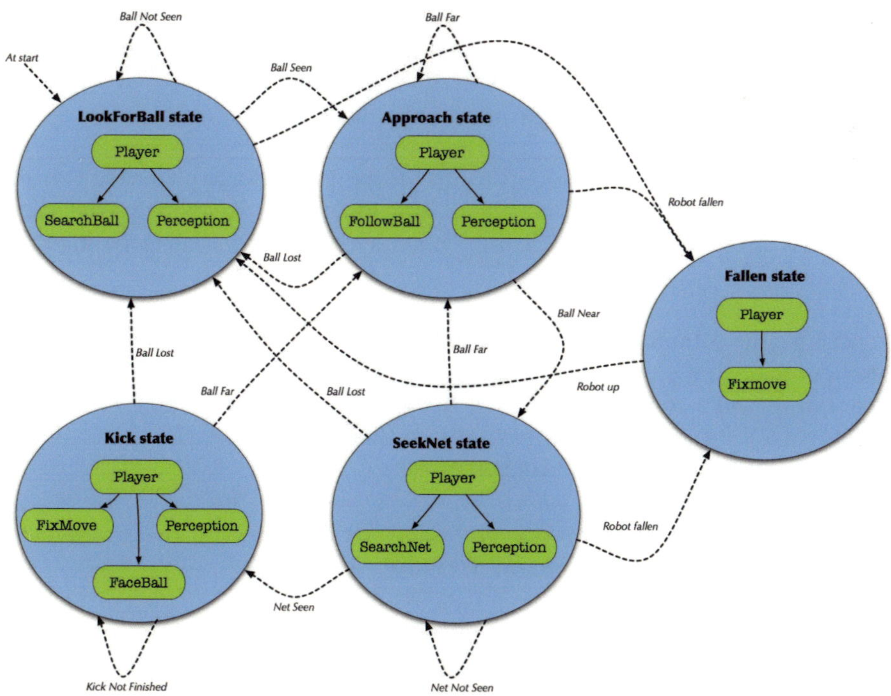

Fig. 25. `Player` component finite state machine with its corresponding activation tree.

When the ball is close to the robot, the robot is ready to kick the ball, but it has to detect the net first in order to select the addecuate movement to score. For this reason, the `Player` component transitates to *SeekNet* state, activating `SearchNet` component.

Once detected the net, the robot must kick the ball in the right direction according to the net position. Player component makes this decission in the *Kick* state. Before activating `FixMove` component, `Player` component ask to `Perception` component the 3D ball position. With this information, it can calculate the displacement needed by the selected kick to perform this kick correctly. Once activated the `FixMove` component, the robot performs the kick.

In this state, this component also activates `FaceBall` component to track the ball while the robot is kicking the ball.

The last state is *Fallen*. This component goes to transitates to this state when the robot falls to the floor. It activates `Fixmove` component and modulates it with the right movement to make it getting up.

These are the components needed for the forward player behavior. In next sections we will explain the tools developed to tune, configurate and debug these components, and also the components developed to make a complete soccer player.

6. Tools

In previous section we described the software that the robot runs onboard. This is the software needed to make the robot perform any task. Actually, this is the only software that is working while the robot is playing soccer in an official match, but some work on calibrating and debugging must be done before a game starts.

We have developed a set of tools useful to do all the previous work needed to make the robot play. *Manager* application contains all the tools used to manage the robot. This application runs in a PC connected to the robot by an ethernet cable or using wireless communications.

To make possible the communication between the robot and the computer we have used the SOAP protocol that NaoQi provides. This simplify the development process because we do not have to implement a socket based communication or anything similar. We only obtain a proxy to the NaoQi module that contains our software, and we make calls to methods to send and receive all the management information.

Next, we will describe the main tools developed inside the *Manager*. These tool let to debug any component, tune the vision filters and the robot movements.

6.1 Component debugging tool

Inside the *Manager* we can debug any component individually. We can activate a component, modulate it and change its frequency. It is also possible to take measures related to the CPU consumption.

In figure 26 we show the GUI of this tool and how the `Turn` component is debugged. We can activate independiently using a checkbox. We can also configurate the frequency, in this case it is set to run at 5 Hz. We use the slider to modulate this component setting its input in the [-1,1] range. In the figure the modulation is 0, so the robot is stopped. Finally, we can obtain the mean, maximum and minimum CPU consumption time in each iteration.

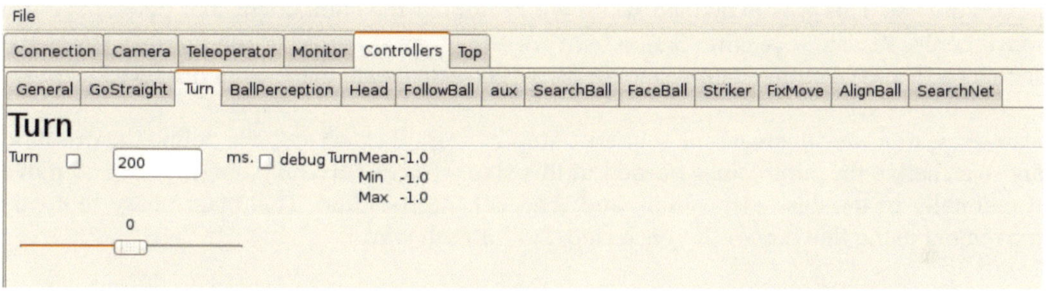

Fig. 26. Component debugging tool and how the `Turn` component is debugged.

6.2 Perception tool

The environment conditions are not similar in every place the robot must work. Even in the same place, the conditions are not similar along the day. For this reason to calibrate the camera characteristics and the color definitions is essential to face these changes in the ligt conditions.

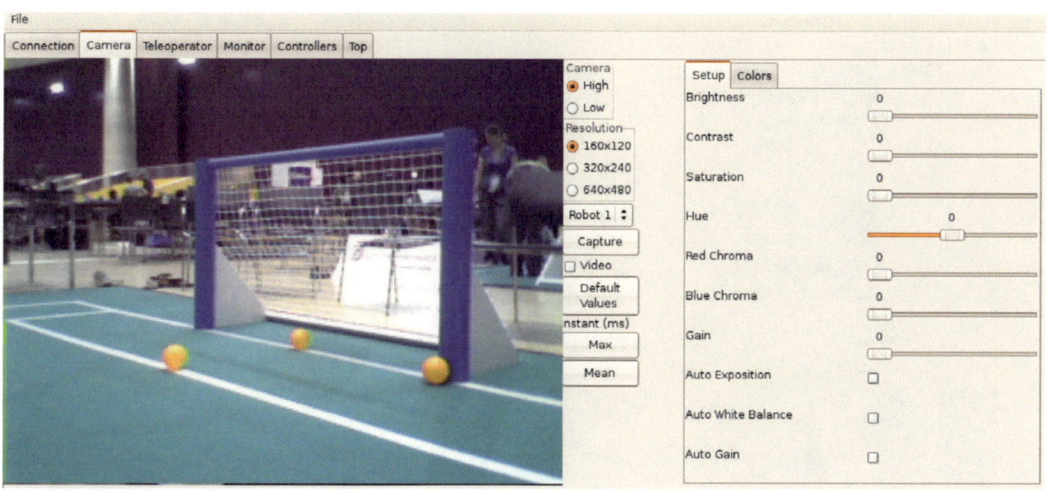

Fig. 27. Camera values tunning.

The *Manager* contains some tools to do this calibratarion. Figure 27 shows the tool used to calibrate the camera values (brightness, contrast, …). Each relevant element to the robot has a different color, as we explained in section 5.1.1. These colors and the recognization values are set using the tools shown previously in figure 15.

6.3 Fixed Movement tool

In some situations the robot must perform a fixed movement. To kick the ball or to get up, the robot needs to follow a sequence of movements that involves many joints. It is difficult to create these movements without a specific tool.

We have implemented a tool to create sequences of movemenet. For each sequence we have to specify the joints involved, the angles that each joint has to be set and the time these angles are set. This was explained when we presented the component that performs these movements, `Fixmove` (section 5.3), where we shown an example of the file that stores the sequence. The goal of this tool, shown in figure 28, is to create these sequence files.

Using this tool we can create the sequence step by step. En each step we define the duration and we change the joint values needed in this step. We can turn off a single joint and move it manually to the desired position, and then get that position. This makes easy to create movement using this process for each step and for each joint.

We have created three types of kicks using this tool. Each kick can be done simetrically with both legs, then we really have six kicks. Also, we have created two movements to make the robot get up from the floor.

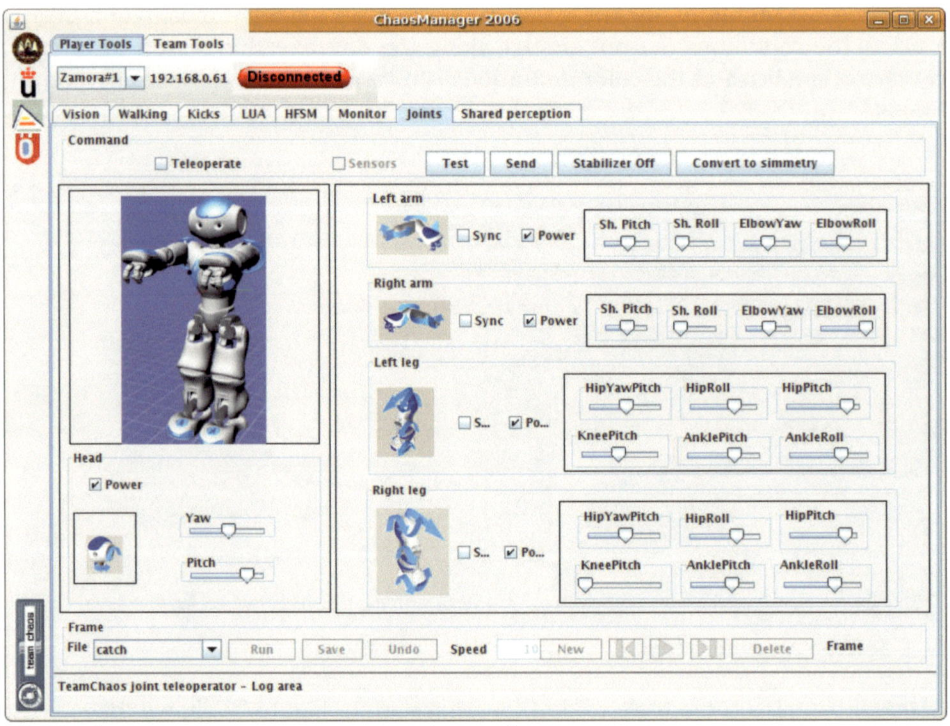

Fig. 28. Fixed movement sequence generation tool.

7. Experiments

In this behavior we have presented our behavior based architecture and a complete soccer player application using it. In this chapter we will show the experiments carried out during and after its development.

7.1 First behavior architecture attempt

Not always the first steps are the right ones. In this architecture design, the proposed solution wasn't the first approximation we took. At initial we tried to exploit all the benefits that NaoQi provides. This software element lets to decompose our application functionality in modules which cooperate among them to achieve a goal. Each module perfoms some processing task and sends data to other modules. This would lets to implement our architecture in a natural way using this approximation. NaoQi has a funtionality to start and stop calling iteratively a method, using a callback to a periodic clock event. This solves the execution cycle to call `step()` method iteratively. Communications among modules are solved by the SOAP messages mechanism that NaoQi provides. We also could use ALMemory as a blackboard where all the information from sensorial components and all the modulations to actuation modules are registered and taken. Even callbacks can be set up in each module to be called each time an interesting value in this blackboard changes. In fact, this was the first approach we took to design our architecture. Unfortunately, and intensive use of these mechanisms had a big impact in NaoQi performance and some real time critical tasks were severely affected. One of these real time critical tasks is movement generation. When the performace in this task was poor, the movement was affected and the robot fallen to floor.

7.2 General behavior creation process using the proposed architecture

The final design tried to use as less NaoQi mechanisms as possible. We use NaoQi to access sensors and actuators, but all the communication via SOAP messages are reduced to the minimun possible. ALMemory as a blackboard was discarded and callbacks to data are used only to the essential information generated by NaoQi that are useful to our tasks. Although we have taken this decission, some of the NaoQi functionality is still essential to us. We use NaoQi for accessing to the camera images or for walking generation. We are not interested in developing our own locomotion mechanism because this is being improved continuosly by the robot manufacturer and we want to concentrate in hich level topics.

With the changes from the initial to the final version we obtained better performance and we avoid to affect NaoQi locomotion mechanism. Althought our software wants to consume too much processing time, only our software will be affected. This is a redical improvement with respect the initial version.

The robot soccer behavior is itself an experiment to test the proposed behavioral architecture. Using this architecture, we have developed a complete behavior able to cope with the requirements that a RoboCup match imposes. We have shown some of the characteristics of this robot architecture during this process:

- FaceBall, SearchNet and SearchBall components reuses the component Head. In this way we have shown how a component can be reused only changing its modulation.
- Only Perception, GoStraight and Turn components face to the complexity of the robot hardware, which let in the other levels to ignore this complexity.
- Component activations and deactivations, for example in section 5.5, let to have diferent behaviors on the robot.

7.3 Forward soccer player test

The final experiment is the real application of this behavior. We have tested it at RoboCup 2009 in Graz. Before the test we had to use the tools described in section 6 to calibrate the colors of the relevant elements in the environment. Once tuned, the robot is ready to work. This sequence has been extracted from a video which full version may be visualized at http://www.teamchaos.es/index.php/URJC#RoboCup-2009.

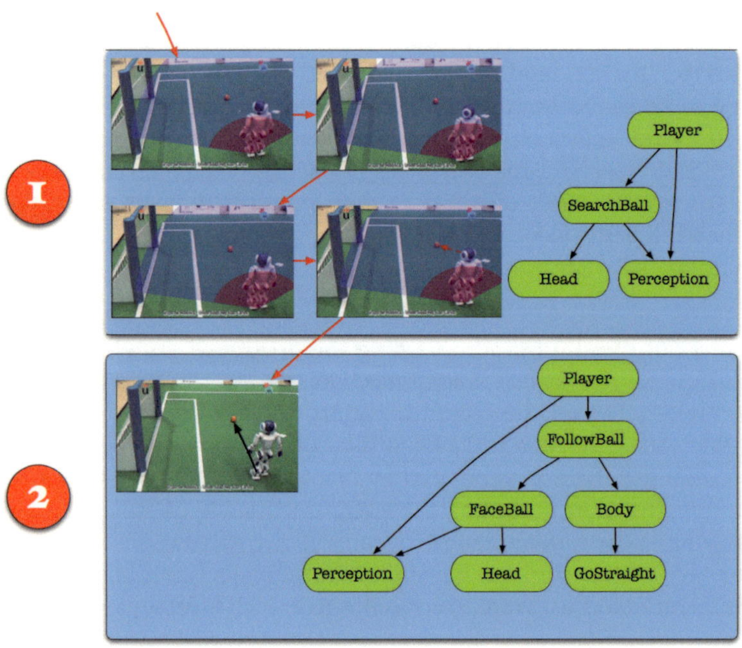

Fig. 29. Ball searching sequence.

Figure 29 shows a piece of an experiment of the soccer player behavior. In this experiment the robot starts with total uncertainty about the ball. Initially, the Player component is in LookForBall state and it has activated the SearchBall component to look for the ball. SearchBall component uses first the bottom camera. The shadow area respresents the camera coverage, where the red area represents the bottom camera coverage and the blue area the upper camera coverage. In the two first images the robot is scanning the nearest space with the bottom camera and it doesn't find any ball. Once completed the near scan, SearchBall component modulates Perception component in order to change the camera, covering the blue marked area. Player component is continously asking Perception component for the ball presence, and when the ball is detected in the image (fourth image in the sequence), SearchBall component is deactivated and FollowBall component is activates, approaching to the ball (last image in the sequence). Take note that in this example, the upper camera is now active.

FollowBall component activates FaceBall component to center the ball in the image while the robot is approaching to the ball. FollowBall activates Body to approach the ball. As the neck angle is less than a fixed value, i.e 35 degrees (the ball is in front of the robot), Body activates GoStraight component in order to make the robot walk straight.

The approaching to the ball, as we said before is made using FaceBall component and Body component. Note that in any moment no distance to the ball is take into account. Only the head pan is used by the body component to approach the ball.

In figure 30, while the robot the robot is approaching to the ball, it has to turn to correct the walk direction. In this situation, the head pan angle is higher than a fixed value (35 degrees, for example) indicating that the ball is not in front of the robot. Immediately, after this condition is true, FollowBall modulates to Body so the angular speed is not null and forward speed is zero. Then, Body component deactivates GoStraight component and activates Turn Components, which makes the robot turn in the desired direction.

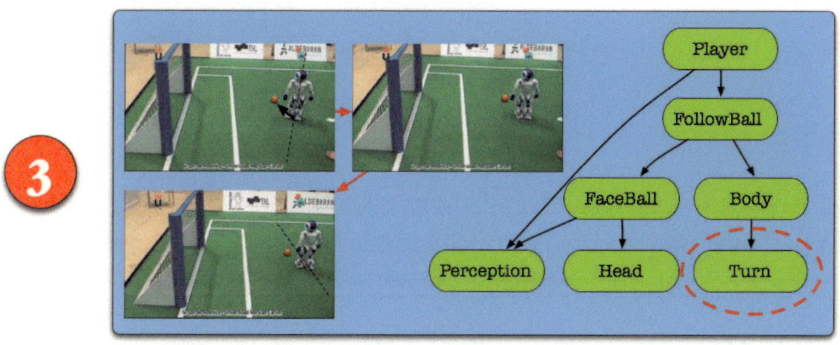

Fig. 30. Ball appraching modulation to make the robot turn.

The robot reached the ball when it is walking to the ball, the bottom camera is active, the head tilt is higher than a threshold, and the head pan is low. This situation is shown in the first image in the figure 31. In that moment, the robot has to decide which kick it has to execute. For this reason, the net has to be detected. In the last image, the conditions to kick the ball are held and the player component deactivates FollowBall component and activates the SearchNet component. The SearchNet component has as output a value that indicates if the scan is complete. The Player component queries in each iteration if the scan is complete. Once completed, depending on the net position (or if it has been detected) a kick is selected. In the second image of the same figure, the blue net is detected at the right of the robot. For this test we have created 3 types of kicks: front, diagonal and lateral. Actually, we have 6 kicks available because each one can be done by both legs. In this situation the robot selects a lateral kick with the right leg to kick the ball.

Before kick the ball, the robot must be aligned in order to situate itself in the right position to do an effective kick. For this purpouse, the player component ask to the Perception module the ball position in 3D with respect the robot. This is the only time the ball position is estimated. The player components activates Fixmove component with the selected kick and a lateral and straight alignement. As we can see in third and fourth image, the robot moves on its left and back to do the kick.

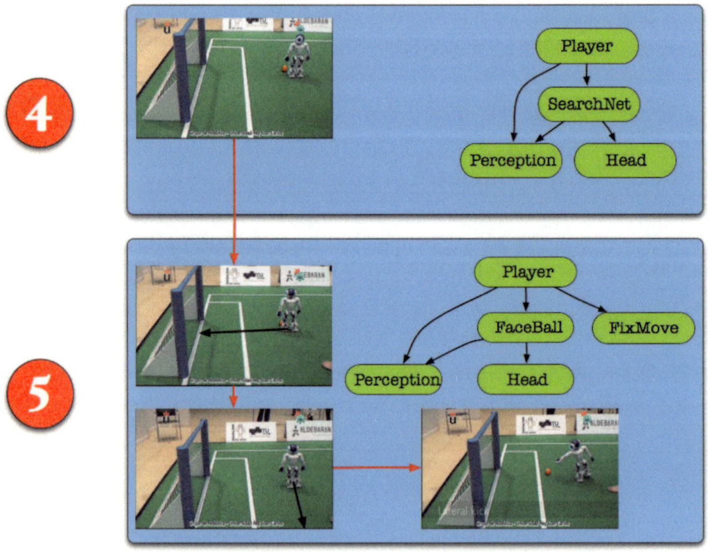

Fig. 31. Search net behavior and kick.

While the kick is performing and after the kick, FaceBall component is activeted to continue traking the ball. This spped up the recovering after the kick and sometimes it is not needed to transitate to the searching ball state, but the approaching to the ball state.

This experiment has been carried out at the RoboCup 2009 in Graz. This behavior was tested in the real competition environment, where the robot operation showed robust to the noise produced by other robots and persons.

7.4 Any Ball challenge

In RoboCup 2009 competitions we also took part in the Any Ball Challenge. The goal was to kick ball differents to the orange official one. To achieve this goal we changed the Perception component to detects the non-green objects under the orizont that seemed like balls. In the figure 32 (and in the video we mentioned before) we can see how the robot was able to kick differents balls.

Fig. 32. Any Ball Challenge. The robot must detect and kick hetereogeneuos balls.

7.5 Camera switching experiment

In the experiment described in section 7.2, between the instant 3 and 4 sequence, a camera switch has made. For clarity, lets use another texperiment to explain this change with another sequence. In figure 33, the robot is approaching to the ball using the upper camera. The Player component has activated FollowBall component and the robot is walking straight to the ball using the upper camera.

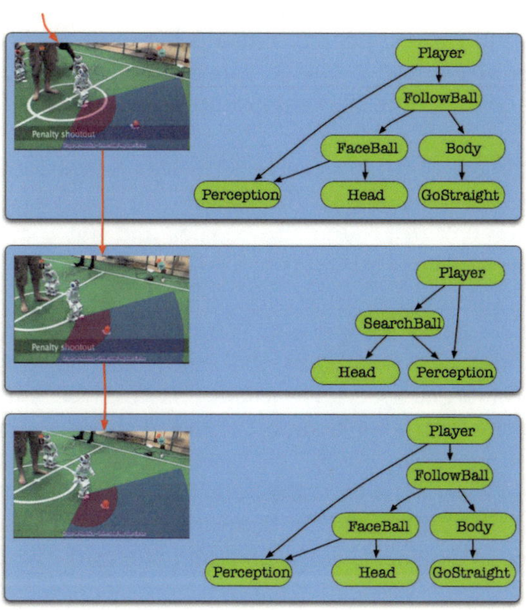

Fig. 33. Camera switching.

When ball enters in the red area and the tilt head has the maximun value, ball dissapears from the image taken by the upper camera. Then, the Player components deactivates FollowBall components (and their activated components in cascade) and activates SearchBall component. SearchBall component always starts with the bottom camera activated and moves the head up. In few steps (maybe 1 or two are enough) ball is detected again. Player component deactivates SearchBall component and activates FollowBall component, that starts with the bottom camera selected. The camera change is made.

8. Conclusions

In this chapter we have proposed a robotic behavior based architecture. With this architecture we can create robotics behaviors. The behavior arises from a cooperative execution of iterative processing units called components. These units are hierarchically organized, where a component may activate and modulate another components. In every moment, there are active components an latent components that are waiting for be activated. This hierarchy is called *activation tree*, and dinamically changes during the robot operation. The components whose ouput is not needed are deactivated in order to save the limited resources of the robot.

We can use this behavior architecture for create any robotic behavior. In this chapter we have shown how the behaviors are implemented within the architecture. We have created a forward player behavior to play soccer in Standar Platform League at RoboCup. This is a dynamic environment where the conditions are very hard. Robots must react very fast to the stimulus in order to play soccer in this league. This is an excellent test to the behaviors created within our architecture.

We have developed a set of component to get a forward soccer player behavior. These components are latent until a component activate it to use it. These component have a standar modulation interface, perfect to be reused by several component without any modification in the source code or to support multiple different interfaces.

Perception is done by the `Perception` component. This component uses the camera to get different stimulus form the environment. The stimulus that it detect are the ball and the net in image coordinates, the ball in ground coordinates and the goal in camera coordinates. This component only perceives one stimulus at the same time, saving the limited resources.

Locomotion is done by the `Body` component, that uses `Turn` or `GoStraight` components alternatively to make the robot walk to the ball. This component is modulated by introducing a lineal or rotational speed. We found this interface is more appropiate that the one provided by the NaoQi high level locomotion API to create our behaviors.

The head movement is not managed from the Body components because it is involved in the perception process and we think that it is better to have a sepparate component for this element. This component is the `Head` component, and moves the robot's neck in pan and tilt frames.

Robot also moves by performing a sequence of basic joint-level movements. This functionality is obtained from the `FixMove` component execution. This is useful to create kicking sequences or getting up sequences.

These components use the NaoQi API directly and are usually in the lower level of the activation tree. They are iterative components and no decision are taken in each step. There are other more complex components. These components activate other components and may vary dinamically the set of components that it activates.

The `FaceBall` component activates `Perception` and `Head` component in order to center the ball in the image. This component is very important, because when it is activated, we can assume that the ball position is where the ball is pointing at. This is used by the `FollowBall` component. This components uses `Body` component to make the robot move ahead when the neck pan angle is small, and turning when it is big. There is not need to know the real distance or angle to the ball, only the neck pan angle.

When the ball is not detected in the image, the `SearchBall` component activates and modulates `Head` and `Body` components to detect the ball. In the same way, the SearchNet component component uses the `Head` component to look for the goals, in order to calculate the right kick to use in order to score.

The higher level component is the `Player` component. This component is implemented as a finite state machine and activates the previously described components in order to obtain the forward player behavior.

This behavior, created with this architecture, has been tested in the RoboCup environment, but it is not limited to it. We want to use this architecture to create robot behaviors to solve another problems out of this environment.

9. References

Thrun, S.; Bennewitz, M.; Burgard, W.; Cremers, A. B.; Dellaert, F.; Fox, D.; Hahnel, D.; Rosenberg, C. R.; Roy, N.; Schulte, J; Schulz, D. (1999). *MINERVA: A Tour-Guide Robot that Learns*. Kunstliche Intelligenz, pp. 14-26. Germany

Reid, S. ; Goodwin, R.; Haigh, K.; Koenig, S.; O'Sullivan, J.; Veloso, M. (1997). *Xavier: Experience with a Layered Robot Architecture*. Agents '97, 1997.

Stoytchev, A.; Arkin, R. (2000). *Combining Deliberation, Reactivity, and Motivation in the Context of a Behavior-Based Robot Architecture*. In Proceedings 2001 IEEE International Symposium on Computational Intelligence in Robotics and Automation. 290-295. Banff, Alberta, Canada. 2000.

Arkin, R. (1989). *Motor Schema Based Mobile Robot Navigation*. The International Journal of Robotics Research, Vol. 8, No. 4, 92-112 (1989).

Saffiotti, A. ; Wasik, Z. (2003). *Using hierarchical fuzzy behaviors in the RoboCup domain*. Autonomous robotic systems: soft computing and hard computing methodologies and applications. pp. 235-262. Physica-Verlag GmbH. Heidelberg, Germany, 2003.

Lenser, S.; Bruce, J.; Veloso, M. (2002). *A Modular Hierarchical Behavior-Based Architecture,* Lecture Notes in Computer Science. RoboCup 2001: Robot Soccer World Cup V. pp. 79-99. Springer Berlin / Heidelberg, 2002.

Röfer, T.; Burkhard, H. ; von Stryk, O. ; Schwiegelshohn, U.; Laue, T.; Weber, M.; Juengel, M.; Gohring D.; Hoffmann, J.; Altmeyer, B.; Krause, T.; Spranger, M.; Brunn, R.; Dassler, M.; Kunz, M.; Oberlies, T.; Risler, M.; Hebbela, M.; Nistico, W.; Czarnetzkia, S.; Kerkhof, T.; Meyer, M.; Rohde, C.; Schmitz, B.; Wachter, M.; Wegner, T.; Zarges. C. (2008). *B-Human. Team Description and code release 2008. Robocup 2008.* Technical report, Germany, 2008.

Calvo, R.; Cañas, J.M.; García-Pérez, L. (2005). *Person following behavior generated with JDE schema hierarchy.* ICINCO 2nd Int. Conf. on Informatics in Control, Automation and Robotics. Barcelona (Spain), sep 14-17, 2005. INSTICC Press, pp 463-466, 2005. ISBN: 972-8865-30-9.

Cañaas, J. M.; and Matellán, V. (2007). *From bio-inspired vs. psycho-inspired to etho-inspired robots.* Robotics and Autonomous Systems, Volume 55, pp 841-850, 2007. ISSN 0921-8890.

Gómez, A.; Martínez, H.; (1997). *Fuzzy Logic Based Intelligent Agents for Reactive Navigation in Autonomous Systems.* Fitth International Conference on Fuzzy Theory and Technology, Raleigh (USA), 1997

Loetzsch, M.; Risler, M.; Jungel, M. (2006). *XABSL - A pragmatic approach to behavior engineering.* In Proceedings of the IEEE/RSJ International Conference on Intelligent Robots and Systems (IROS 2006), pages 5124-5129, Beijing, October 2006.

Denavit, J. (1955). *Hartenberg RS. A kinematic notation for lower-pair mechanisms based on matrices.* Transactions of ASME 1955;77: 215–221 Journal of Applied Mechanics, 2006.

Herrero, D. ; Martínez, H. (2008). *Embedded Behavioral Control of Four-legged Robots.* RoboCup Symposium 2008. Suzhou (China), 2008.

Akin, H.L.; Meriçli, Ç.; Meriçli, T.; Gökçe, B.; Özkucur, E.; Kavakhoglu, C.; Yildiz, O.T. (2008). *Cerberus'08 Team Report.* Technical Report. Turkey, 2008.

Chown, E.; Fishman, J.; Strom, J.; Slavov, G.; Hermans T.; Dunn, N.; Lawrence, A.; Morrison, J.; Krob, E. (2008). *The Northern Bites 2008 Standard Platform Robot Team.* Technical Report. USA, 2008.

5

Robot soccer educational courses

Hrvoje Turić, Vladimir Pleština, Vladan Papić and Ante Krolo
University of Split
Croatia

1. Introduction

Robotics encompasses multiple disciplines, including mechanical engineering, software programming, electronics and even human psychology. Robot soccer is an international project intended to promote these disciplines as well as other related fields due to increasing demand for the properly educated engineers. Basically, it is an attempt to foster AI and intelligent robotics research by providing a standard problem where wide range of technologies can be integrated and examined. The idea of introducing robot soccer and robot soccer league turned out as a great success in popularization of robotics and AI but also the other fields such as mechanical engineering and electronics.

Practical courses for the undergraduate and graduate students can be domain - focused towards a particular research field such as intelligent agents (Coradeschi & Malec, 1999; Anderson & Baltes, 2006), computer vision, artificial intelligence (Riley, 2007), control (Bushnell & Crick, 2003), etc. During the courses students are either constructing the robots, developing software for the robots or doing the both things (Beard et al., 2002; Nagasaka et al., 2006), usually divided into different research teams (Archibald & Beard, 2002, Cornell). Anyhow, they are presented with the real-life problems and have the opportunity to work on challenging project that has a motivating goal. Learning theory that supports this approach is constructionism (Piaget & Inhelder, 1966; Papert, 1980; Papert, 1986). Constructionism holds that learning can happen most effectively when people are also active in making tangible objects in the real world so we can say that experiential learning is optimal for adoption of new knowledge.

Even early works have acknowledged the need to divide robotics courses into different groups depending on their educational goal and complexity (prerequested knowledge). In his survey, Lund presented three set-ups that have been designed as a three step educational process (Lund, 1999). He considers his approach as a guided constructionism because, unlike unguided constructionism approach, it combines the constructionism approach with other methods (guidance) in order to allow the students to acquire knowledge in the most profound way. Strengthening of Educational Robotics as a pedagogic tool and integration of the Educational Robotics into the Curriculum has been subject of investigation for several years (Bruder & Wedeward, 2003; Novales et al., 2006).

Although practical courses for the university students are the most obvious choice because of generally high prerequested domain knowledge, children in elementary and secondary schools (K-12) are also targeted audience. With the current negative enrolment trends at the

technical universities and the increasing demands on the labour market, early days popularization of the technical sciences is necessary in order to provide better and more massive input for the technical sciences oriented studies. Robot soccer has the capability of attracting attention of younger population because it provides both fun and educational experiences. The term sometimes used is 'edutainment robotics' (Miglino et al., 1999; Lund, 2001). Of course, robotic soccer is not the only approach in motivating children for the robotics (McComb, 2008), but it is one of the most popular and perhaps the most comprehensive one. Choice of the platform for children and process adopting courses for their education is certainly interesting and demanding task (Baltes & Anderson, 2005). Various researchers present different modular concepts for the introduction of robotics and computer science education in high schools (Verner & Hershko, 2003; Nourbakhsh et al., 2004; Henkel et al., 2009). In fact, robot design is considered as the suitable school graduation project (Verner & Hershko, 2003). Even very young children (8 to 9 years of age) can be included in courses that can change their way of thinking and teach them basics of robotic technology as well as team work (Chambers et al., 2008).
Introduction of robot soccer courses into K-12 education has some other issues to be solved other than only adopting course difficulty level. One of the most important issues that have to be solved (other than finances) is proper education of the school teachers because they have a broad range of educational backgrounds (Matarić, 2004). Proper documentation and hardware should be available to the teachers because they are best prepared to innovate when working from a solid foundation, prepared by robotics educator, not when starting from the beginning (Wedeward & Bruder, 2002; Matarić et al., 2007). An interesting project that should be mentioned is the TERECOP project (Teacher Education on Robotics-Enhanced Constructivist Pedagogical Methods). It's overall aim is to develop a framework for teacher education courses in order to enable teachers to implement the robotics-enhanced constructivist learning in school classrooms (Arlegui et al., 2008).
As it has already been said, different age groups require an adaptive and modular approach and that was the main idea behind the concept that will be presented here. Short practical courses for three age groups have been developed and proposed: 1) younger K-12 school children (ages 13-15), 2) senior K-12 school children (ages 15-18) and 3) university students (ages > 18).
In this chapter, different modules incorporated in the practical courses are explained and curriculum, aims and tasks for each course level is described. The attention is focused on the modules integration. Although there are some commercial systems available at the market (Lund & Pagliarini, 1999; Gage, 2003; Baltes, J. et al., 2004), in order to provide courses with full range of possible learning themes as a basis for the proposed constructive education courses, development of cheap and simple robots for the robot soccer team is explained in detail.

2. Robot design

First, it should be stated that an inspiration and great help in understanding the most important problems and issues that have to be resolved during design process of a soccer robot was very detailed documentation that can be found on the Cornell University web site (Cornell). Because the robots presented in mentioned documentation are state of the art, development of similar robot for the purpose of an educational course would be too expensive and complicated. The robot presented here is much cheaper but still, it has all main components and functionality in order to fulfil educational goals and learning outcomes. Basic parts of designed robot soccer player (Figure 1.) are:

- RF two channel communication on 433/434 MHz
- 4 DC electro motors (12 V)
- Solenoid kicker which operates on 6 V
- Special made battery charger
- Microcontroller AT89C4051 with electronics
- 7.2V battery for DC electro motors and solenoid
- 4.8V battery for electronics

Fig. 1. Robot soccer player

Robot consists of four levels. The first three levels represent drive and the fourth is the control level. At the first level there are four DC electro motors with gearbox and wheels. At the same level there is also a solenoid which is used to kick the ball. At the second level the two batteries are placed. The first, 7.2 V battery, supplies four DC electro motors and the second, 4.8 V battery, supplies the electronics. Battery charger is placed on the third level. At the uppermost, fourth level, there is a microcontroller and a RF receiver. This level is also known as control level, because it contains electronics for managing electro motors. It controls their speed and orientation depending of signals received from RF receiver. This electronics also controls the kick of solenoid. RF signal is sent by RF transceiver which is connected to personal computer.

The main feature of this robot is the use of a global vision system (Figure 2). Local vision is also possible, but in order to simplify the solution, global vision option was chosen which means that only one camera is used and placed over the soccer field. The camera is connected to computer which is used for image processing and strategy planning. Computer acquires picture from the camera and recognizes the field ground, the robots and the ball. Recognition is mainly based on color segmentation because the colors of the robot teams are predefined (yellow and blue circles placed in the centre on the robot's top plate) and the ball color is predefined also (orange golf ball). Depending on the number of robots in each team, robots have additional color marks on their top so they can be distinguished by the strategy planning program. Picture background has to be green as it is the field ground. This "color rules" are set according to the rules of "small robot league" (Robocup Official Page). Software package used for the image processing and communication with the serial port is MATLAB. Image processing has to be fast and in real-time.

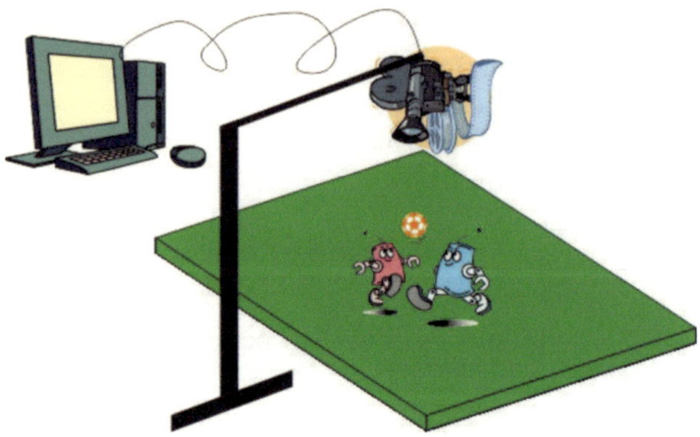

Fig. 2. Robot soccer player system using the global vision

Depending on the positions of the recognized objects, computer sends signals to the microcontrollers in robots. These signals are the upper level instructions that will be translated by the microcontrollers to the lower level instructions for the actuators.
Processed results are sent through the RF transmitter to the robot's RF receiver. Actually, RF transceivers were used, so the communication could be done in both directions. Detailed explanation of all the basic modules will be given in the following sections.

3. Robot modules

3.1 Operative module

Operative module consists of four DC motors that drive the robot and one solenoid. Motors themselves have gearboxes that reduce motor speed but also increase the torque. When choosing motor, the price was crucial and this is because the professional micro-motors for robotics are expensive. Table 1 shows the characteristics of the chosen motor. Figure 3 shows the appearance of the motor with gearbox and motor characteristics.

	No Load		Max efficiency							Stall			
Voltage	Speed	Current	Speed	Torque			Current	Output	Eff.	Torque			Current
V	RPM	A	RPM	mN-m	g-cm	oz-in	A	W	%	mN-m	g-cm	oz-in	A
12	94.7	0.023	69	28.8	294	4.08	0.06	0.21	28.0	106	1080	15.0	0.17

Table 1. Characteristics of the chosen motor

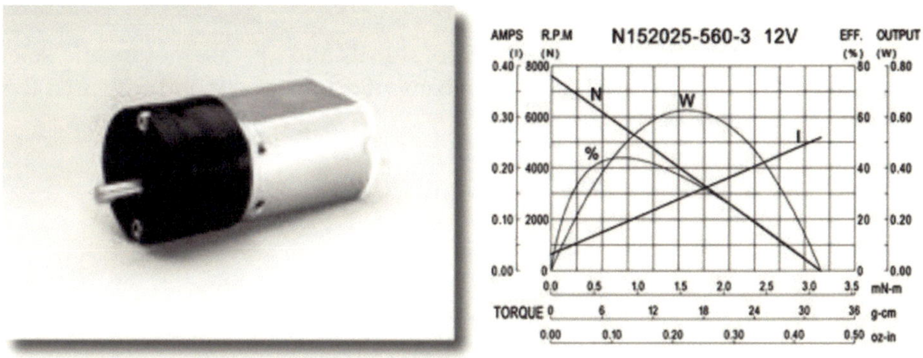

Fig. 3. Appearance of the motor with gearbox and motor caracteristis.

3.2.1. Drive motor and wheel layout

Drive motor and therefore wheel layout as in Figure 4 was chosen because the robot has to be agile and able to turn quickly around its axis. To increase the speed of robot motion forward, the front motors are set at the angle of 33 degrees.

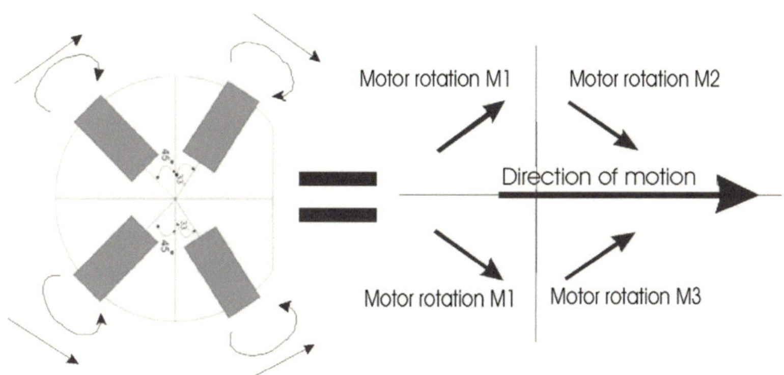

Fig. 4. Drive motor layout

Due to possibility of robot motion linearly with this set of wheels, it is necessary to use special wheels called omniwheel wheels. Although some robot soccer teams develop their own 'home-made' multidirectional wheels (Brumhorn et al., 2007), a commercial solution presented in Figure 5 is used. Wheels have transverse rollers that can rotate, so the wheel, without creating high resistance, can move vertically according to the first direction of motion.

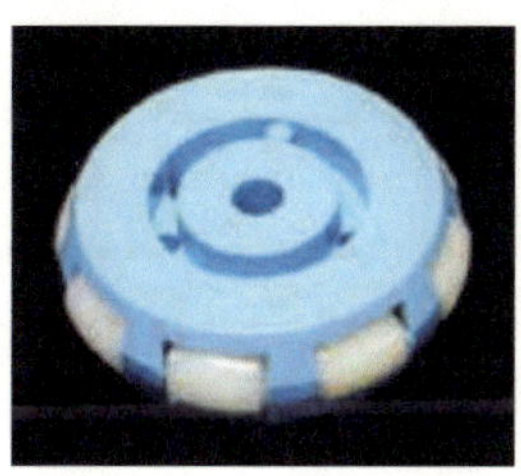

Fig. 5. Omniwheel.

3.2. Electronics module

Electronics module has function to receive control signals and operate motors and solenoid. Receiver placed on robot receives signals from computer transmitter and forwards them to the microcontroller (Figure 6).

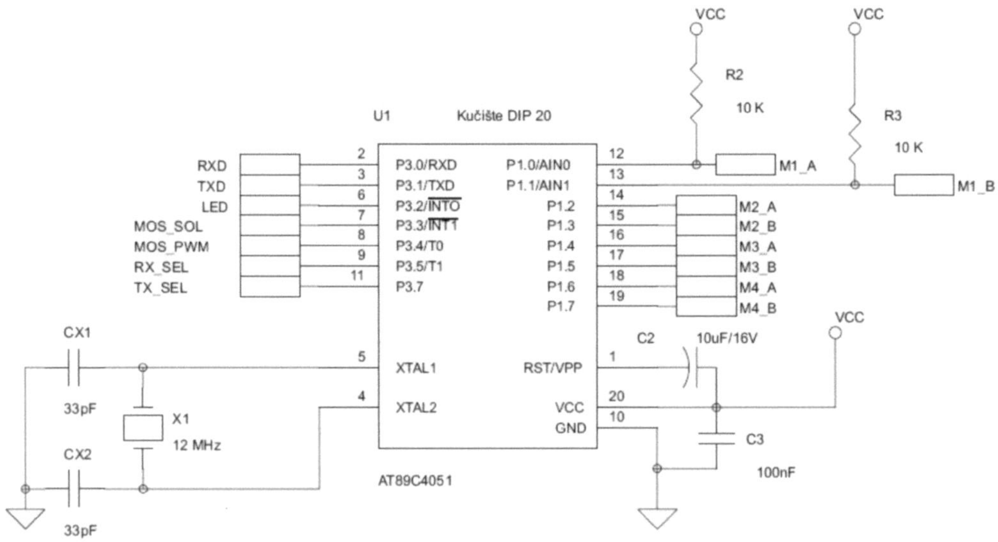

Fig. 6. Electric scheme – microcontroller

For this operation an AT89C4051 microcontroller is used and programmed. Control signals are commands for turning motors on and off in order to settle the direction of the robot motion. Input pins 2, 3, 9, 11 (Figure 6) are connected to the receiver pins 12, 14, 15, 16 (Figure 20). Microcontroller operates four motors (M1, M2, M3, and M4).

3.2.1. Motor controllers

TA7288P drivers are used to control motors speed and direction. There are four drivers, one for each motor. Electric scheme for control of one motor is shown in Figure 7.

Motor management is quite simple. Combination of A and B pins from microcontroller as a result has three functions (Table 2):

- rotate motor left
- rotate motor right
- stop motor rotation

	A	B
Rotate left	0	1
Rotate right	1	0
Stop	0	0

Table 2. Motor control function

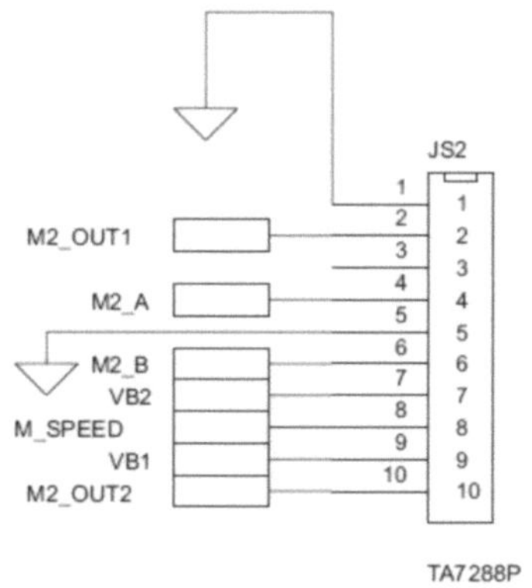

Fig. 7. M2 Motor control

Motor speed is controlled by transistor switch shown in Figure 8. Robot motion is defined by synchronized movement of all four motors. If the robot receives command "move left" or "move right", then the motors M1, M2, M3 and M4, each through its driver (Figure 7) receive orders from the table 2. Depending on these commands motors rotate and bring the robot into desired position.

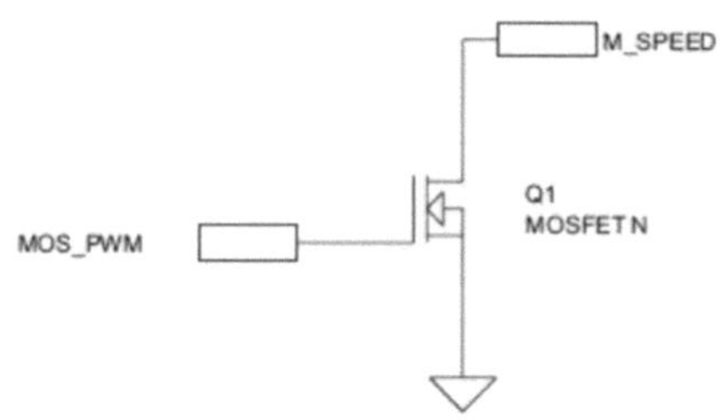

Fig. 8. Motor speed control

Motor speed (over M_SPEED pin) is common for all motors. Applying digital logic on MOS_PWM pin performs speed control. Depending on frequency of digital signal, voltage on M_SPEED pin changes. That results in greater or lesser number of revolutions of the motor.

Presented robot soccer player can move with four different speeds. Speed variation is achieved by sending four different rectangular signal frequencies on MOS_PWM. Also, robot speed depends on distance between robot and ball. If robot is far away from the ball it moves faster. In the immediate vicinity of the ball, the robot is moving slowly to be as accurate as possible.

3.2.2. Solenoid control

Solenoid is electromagnet used to hit the ball. In the immediate vicinity of the ball, the robot slows down and tries to kick ball with solenoid.

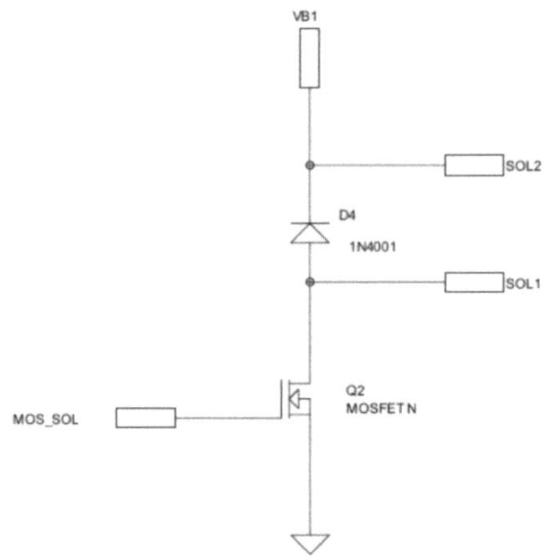

Fig. 9. Solenoid control

When the control software estimates that the robot is near the ball, it sends commands to kick it (optional). Command signal is applied on MOS_SOL pin of solenoid control. Electromagnet is activated and throws the solenoid armature; a little spring provides that solenoid return to its original position. Figure 9 shows the transistor switch that controls the solenoid. SOL1 and SOL2 are connected directly to the solenoid.

3.2.3. Power supply

Designed robot uses two battery power systems. 7.2 V battery is used for motors and solenoid and four 1.2 V batteries are connected in series and power electronics. In Figure 10 is shown the voltage stabilizer.

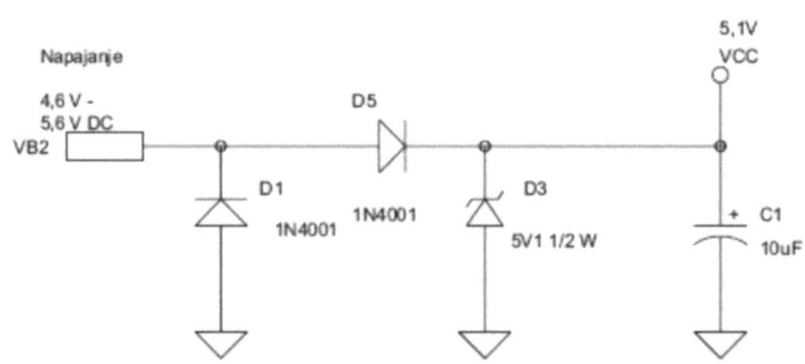

Fig. 10. Voltage stabilizer

Batteries (Figure 11) are rated at 1.2 V, four serial connected should give a voltage of 4.8 V. When batteries are full, they give the slightly higher voltage. In this case measured voltage is 5.6 V. Therefore, the stabilizer shown in Figure 10 is used. Battery charger (Figure 12 and 13) is specifically designed for this robot. There are two ways of charging. Quick 3 hours charge and slow 12 hours charge. Slow charging is safer but fast charging is option for special circumstances. This charger, of course, doesn't need to be a part of the robot, but, if it is included as presented in our schematics, battery changing which can be quite tricky is avoided.

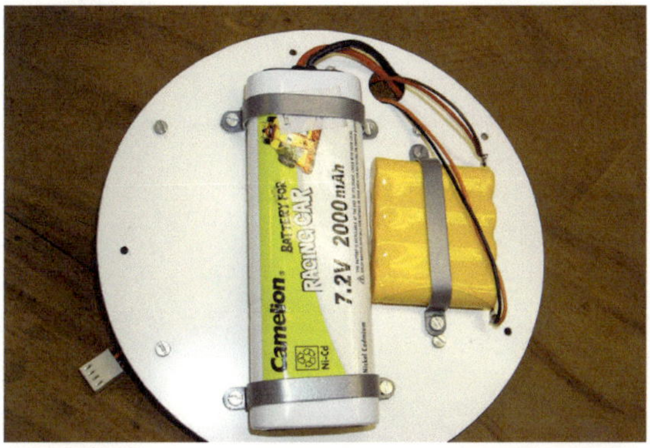

Fig. 11. Batteries

Fig. 12. Battery charger

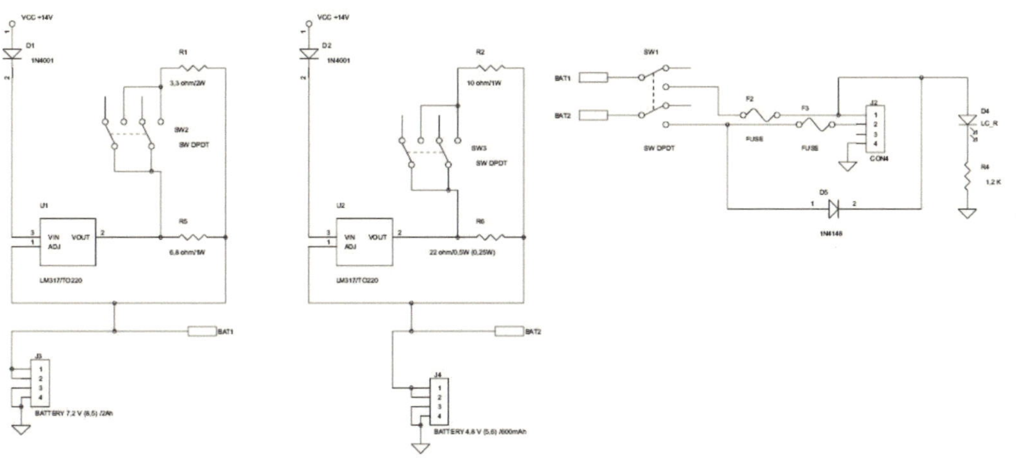

Fig. 13. Battery charger electronic scheme

3.3. Vision module

Simple local vision scheme of our vision module that uses MATLAB software package for the image processing on the central processor is presented in Figure 14.

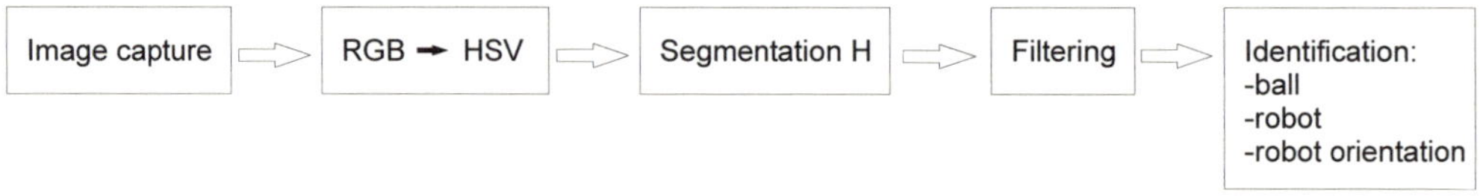

Fig. 14. robot soccer vision module scheme

3.3.1. Image capture

Image capture is the process where a color image is grabbed from the camera and placed into a memory location (buffer) on the host computer. This image must be transformed into a packed RGB image before it can be processed. Figure 15. shows simple example.

Fig. 15. Image capture example

Fig. 16. H component

3.3.2. Color model transformation

Before the segmentation, image has to be converted from RGB color model into HSV color model. Generally, it can be stated that traditional RGB color space is not convenient for this kind of applications due to the high correlation between color components. Although HSI (Hue, Saturation, Intensity) as well as HSV (Hue, Saturation, Value) color spaces has also some problems especially with the low saturation images, they are better choice for wide range of Computer Vision applications (Cheng et al., 2001; Barišić et al., 2008). After that, component H has been isolated. Component H represents hue, i.e. wavelength color. Figure 16. shows the isolated H component in gray scale.

3.3.3. Segmentation

Segmentation refers to the process of partitioning an image into multiple segments. The goal of segmentation is to simplify the representation of an image into something that is more meaningful and easier to analyze. Image segmentation is used to locate objects in images.
In the case example, objects are the robot and the ball. Robot has two markers placed on its top plate. One marker indicates the robot while the other one is used to obtain information on its orientation. In Figure 17, result of region separation is shown.

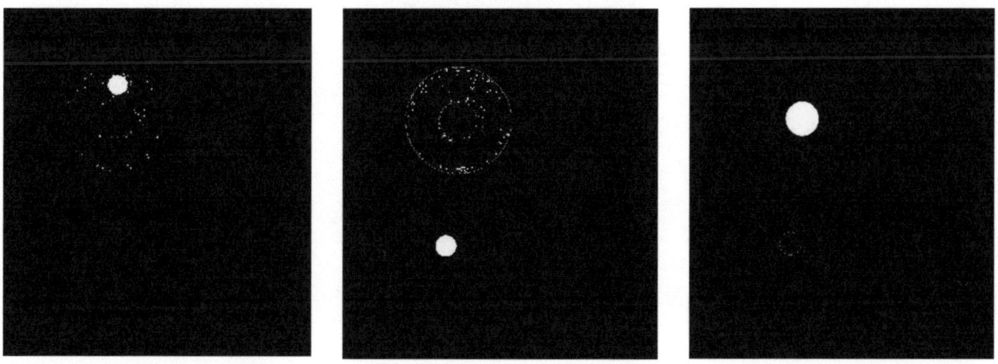

a) b) c)

Fig. 17. Separated regions: a) red regions; b) ball - orange regions; c) yellow regions

a) b) c)

Fig. 18. Filtered regions: a) red regions; b) ball - orange regions; c) yellow regions

3.3.4. Image filtering

Image filtering is a process by which we can enhance images. The first step in objects analysis process is erosion. Erosion gets rid of most of the image noise and reduces the objects to those used to identify the robots and the ball. Figure 18 shows filtered regions.

3.3.5. Identification and orientation

The ball is easily identified because it has distinguishing orange color. If no ball is located it is considered missing and no location is sent to the computer.

Determining the general location of our robots is done by locating the center marker. In our case example that is the yellow regions. Red region is used to determine the orientation of the robot in relation to the ball. Figure 19 shows the position and orientation of robot and the ball.

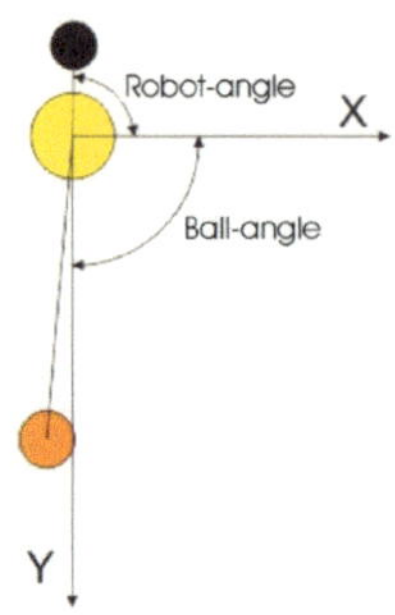

Fig. 19. Position and orientation of robot and the ball

3.4. Communication module

When robot position, orientation and distance from the ball are known, software determines what the robot should do. For example, rotate it, move forward to the ball, or kick the ball. In order to make robot do this operations, it is needed to receive predefined control commands. Although it is possible to apply Bluetooth-based control as well, in our example control commands are sent by radio transmitter and received by receiver that works on 433/434 MHz. The transmitter is connected to a computer via serial port.

Image that is obtained from the camera is processed using the image processing software on PC. Results of the image analysis are used to produce commands that are sent via the RF transmitter. RF receiver located on robot receives commands and forwards them to microcontroller. Microcontroller manages four motors and the solenoid according to the received commands. Table 3 contains set of all commands used to control designed robot soccer player and their meanings.

In Figure 20 transmitter/receiver electrical scheme is shown. For transmitter RXD, TXD, TX_SEL and RX_SEL are connected to the computer RS232 (serial) port through which computer sent commands. Same device is on robot and works as receiver. RXD, TXD, TX_SEL and RX_SEL are connected to microcontroller.

Command	Command meaning
Rotol	Rotate left
Rotor	Rotate right
Pravo	Go straight
Nazad	Go back
Stani	Stop
Udari	Hit the ball
Brzi1	Speed 1
Brzi2	Speed 2
Brzi3	Speed 3
Brzi4	Speed 4
Brzi5	Speed 5

Table 3. Commands

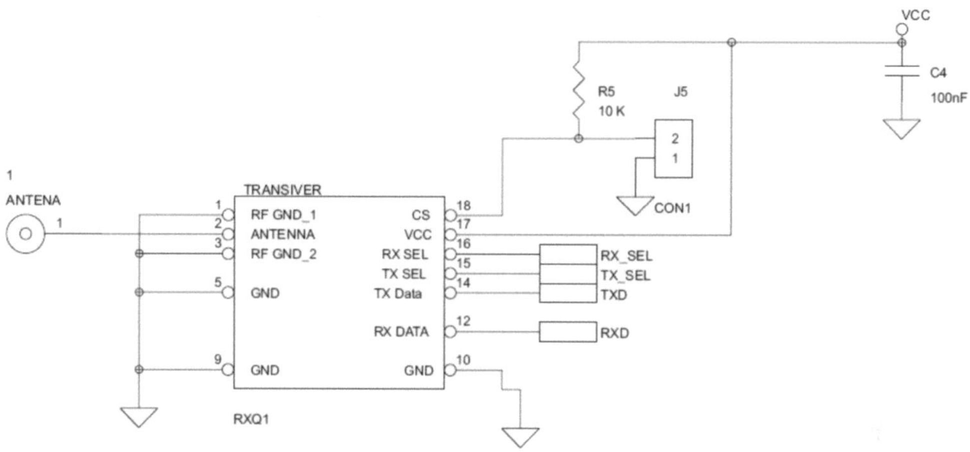

Fig. 20. Electrical scheme transmitter / receiver

4. Curriculum, aim and tasks

Because of wide educational scope provided with the robot soccer idea, instructor and designer of a particular course should be aware of various possibilities available for different groups of students. A simplified overview of the most dominant educational areas is shown in Figure 21. It should be stressed that presented schema does not include all possible areas of investigation and education as well as all possible connections between presented areas. Complexity level should be taken as provisional information because upper complexity boundaries are almost infinite. Only lover boundary of the position at which certain term occur in the figure roughly correspond to the suggested level of needed student previous education in order to attend a course.

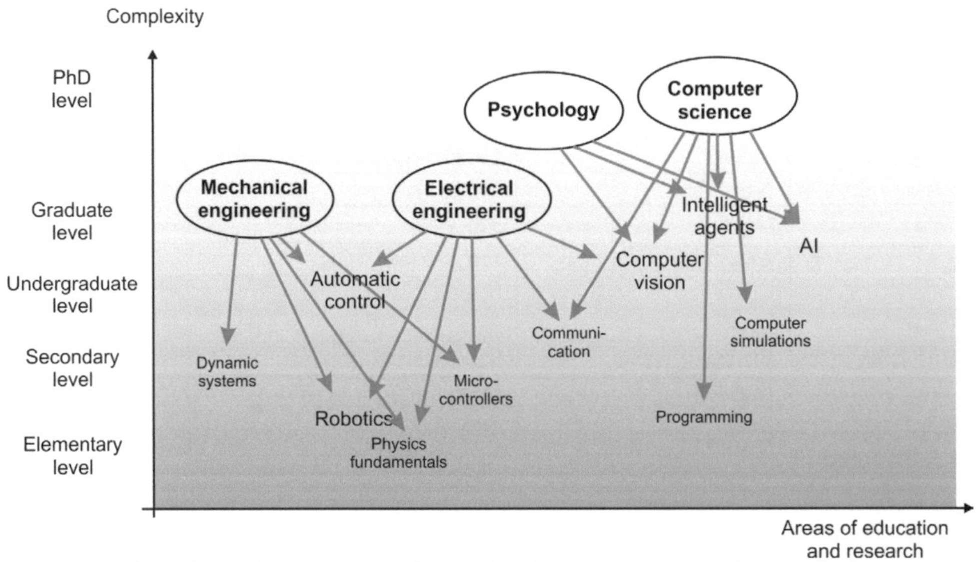

Fig. 21. Simplified chart of course complexity, level of education and area of education

Some terms in the Figure 21 overlap or can be regarded as a part of some other but the idea is to accentuate possible autonomous courses and modules that can also be further combined in order to achieve desired educational goal.

4.1 Curriculum aim

Combine acquired mechanics, electronics, informatics and programming knowledge through autonomous construction of robot soccer player.

4.2 Curriculum tasks

4.2.1 Elementary school (K-12, age 13-15)

As we already mentioned, the course is different for three age groups. The final result is the same - construction of a robot soccer player. The difference is in the amount of autonomous work, which, of course, depends on educational level of certain groups. Therefore, the elementary school students will get ready made modules that they will have to assemble in the one unit. The aim is the same, but their knowledge from this field is lower, therefore their tasks through this course will be simpler.

Educational tasks

To enumerate all the robot modules
To explain the working principle of every single module
To explain the working principle of the entire robot system

Functional tasks

To identify the each part of robot
To identify each module
To assemble robot modules into one unit

Pedagogical tasks
To develop the culture of communication and to express their own views
To acquire the habit of tidiness in the work room

4.2.2. Secondary school (K-12, age 15-18)

Secondary school students will have more complex tasks. Through those tasks, they will manufacture the certain modules by themselves. However, the programming of control functions, just like the manufacturing of image recognition software is not the main task for the students and it is playing just an informative role. In this case, students will just have to change the certain parameters inside of the computer software.

Educational tasks
To enumerate all the robot modules
To explain the working principle of every single module
To explain the working principle of the entire robot system
To explain the particular parts of the module and their role (in it)

Functional tasks
To braze the elements on ready made circuit board individually
To identify each part of the robot
To identify each part of the modules
To identify the certain modules
To assemble robot modules into one unit
To change the parameters of image processing computer software individually

Pedagogical tasks
To develop the culture of communication and to express their own views
To acquire the habit of tidiness in the work room

4.2.3. Students

The students will reach the aim by themselves. Their tasks are the most complex ones. The students will have all the required electrical and mechanical schemes, plans of robot and the prepared material. They will individually manufacture the certain module and finally they will develop the image recognition computer software.

Educational tasks
To enumerate all the robot modules
To explain the working principle of every single module
To explain the working principle of the entire robot system
To explain the certain parts of the module and their role in it
To explain the certain function of each element inside of the module

Functional tasks
To manufacture the module circuit board according to electrical scheme by themselves
To braze the elements on already made module circuit board by themselves
To identify each part of the robot

To identify each part of the modules
To identify the particular modules
To assemble robot modules into one unit
To manufacture the image processing computer software by themselves

Pedagogical tasks
To develop the culture of communication and to express their own views
To acquire the habit of tidiness in the work room

5. Curriculum schedule

Day 1:
Introducing robotics and robot soccer players to the students. Introducing to students the tasks they will need to accomplish. After the short introduction with the tasks, we start with manufacturing of the actuating module of the robot soccer player. The actuating module is the first level of the robot. It consists of a platform with four motors (with gearboxes), the wheels and a solenoid shooter.
Day 2:
Manufacturing the second and the third level of the robot.
Elementary school students: They connect batteries with ready made rechargeable circuit and assemble it all together on the metal platform which makes the second and the third level of the robot.
Secondary school students: They braze the elements of rechargeable circuit on ready made circuit board. Then they connect the batteries with rechargeable circuit and assemble it all together on the metal platform.
University students: They manufacture the module circuit board according to electrical scheme. They braze the elements of rechargeable circuit on the circuit board. Then they connect the batteries with rechargeable circuit and assemble it all together on the metal platform.
Day 3:
Manufacturing the last (fourth) level. This is the most complex level. At this level there is a controlling and receiving circuit.
Elementary school students: They get ready made module. The module is explained and described to them in detail. They assemble the module on the metal platform and put them all together in the one unit (robot).
Secondary school students: They braze the elements of the controlling and receiving circuit on ready made circuit board. Then they assemble the module on the metal platform and put them all together in the one unit (robot).
University students: They manufacture the module circuit board according to electrical scheme. They braze the elements of the controlling and receiving circuit on the circuit board. Finally, they assemble the module on the metal platform and put them all together in the one unit (robot).
Day 4:
Manufacturing the computer software or explaining the way of working for the lower levels.
Elementary school students: Explaining the way of working of the image processing and the computer vision software.

Secondary school students: Explain them the way of working and the principles of the computer vision. They autonomously change the parameters inside of the image processing computer software.
University students: With teacher's assistance, they manufacture the computer software using Matlab software package.
Day 5:
Robot activation. Synchronization and adjusting of the computer software. Demonstration.

6. Discussion

It is important to accentuate that presented courses are short-termed. Longer courses that last for one semester or even longer can handle much more topics and go in more details (Archibald & Beard, 2002; Bushnell & Crick, 2003; Baltes et al., 2004; Hill & van den Hengel, 2005) or can even allow students to manage complete robot building project by themselves (Pucher et al., 2005). If the focus of the course is shifted from building a robot towards artificial intelligence, an approach using the commercial solution of already made robots such as Khepera or LEGO Mindstorms can be used (Miglino et al., 1999; Lund, 2001). Also, size of the student group, number of available teachers must be considered before presenting course plan in order to set achievable goals.
As for the proposed and presented courses, they have been conducted for K-12 children and undergraduate and graduate students as well. In all of these courses, final result of this age levels courses are fully operational soccer robots. Each robot is made by team of 4-5 students with their mentor.
Computer vision education modules were already included as a part of the present Image processing and Computer vision course for the undergraduate and graduate students of informatics and technics at the University of Split. Another module that has already been included as a part of the Computers in technical systems course is the microcontroller programming module.
Preliminary results are encouraging, scholars are highly motivated and the response and working atmosphere is great. Members of the K-12 groups are, according to their statements, very keen on continuing with the work related with the presented problem. University students also showed great interest in the presented course modules. During the development of the system occurred some real life problems such as light and vision, noise in communication, speed of communication with PC port. Students had to deal with them or were given an insight because these are the problems that are preparing them for the actual work after graduation.
So far, curriculum described in Section 3 and 4 did not included collaboration between robots and global strategy planning. This option should be added soon especially for the undergraduate and graduate students in computer science that are listening some of the AI courses in their regular curriculum. Also, robot simulation software has not been developed yet so it has not been included here although simulators are providing significant possibilities in investigation of artificial intelligence.

7. Conclusion

In this chapter we have presented a framework for modular practical courses using robotics and robot soccer problem as an educational tool. Presented approach is flexible so it can be adapted for various age groups with different scopes of interest and previous knowledge. Some case examples have been shown. Also, we have described the robot electronics, mechanics and the whole global vision system that was developed for this purpose. Main requirements for the system and robots were: simplicity, overall price (around 300 EUR/robot + PC and a camera) and openness for further improvements.

Integration and need of interdisciplinary knowledge and its application for the successful completion of the project defined by the course aims provides possibility of collaboration between different departments and teachers. It offers the possibility to apply slightly adopted courses to different age groups. Constructive education approach and the possibility of using the presented practical course modules as a support for wide range of existing engineering courses along with a great first response from the scholars motivates authors to continue with the development of the course and its modules. Further development of the courses for the youngest children, as well as specialized AI courses, is expected in the future.

8. References

Anderson, J. & Baltes, J. (2006). An agent-based approach to introductory robotics using robotic soccer. *International Journal of Robotics and Automation*, Vol. 21, Issue 2 (April 2006), pp. 141 - 152, ISSN 0826-8185, ACTA Press Anaheim, CA, USA.

Archibald, J. K. & Beard, R. W. (2002). Competitive robot soccer: a design experience for undergraduate students, *Proceedings of the 32nd Annual Frontiers in Education*, fir, Vol. 3., pp. F3D14-19, Boston, MA, USA, November, 2002.

Arlegui, J.; Fava, N.; Menegatti, E.; Monfalcon, S.; Moro, M. & Pina, A. (2008). Robotics at primary and secondary education levels: technology, methodology, curriculum and science, *Proceedings of 3rd International Conference ISSEP Informatics in Secondary Schools Evolution and Perspectives, July, 2008, Torun, Poland.*

Beard, R.W.; Archibald, J.K. & Olson, S.A. (2002). Robot soccer as a culminating design project for undergraduates, *Proceedings of the 2002 American Control Conference*, Vol. 2, pp. 1086-1091, ISBN 978-0780372986, Anchorage, Alaska, USA, May, 2002, IEEE, Los Alamitos, CA, USA.

Baltes, J. & Anderson, J. (2005). Introductory programming workshop for children using robotics. *International Journal of Human-Friendly Welfare Robotic Systems*, Vol.6, No.2, 17-26, ISSN 0929-5593.

Baltes, J.; Sklar, E. & Anderson, J. (2004). Teaching with robocup, *Proceedings of the AAAI Spring Symposium on Accessible Hands-on Artificial Intelligence and Robotics Education*, pp. 146-152, Stanford, CA, March, 2004.

Barišić, B.; Bonković, M. & Papić, V. (2008). Evaluation of fuzzy clustering methods for segmentation of environmental images, *Proceedings of 2008 International Conference on Software, Telecommunications and Computer Networks*, ISBN 978-953-290-009-5, Split-Dubrovnik, Croatia, September, 2008, FESB, University of Split.

Bruder, S. & Wedeward, K. (2003). An Outreach Program to Integrate Robotics into Secondary Education. *IEEE Robotics & Automation Magazine,* Vol. 10, September, 2003, pp. 25-29, ISSN 1070-9932.

Brumhorn, J.; Tenechio, O. & Rojas, R. (2007). A Novel Omnidirectional Wheel Based on Reuleaux-Triangles. In: *RoboCup* 2006 : Robot Soccer World Cup X. Lakemeyer, G.; Sklar, E.; Sorrenti, D. G.; Takahashi, T. (Eds.), 516-522, Springer Verlag LNAI 4434, ISBN 978-3-540-74023-0.

Bushnell, L.G. & Crick, A.P. (2003). Control Education via Autonomous Robotics, *Proceeding of 42nd IEEE Conference on Decision and Control,* Vol.3, pp.3011-3017, ISBN 0-7803-7924-1, Maui, Hawaii, USA, December, 2003, IEEE Control Systems Society.

Chambers, J.M.; Carbonaro, M. & Murray, H. (2008). Developing conceptual understanding of mechanical advantage through the use of Lego robotic technology. *Australasian Journal of Educational Technology,* Vol. 24(4), pp. 387-401, ISSN 1449-3098.

Cheng, H. D.; Jiang, X. H.; Sun, Y. & Wang, J. L. (2001). Color Image Segmentation: Advances & Prospects, *Pattern Recognition,* Vol. 34(12), pp. 2259-2281, ISSN 0031-3203.

Coradeschi, S. & Malec, J. (1999). How to make a challenging AI course enjoyable using the RoboCup soccer simulation system. In : *RoboCup98: The Second Robot World Cup Soccer Games and Conferences.* Asada, M. & Kitano, H. (Eds.), 120-124, Springer Verlag LNAI, ISBN 978-3-540-66320-1, Berlin / Heidelberg.

Cornell RoboCup Team documentation. (31.08.2009.). http://www.cis.cornell.edu/boom/2005/ProjectArchive/robocup/documentation.php.

Gage, A. & Murphy, R. R. (2003). Principles and Experiences in Using Legos to Teach Behavioral Robotics, *Proceedings of 33rd ASEE/IEEE Frontiers in Education Conference,* pp. 1-6, ISBN 0-7803-7961-6, November 5-8, 2003, Boulder, CO, IEEE.

Hill, R. & van den Hengel, A. (2005). Experiences with Simulated Robot Soccer as a Teaching Tool, *Proceedings of the Third International Conference on Information Technology and Applications (ICITA'05),* Vol.1, pp. 387-390, ISBN 0-7695-2316-1, Sydney, Australia July, 2005, IEEE Computer Society, Los Alamitos, California, USA.

Henkel, Z.; Doerschuk, P. & Mann, J. (2009). Exploring Computer Science through Autonomous Robotics, *Proceedings of 39th ASEE/IEEE Frontiers in Education Conference,* October, 2009, San Antonio, USA.

Lund, H. H. (1999). Robot soccer in education. *Advanced Robotics,* Vol. 13, No.6-8, 1999, pp. 737-752(16), VSP, an imprint of Brill, ISSN 0169-1864.

Lund, H. H. & Pagliarini, L. (1999). Robot soccer with lego mindstorms. In : *RoboCup98: The Second Robot World Cup Soccer Games and Conferences.* Asada, M. & Kitano, H. (Eds.), 141-151, Springer Verlag LNAI, ISBN 978-3-540-66320-1, Berlin / Heidelberg.

Lund, H. H. (2001). Adaptive Robotics in Entertainment. *Applied Soft Computing,* Vol.1, pp. 3-20, Elsevier, ISSN 1568-4946.

Matarić, M. (2004). Robotics Education for All Ages, *Proceedings of the AAAI Spring Symposium on Accessible Hands-on Artificial Intelligence and Robotics Education,* Stanford, CA, March, 2004.

Matarić, M.J.; Koenig, N. & Feil-Seifer, D.J. (2007). Materials for Enabling Hands-On Robotics and STEM Education, *Papers from the AAAI Spring Symposium on Robots and Robot Venues: Resources for AI Education,* 2007, pp. 99-102, ISBN 9781577353171, March, 2007, Stanford University, Stanford, CA, USA, AAAI Press, Stanford.

McComb, G. (2008). Getting kids into Robotics. *Servo Magazine*, October, 2008, pp. 73-75, T&L Publications, Inc., North Hollywood, CA, USA, ISSN 1546-0592.

Miglino, O.; Lund, H. H. & Cardaci, M. (1999). Robotics as an educational tool. *Journal of Interactive Learning Research*, Vol.10, Issue 1 (April 1999), pp. 25-47, ISSN 1093-023X, Association for the Advancement of Computing in Education, USA

Nagasaka, Y. ; Saeki, M.; Shibata, S. ; Fujiyoshi, H.; Fujii, T. & Sakata. T. (2006). A New Practice Course for Freshmen Using RoboCup Based Small Robots. In: *RoboCup* 2005 : Robot Soccer World Cup IX. Bredenfeld, A.; Jacoff, A.; Noda, I.; Takahashi, Y. (Eds.), 428-435, Springer Verlag LNAI 4020, ISBN 978-3-540-35437-6.

Nourbakhsh, I.R.; Hamner, E.; Crowley, K. & Wilkinson, K. (2004). The educational impact of the Robotic Autonomy mobile robotics course, *Proceedings of 2004 IEEE International Conference on Robotics and Automation*, Vol.2, pp. 1831-1836, April-May, 2004, New Orleans, LA, USA, IEEE, USA.

Novales, M.R.; Zapata, N.G. & Chandia, S.M. (2006). A strategy of an Introduction of Educational Robotics in the School System. In: *Current Developments in Technology-Assisted Education*, Vol.2, Méndez-Vilas, A.; Martín, A.S.; González, J.A.M. & González, J.M (Eds.), pp. 752-756, Formatex, ISBN 978-84-690-2472-8, Badajoz, Spain.

Papert, S. (1980). *Mindstorms: Children, Computers, and Powerful Ideas*. NY, New York: Basic Books.

Papert, S. (1986). *Constructionism: A New Opportunity for Elementary Science Education*. A MIT proposal to the National Science Foundation.

Piaget, J. & Inhelder, B. (1966). *La psychologie de L'enfant*. Paris: P.U.F.

Pucher, R.K.; Wahl, H.; Hofmann, A. & Schmöllebeck, F. (2005). Managing large projects with changing students – the example of the roboter soccer team "Vienna Cubes", *Proceedings of the 22nd ASCILITE Conference*, Vol. 2, pp. 561-567, ISBN 0975709313, Brisbane, December, 2005, Australasian Society for Computers in Learning in Tertiary Education, Figtree, NSW, Australia.

Riley, J. (2007). Learning to Play Soccer with the Simple Soccer Robot Soccer Simulator, In : *Robotic Soccer*, Lima, P. (Ed.), pp. 281-306, Itech Education and Publishing, ISBN 978-3-902613-21-9, Vienna, Austria.

Robocup official page. (31.08.2009.). http://www.robocup.org.

Verner, I.M. & Hershko, E. (2003). School Graduation Project in Robot Design: A Case Study of Team Learning Experiences and Outcomes. *Journal of Technology Education*, Vol. 14, No. 2, pp. 40-55, ISSN 1045-1064.

Wedeward, K. & Bruder, S. (2002). Incorporating robotics into secondary education, *Proceedings of the 5th Biannual World Automation Congress (WAC 2002)*, Vol. 14, pp. 411-416, ISBN 1-889335-18-5, Orlando, USA. June, 2002, Albuquerque, New Mexico : TSI Press.

6

Distributed Architecture for Dynamic Role Behaviour in Humanoid Soccer Robots

Carlos Antonio Acosta Calderon, Rajesh Elara Mohan and Changjiu Zhou
Advanced Robotics and Intelligent Control Centre, Singapore Polytechnic
500 Dover Road, Singapore 139651

1. Introduction

In recent years, robotics competitions have flourished all over the world. These competitions have been accepted among the scientific community because of their roles in the in the advancement of science. Roboticists have understood that competitions do not only nurture innovative ideas, but they also serve as a common testbed, where approaches, algorithms, and hardware devices could be compared by evaluating them in the same environment and under identical conditions. Competitions also motivate students to be involved in robotics to acquire new technological and problem solving skills. Robot soccer has proved to be a challenging and inspiring benchmark problem for artificial intelligence and robotics research. In a soccer game, one team of multiple players must cooperate in a dynamic environment and sensory signals must be interpreted in real time to take appropriate actions. The soccer competitions test two multi-robot systems competing with each other. The presence of opponent teams, which continuously improve their systems, makes the problem harder every year. The number of goals scored is an objective performance metric that allows a comparison of the systems.

Two of the most successful competitions for robot soccer are FIRA (Federation of International Robot-soccer Association) and RoboCup. Both FIRA and RoboCup have their international conferences co-located with the games to promote scientific dissemination of the novel ideas and solutions proposed by the teams. Currently, there is a number of different soccer leagues in RoboCup and FIRA focusing on different aspects of the soccer challenge. Both competitions have a humanoid league, where autonomous robots with a human-like body and human-like senses play soccer against each other. RoboCup has set the final target of the humanoid robot soccer competitions for being able to develop a team of humanoid soccer robots capable of defeating the human world champion team by 2050 (Kitano & Asada, 2000). Although humanoid soccer robots are far from human performance, their progress is particularly visible. Nowadays, the robots manage basic soccer skills like walking, kicking, getting-up, dribbling, and passing.

The early stages of these competitions consisted only of basic robotic soccer skills, such as walking, getting-up, penalty kick, and obstacle avoidance. In 2005 RoboCup introduced the 2 vs. 2 soccer games for the Humanoid League, which became 3 vs. 3 soccer games in 2008. It is planned to increase the number of robot players in the games until eventually it reaches

11. Raising the number of players poses new challenges for the roboticists and further expands the possibilities of team play. The increased complexity of soccer games with more players will make structured behaviour a key factor for a good humanoid soccer team. Cooperative behaviour of the humanoid robots would give advantage to the team to achieve its ultimate goal, to win the game.

In soccer, cooperative behaviour is displayed as coordinated passing, role playing, and game strategy. Wheeled robots with global vision systems and centralized control have achieved such behaviours; example of this is the RoboCup Small-Size League. In the RoboCup Small- Size League, teams are able to display formations according to the strategy of the team. In addition, the robots show a well-defined behaviour according to their assigned roles during the game. The robot's role would determine the type of contribution of a robot to the strategy of the team. Passing is therefore a consequence of the behaviour for the roles and the support for the strategy. In RoboCup Small-Size, a central computer is responsible for deciding the roles, positions and behaviour of the five robots of the team. The central computer receives a complete image of the soccer field from an overhead camera; this image is then processed and used to calculate new positions and states for each robot. Finally, the position and states are sent to the robots. In the humanoid soccer games, there is no central computer; each robot is meant to be autonomous and is equipped with a camera as a main source of information about its environment. The partial information about the environment that the humanoid robot can collect with the camera, along with received information from the team-mates, is the information used to determine its behaviour. As it could be guessed, the partial information about the environment and other robots makes the problem of behaviour control quite challenging. Most of the teams in the RoboCup Humanoid League have identified the need to have different roles for the robots in the team. This role assignation is then useful to specify particular behaviours that must be unique to the role of the robot, e.g. the goalie is the only robot that is allowed to dive to block the ball when it is drawing near the goal. Despite the obvious advantages to the static role assignation, some drawbacks are still observed. The roles can only be changed before or after the game, i.e. during a normal game, the roles of the robots are not allowed to change. If a robot is damaged, and cannot continue the game, the other robots could not adjust their roles to compensate for the team's disadvantage.

This Chapter describes the approach developed at the Advanced Robotics and Intelligent Control Centre (ARICC) of the Singapore Polytechnic with the team of humanoid robots named Robo-Erectus. Robo-Erectus team has taken part in the RoboCup Humanoid League since 2002 (Zhou & Yue, 2004). The work described here deals with role assignation for the robots, team formation and the relation of strategy and behaviour for the humanoid robots.

The rest of the Chapter is organized as follows. First, a review of related works are presented; part of this work has been done in different robotic platforms. After reviewing these approaches, the proposed method is then presented. Next is an introduction of the Robo-Erectus humanoid robot, the hardware, the control, and the software architecture used for the experiments. Section 5 presents the experiments and results obtained with the proposed approach. These experiments were conducted on a simulator as well as the actual robots. Finally, Section 6 provides concluding remarks about this approach and future work.

2. Literature Review

Cooperative behaviour is an intrinsic feature of the soccer robot competitions. It has been addressed from different points of view; these approaches are based on the capabilities of the robots to perceive the world. The perception of the system would bring advantages and disadvantages and not all these approaches can be shared among the different soccer leagues. For example, the Small Size League of the RoboCup and FIRA use a global vision system that obtains the positions of the robots and the ball from the images. This information is used by a server computer to calculate and send the next positions of all the robots. In this scenario, the cooperative behaviour of the system is conceived by one central orchestrator, who has a full picture of the game. In this work, despite the main focus are humanoid robots, relevant approaches of other leagues are also discussed.

The RoboCup Four-Legged League successfully addressed many aspects of the soccer problem. In this league, four autonomous Sony AIBO robots play in each team. Each robot perceives the environment from a camera and tries to estimate its position as well as the positions of its teammates, foes, and the ball. Cooperative behaviour in this league is an emergent behaviour achieved by all the independent robots in the team. The challenge is to have the group of robots working together towards the same goal, without interfering among themselves, but also supporting their roles.

Previous work in the RoboCup Four-Legged League had addressed the cooperative behaviour problem. Phillips and Veloso presented an approach to coordinate two robots for supporting the team attack. The robots achieved this behaviour by assigning roles to the robots, i.e. attacker and supporter. The supporter robot will position itself to not interfere with the attacker, yet able to receive a pass or recover a lost ball (Phillips & Veloso, 2009). Other researchers have focused on the autonomous positioning of the robots at the beginning of a game or after a goal. Most of the leagues consider it a foul to manually place the robots to resume a game, and each league would assign some kind of penalty to the team that incurs in such situation. In addition, after a goal the robots would have a certain amount of time to self-position themselves before the game is resumed. It is quite common to have robots that played as defenders located in a different region than a defender should be located. If that is the case it might be more meaningful to place the robot behaving as another role just for that specific period of time, instead of having the robots moving back to its defending area, which would require more time than the allowed. Work et. al. proposed a method for player positioning based on potential fields. The method relies on roles that are assigned by a given strategy for the game. The potential fields calculate the shortest paths for the robot to self-position after a goal (Work et al., 2009). Zickler and Veloso proposed random behaviour tactics for the robots in the team. The proposed method can be used to generate a shorter plan in contrast with plans which are too far in the future. The main advantage of the method is the ability of re-planning short plans (Zickler & Veloso, 2009).

Teams in the RoboCup Four-Legged League, Standard Platform League, and Humanoid League have studied the problem of cooperative behaviour from the point of view of player role. A robot with a specific role in the game would contribute to the final objective of the team in a different way. Some teams have addressed the role assignation as a static problem (Acosta et al., 2007) others have addressed the problem as a dynamic assignation. An example of role assignment concept in the planning layer is used in NimbRo humanoid robots (Behnke & Stueckler, 2008). It implements default role negotiation and role switching.

A player can be assigned as a striker, a defender, or a goalkeeper. If only one player is on the field, it plays offensive. When the team consists of more than one field player, the players negotiate roles by claiming ball control. As long as no player is in control of the ball, all players attack. If one of the players takes control, the other player switches to the defensive role. Another application of the role concept is goal clearance by the goalkeeper. The goalkeeper switches its role to field player when the ball gets closer than a certain distance. In this case, it starts negotiating roles with other field players like a standard field player. Thus, the goalie might walk toward the ball in order to kick it across the field.

Coelho et. al. approached the coordination and the behaviour of the robots in a team from the point of view of genetic algorithm (Coelho et al., 2001). They use the genetic algorithms to optimize the coordination of the team. The fitness of the objective function is associated with the solution of the soccer problem as a team, not just as player of the team. The algorithm is flexible in the sense that we can produce different configurations of the team. One drawback of the method is that this process must be done offline and it does not permit online adjustments.

Communication between the Darmstadt Dribblers humanoid robots is used for modelling and behaviour planning (Friedmann et al., 2006). The change of positions of opponents or team members can be realized. A dynamical behaviour assignment is implemented on the robots in such a way that several field players can change their player roles between striker and supporter in a two on two humanoid robot soccer game. This change is based on their absolute field position and relative ball pose.

Risler and von Strik presented an approach based on hierarchical state machines that can be used to define behaviour of the robots and specify how and when the robots would coordinate (Risler & von Strik, 2009). The roles of the robots are assigned dynamically according to the definition given inside the state machines. For example, if a robot is approaching towards the ball with a good chance to score, the robot's role would become striker, and if the state machine defines that only one striker should be in the team, the previous striker would negotiate for another role.

Previous works on cooperation of the behaviour are mainly based on the role of the player. As presented, some works focus on static roles of the players that cannot change during the execution of the game. Others use dynamic assignation of the roles during a game, the criteria are based on the position of the players. The work presented here uses a dynamic role assignation based on the strategy that the team has for the game. Other factors like remaining time and goal difference are used to determine the new formation of the team.

3. Dynamic Role Assignation

On a three versus three humanoid robots game, each robot has assigned a role, e.g. goalie, defender, striker. The fixed or static roles of the robots do not change throughout the whole duration of a robot soccer game, regardless of the game time and goal difference. This method does not cater for scenarios when the team needs to win the game to qualify for the next round, or a premeditated draw game as part of a league game strategy. In some cases the roles of robots are not bounded or limited to an area of the soccer field, where the robots are free to roam in the field. This will cause unnecessary and inefficient movement of the robots.

In a robot soccer game, the environment is highly competitive and dynamic. In order to work in the dynamically changing environment, the decision-making system of a soccer robot system should have the flexibility and online adaptation. Thus, fixed roles are not the best approach, even though it is possible to display some level of cooperative behaviour, the system lacks the flexibility to adapt to unforeseen situations. A solution to the fixed roles of the robots is to allow a flexible change of strategies and roles of the robots in a soccer game, according to the game time and goal difference (Acosta et al., 2008). The robot's area of coverage may be limited according to their current roles. This is to allow better deployment and efficiency of the robots movement.
The proposed approach conceived the team as a self-organizing strategy-based decision-making system, in which the robots are able to perform a dynamic switching of roles in the soccer game.

3.1 Cooperative behaviour

In order to achieve a designated goal, agents must work together as a team. However, one thing to remember is that each agent has a different role, and that performing the role is crucial to achieve the desire goal. The changing of roles will be filtered based on three criteria; Strategy, Game Time and Goal Difference respectively.

- Strategy, the strategy to be used for the game will be selected before kick-off and half time of the game. The strategy defines the final objective of the team, and specifies if team should be more offensive or defensive in their play.
- Game Time, the time of the game, 10 minutes for each half of the normal game, and 5 minutes for each half of extra time when required.
- Goal Difference, defined as the difference of own team goal score and opponent team goal score.

The above criteria would then determine a particular formation for the team. Formation here is the number of players for particular roles, not directly the position of the robots on the field. For example, a formation for a team of three robots could be one goalie, one defender, and one striker; another formation could be one goalie and two strikers. In this regard, the formation would specify the number of players with different roles that are required at that moment.
Each role would specify the behaviour of the robot and the region where the robot should be located. However, the robots are free to move around the field, but the specified regions are used as reference for the robots and also for some other behaviour like self-positioning.
The change of formations based on the three criteria has been implemented as a finite state machine on all the robots. The three criteria are updated as follows:

- Strategy, the strategy can only be selected before kick-off and half time of game. The strategy would remain the same for as long as the game last.
- Game Time, the time of the game is kept by each robot, and it is updated when a signal is received from the computer that provides messages like kickoff, stop, etc.
- Goal Difference, it is updated when receive signals from the computer that provides the messages.

The computer that sends the signals for kickoff, stop, resume, etc is known as a referee box. Many leagues in the RoboCup have implemented the use of the referee box with two purposes. First, to standardize the signals that are sent to the teams, the referee box broadcasts the signals to the wireless networks of both teams; and second, to reduce the human interference in the game and to increase the autonomy of the system. The referee box sends signal only when the referee wants to let the robots know about particular situation e.g. a free kick. Most of the referee boxes include some other kind of information that robots are not able to perceive just yet, information such as time and goal difference. In our proposal, we have included both into the referee box messages.

3.2 Strategies

The strategy of a team will define the main objective of the team for a particular game. The strategy is fixed for our approach, which means that it can only be changed if the game stops i.e. half time or of full time (when playing extra time). However, from our experiments we have discovered that it is more meaningful not to change the strategy during a game, unless it is really necessary. While other works have defined some strategies for the games, we have defined four strategies that embrace, in our opinion, all the possibilities of the soccer games.

3.2.1 Normal Game Strategy

The Normal Game strategy is used when the team does not have a specific agenda, which may be used in a friendly game so as not to reveal unnecessary strategies. This strategy uses a Normal Formation throughout the whole game including extra time, regardless of game time and goal difference. Figure 3.1 below illustrates the Normal Game Strategy.

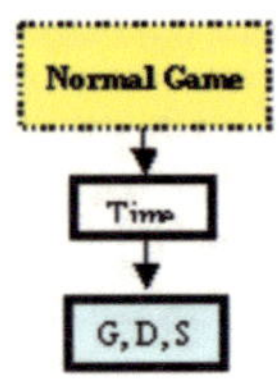

Fig. 3.1 The finite state machine for the Normal Game Strategy.

3.2.2 Must Win Strategy

The Must Win strategy is used when the team has to win the game. The nature of this strategy is more aggressive, with Offensive Formation implemented during the second half of normal game time, and All Out Formation implemented during the second half of extra time, if the team is still losing or draw. When the team is winning, the formations will change to defensive mode to maintain the lead. Figure 3.2 below illustrates the Must Win Strategy.

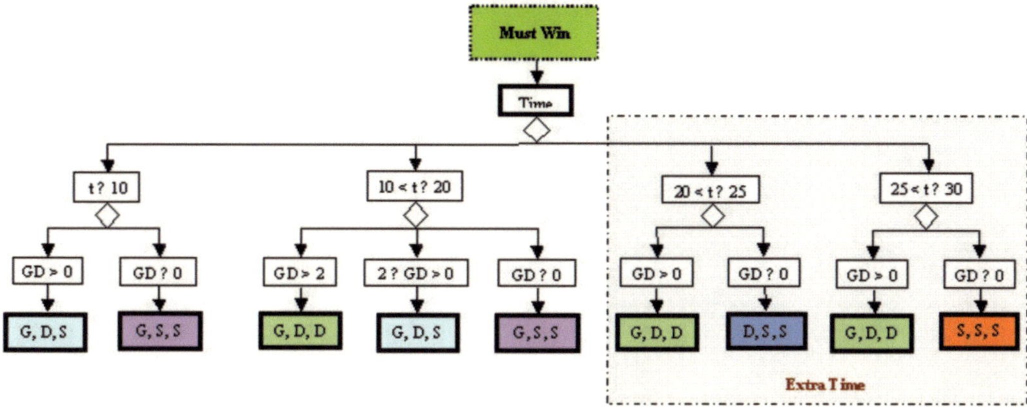

Fig. 3.2 The finite state machine for the Must Win Strategy.

3.2.3 At Least Draw Strategy

The At Least Draw strategy is used when the game strategy is just to aim for a draw or a marginal win. This strategy can be used as a part of first round game, when the team does not want to unnecessarily reveal the full potential of the robots to rival teams. This strategy will implement a normal formation when draw, and a defensive formation when the team is leading. Figure 3.3 below illustrates the At Least Draw Strategy.

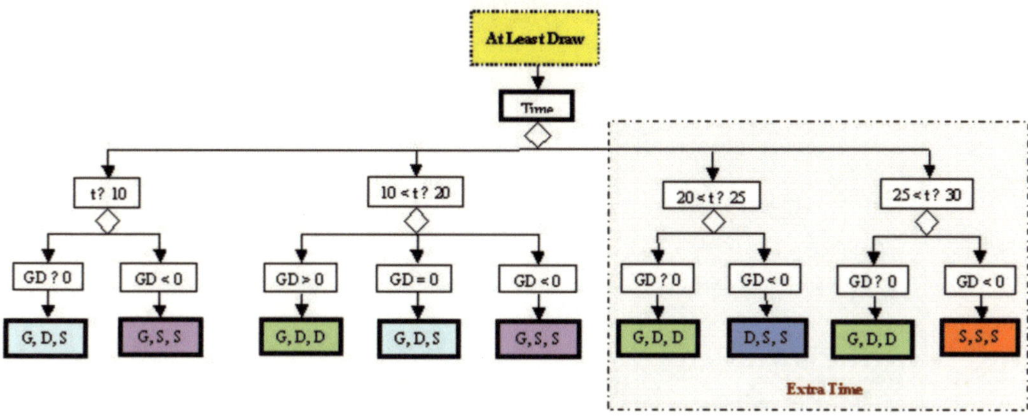

Fig. 3.3 The finite state machine for the At Least Draw Strategy.

3.2.4 Close-Up Strategy

The Close-Up strategy is used to narrow the goal difference when the team is losing to the opponent team. For example, when opponent team scores 10 goals and own team scores 3 goals, this strategy will try to narrow the goal difference to 10 goals versus 6 goals. Figure 3.4 below illustrates the Close-Up Goals Strategy.

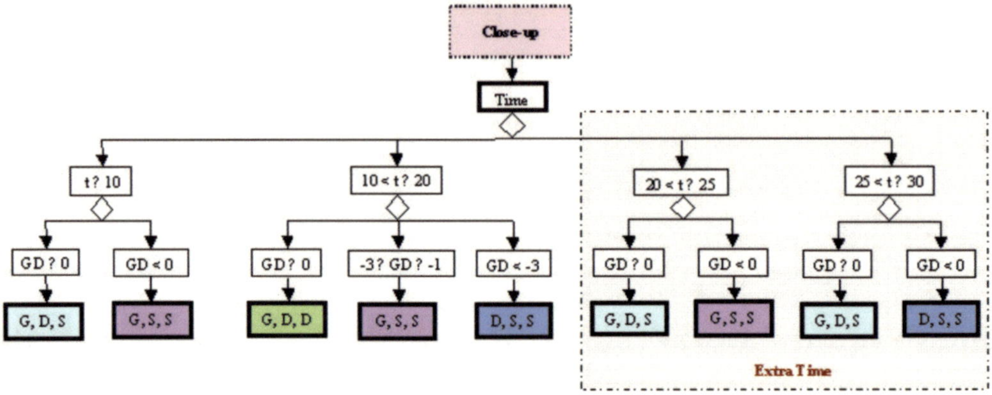

Fig. 3.4 The finite state machine for the Close-up Strategy.

3.3 Negotiation of roles

At several points during the robot soccer game, the robots need to communicate with one another. This may involve informing agents of events or responses, asking for help or information, and negotiating to iron out inconsistencies in information or to agree on a course of action. Negotiation will be necessary when changing of roles is required. This process is complicated since each robot must have its own decisions and there is no central coordinator to assign the roles to each robot.

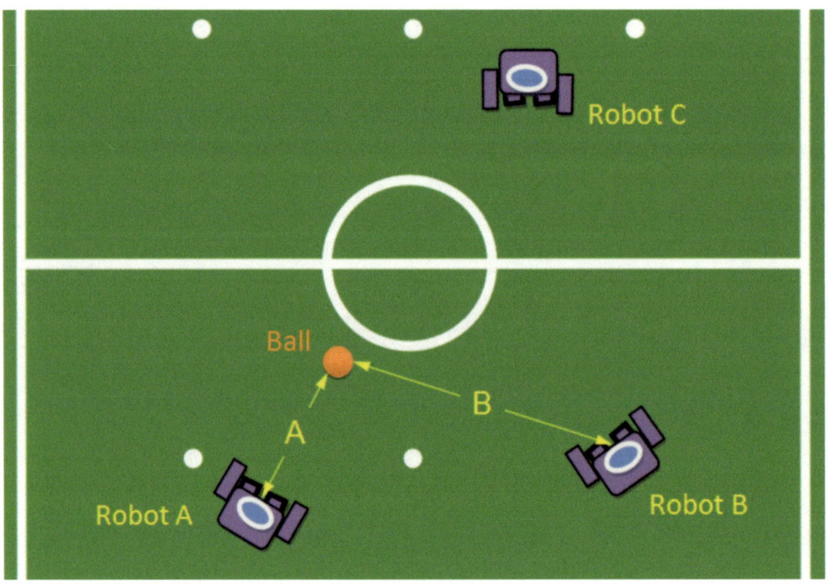

Fig. 3.5 Robots approach through ball approximation

In review of the dynamic roles design, the ball approach method should be taken into consideration. This is for better robot deployment and efficiency during dynamic role change. Team formation change may be based on the positions of the three robots. This could be useful when two robots required negotiating for a particular role. The proximity and approach to the ball could be used to determine which robot would get the role. This means that if we have a situation as the one presented in Figure 3.5, where robot A and robot B are disputing for the role of striker and defender, how should they solve this situation? There are two criteria employed to solve this problem. First, the robots should evaluate if they are attacking or defending. This is evaluated by determining if the opponents are close to the ball, in that case it is considered that the team is defending; otherwise, it is considered that the team is attacking. Second, the robots will evaluate the distance to the ball. If the robot believes that its distance is shorter than that of the other robot, it will approach to the ball and win the role, i.e. defender when the team is defending or striker when attacking.

The robots in the field can approach the ball based on proximity, the robot nearer to the ball will move forward. As illustrated in Figure 3.5, Robot A will approach the ball rather than Robot B due to its proximity, as distance A is shorter compared to distance B. This is taken into consideration that both Robot A and B heading is facing the ball. As each robot has the ability of knowing its own heading orientation, the ball approach method considers the heading of the robot. Illustrated in Figure 3.6 below, Robot A is heading away from the ball, while Robot B is heading towards the ball. Although Robot A distance is nearer compared to Robot B, however it will take time for Robot A to turn its heading position towards the ball. Hence Robot B will approach the ball instead, while Robot A will proceed with heading adjustment. This method is limited to a difference in distances, defined by a threshold, in comparison to the other nearby robots. If the ball distance for Robot A is less than the threshold difference compared to Robot B, then Robot A will still adjust its heading and approach the ball.

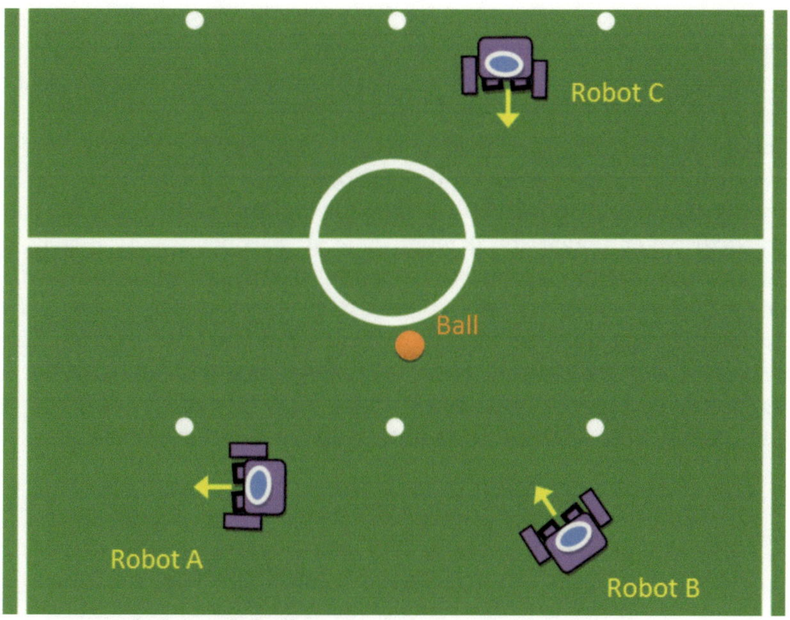

Fig. 3.6 Robots approach through ball approximation and heading.

Figure 3.7 below illustrates the close-up view of the robots approach through ball approximation and heading. With the robot heading perpendicular to the ball as the 90°

heading reference, robot heading 181° to 359° will not initiate the approach. As illustrated below, Robot A heading falls out of the 0° to 180° heading range. While Robot B distance is less or equal to the threshold, therefore Robot B will proceed to approach the ball, and will change its role accordingly e.g. from defender to striker.

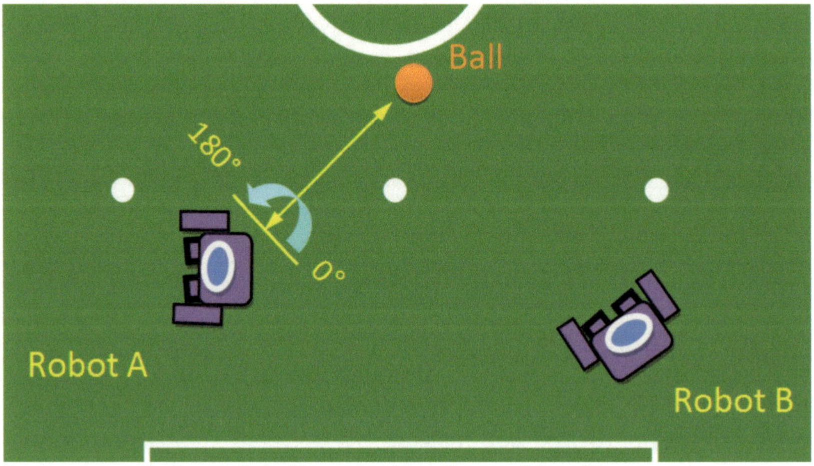

Fig. 3.7 Robots approach through ball approximation and heading.

3.4 Formations, roles, and players

During a game there could be situations where a team must play with substitute robots, and occasionally the team must play with fewer players. To deal with these situations, the proposed formations have roles with priorities. These priorities indicate which roles must be filled first, and which roles must always remain filled. Each robot keeps a list of the other robots broadcasting in its network; with this information each robot is aware of the number of team members of the team. However, when a robot that is currently playing fails and needs to be replaced by another robot. This new robot will be added to the list, the previous robot will identify that there are four robots in the list when this happen, the robots monitor the messages to discover which robot went dead. If a robot is not able to discover the missing robot, it will broadcast a request, so that the playing robots will reply. This request is also broadcasted if one robot does not broadcast any message for a period of time. This is done to try to identify any dead robot. When a robot is faulty and its role is a priority one after the replacement is in the field the robot will renegotiate their roles. In the scenario that there is no replacement, a robot with a non-priority role will switch to the priority role. The Table 3.1 below presents the different formations and the priority of the roles.

Formation	Highest	High	Low
Defensive	Goalie	Defender	Defender
Normal	Goalie	Defender	Striker
Offensive	Goalie	Striker	Striker
Super Offensive	Defender	Striker	Striker
All Out	Striker	Striker	Striker

Table 3.1 Formations and priorities of the roles per formation.

3.5 Area of Coverage

The proposed method also defines the area of coverage or region where the robots should be limited according to their roles. This is to facilitate effectiveness and prevent unnecessary movement during the role changing. This roles area of coverage proposal is to enhance the ball approach criteria described in the previous Section 3.4. Figure 3.8 shows the areas of coverage for the roles of goalie, defender and striker. These areas are also used for the self-positioning behaviour, the robots defined points on the field as starting positions, but they have a higher priority to be inside the area rather than to reach the point. Only during the self-positioning behaviour the area of coverage of a striker becomes the same as that of the defender, but the attraction points are different; i.e. for the striker is closer to the half line.

- Goalie - Movement limited to the Goal Area. This is to cater for cases when the ball stops just at the goal line, and the goalie has to move behind the goal line to kick out the ball, or else the goalie would not know how to react in a situation when ball stops just at the goal line.
- Defender - Movement limited to the own half of the field, as illustrated in Figure 3.8(b) below. The defender's area also includes the goal area.
- Striker - Movement limited from the middle of lower half of field, to the opponent's goal, as illustrated in Figure 3.8(c). However, when the formation has more than one striker, the striker's area becomes the whole field.

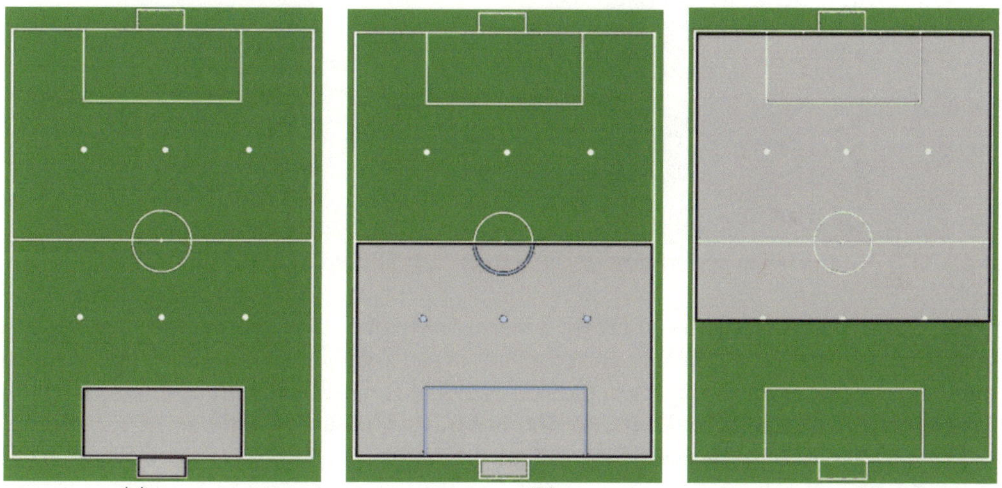

(a) (b) (c)

Fig. 3.8 Area of coverage for roles of (a) goalie, (b) defender, and (c) striker.

4. Robo-Erectus, The Humanoid Soccer Robot

The Robo-Erectus project (www.robo-erectus.org) has been developed in the Advanced Robotics and Intelligent Control Centre (ARICC) of Singapore Polytechnic. The humanoid robot Robo-Erectus is one of the pioneering soccer-playing humanoid robots in the RoboCup Humanoid League (Acosta et al., 2007). Robo-Erectus has collected several awards since its first participation in the Humanoid League of RoboCup in 2002. Robo-Erectus won the 2nd place in the Humanoid Walk competition at the RoboCup 2002 and got 1st place in the Humanoid Free Performance competition at the RoboCup 2003. In 2004, Robo-Erectus won the 2nd place in Humanoid Walk, Penalty Kick, and Free Performance. In 2007, it finished 6th in the 2 vs. 2 games, and 3rd in the technical challenge. In the RoboCup 2009, it qualified to the second round of round robin of 3 vs. 3 games.

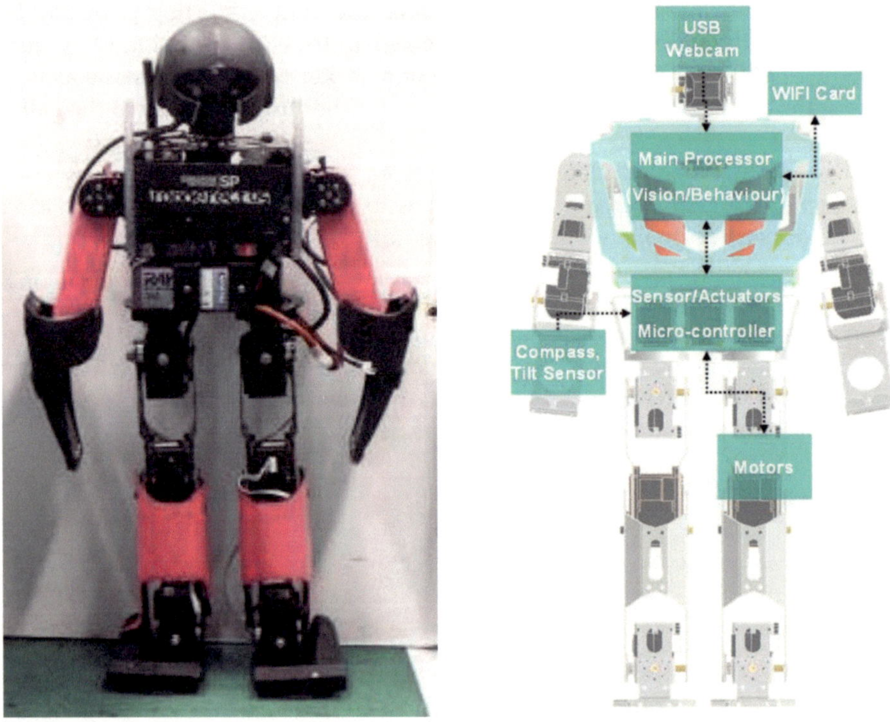

Fig. 4.1 REJr-X1, the latest generation of the family Robo-Erectus.

The Robo-Erectus project aims to create a humanoid robot that can be used for teaching, research and competition. After the introduction of the TeenSize to the RoboCup Humanoid League, the team has developed two parallel types of Robo-Erectus. Robo-Erectus Junior is the name for the robots that take part in the RoboCup Humanoid KidSize, as a group of at least three robots, where the restriction is that the robots must be less than 60cm in height. The second type of Robo-Erectus is called Robo-Erectus Senior and this robot competes in the RoboCup Humanoid TeenSize, in which only one robot takes part, whereby the restriction of the league is that the robots should be more than 1m tall.

The proposed method in this Chapter applies to a team of humanoid kidsize robots. Therefore, the rest of this Section will focus on the Robo-Erectus Junior. Each robot in the team has the same hardware and software features as described here.

4.1 Robo-Erectus Junior

The latest version of Robo-Erectus named Robo-Erectus Junior-X1 (REJr-X1), as shown in Figure 3, has been designed to be fully autonomous and to deal with the challenges of the 3 vs. 3 games.

Figure 4.1 shows the design of the humanoid robot REJr-X1. The skeleton of the robot is constructed with aluminium braces, the head and arms of the robot are made of plastic. Despite its simplicity, the mechanical design of the robot is robust and lighter than their predecessors. Its human-like body has a height of 52cm and weight of just 3.3kg, including batteries.

Robo-Erectus Junior has a total of 21 degrees of freedom. Table 4.1 shows the body parts and their associated degrees of freedom. Each degree of freedom uses as actuator a Dynamixel DX-117 Digital Servomotor. These servomotors have a typical torque of 28.89kg·cm and a speed of 0.172sec/60°. Each knee joint uses a Dynamixel RX-64 Digital Servomotor that provides a higher torque than that of DX-117. Each smart actuator has a micro-controller in charge of receiving commands and monitoring the performance of the actual motor. An RS485 serial network connects all the servomotors to a host processor, which sends positions and receives the current data (angular positions, speed, voltage, and temperature) of each actuator.

Body Part	Roll	Pitch	Yaw
Head		✓	✓
Body			✓
Shoulder	✓	✓	
Elbow		✓	
Hip	✓	✓	✓
Knee		✓	
Ankle	✓	✓	

Table 4.1 List of Degrees of Freedom for the humanoid robot Robo-Erectus Jr.

The control system of the Robo-Erectus Jr-X1 consists of a network with two micro-processors and several devices. Instead of using a single processor to deal with sensory data, processing and actuators control, Robo-Erectus uses dedicated processors for particular tasks. This design was implemented firstly in the Robo-Erectus Junior-AX and it has proven to improve the performance of the system (Acosta et al., 2008). Below are listed the tasks of each processor:

1. The main processor is responsible to coordinate all behaviours of the robot. It receives the processed sensorial information from the sensor-motor processor, which is used to take decisions and finally send the motors commands back to the sensor-motor processor. This processor has a wireless network interface that allows communication with the other robots and a computer that sends game signals. This processor also processes the images from an USB camera.
2. The sensor-motor processor is a dual DSP microcontroller, which receives the motor commands from the main processor. These commands are parameters for the gait generator. The gaits of the humanoid robot are generated online and finally sent to the servo-motors by a RS485 bus. The motor feedback is collected, processed, and sent back to the main processor by a RS232 bus. The data of servo-motors is updated every 16.6 ms. Along with the motor information, this processor also collects data from the accelerometers and gyroscopes before passing these data to a Kaman filter to produce more relevant information, about the state of the robot.

4.2 Virtual RE, The Robo-Erectus Junior Simulator

Robo-Erectus is equipped with sensors and actuators that allow it to navigate autonomously and display an intelligent behaviour. The robot uses a camera as main sensor, and it is able to perceive different colours and to track them. It also contains a dedicated processor to control the behaviour of the robot, wireless communication with the control PC and the teammates, and a sub-system to control sensors and actuators. A simulator was developed to simulate all the sensor of the robot. In addition the simulator is able to simulate two teams with a number variable of robots to reproduce a game (see Figure 4.2). The simulator uses the server-client framework, where each robot is considered a server and a single client could connect to it and control it.

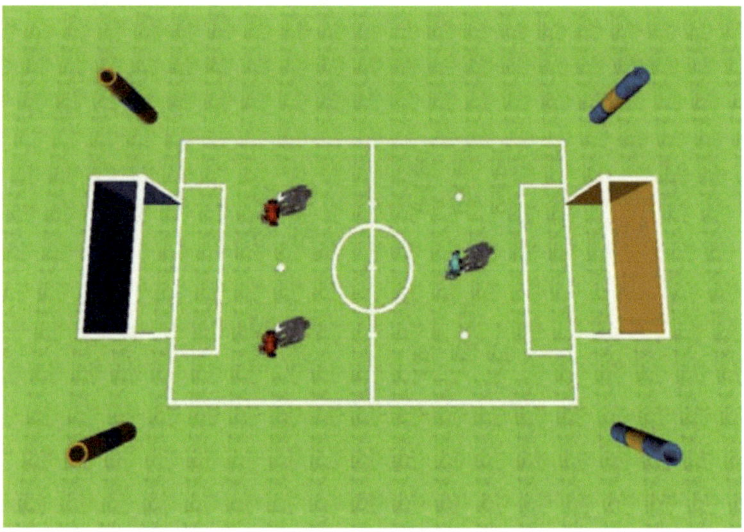

Fig. 4.2 Virtual RE the simulator for the Robo-Erectus Jr.

The simulator of Robo-Erectus Junior-AX, called Virtual-RE, was also used during these experiments. Virtual-RE provides several possibilities of visualization and interaction with the simulated worlds (Acosta et al., 2007). To simulate rigid body dynamics, Virtual-RE uses the Open Dynamics Engine (ODE), which has a wide variety of features and has been used successfully in many other projects. The visualization, as well as the computation of imaging sensor data is based on OpenGL libraries, because this standard offers the best performance with modern hardware on different platforms.

5. Experimental results

In order to test the effectiveness of our algorithms, we performed two sets of experiments: one with the simulator Virtual-RE and the other with the actual Robo-Erectus Jr robots. All the sets involve the four strategies as well as winning and losing scores for both the regular time and the extra time. In addition, each set contains games with faulty players, some with substitute robot and others without.

The behaviours of each role have been implemented in a finite state machine, with simple actions for each robot and self-positioning. The opponents for all the sets played with static formation: goalie, defender, striker.
Table 5.1 shows the resulting analysis of the data. The data presented is the percentage of wrong formation during the possible changes in the game. The possible changes happen every five minutes or after a goal. The 7% of wrong role selection in the simulation is due to situation where two robots decided to have the same role. In all these cases, the role was successfully corrected after one or two seconds to conclude their negotiation. In the case of the Robo-Erectus Jr robots the percentage was higher, because of some unexpected situations with a few robots and they have to be taken out, serviced and their computer rebooted. This situation caused a wrong role assignment, however the robots corrected their roles after negotiation. In a similar way, the wrong formation is due to the faulty robot.

Set	Wrong Formation	Wrong Role
Simulation	0%	7%
Robo-Erectus Jr	5%	18%

Table 5.1 Percentage of wrong formation and wrong role selection for possible changes without faulty robots.

Table 5.2 shows the wrong formation and wrong role selection with faulty robot with and without replacement. For the replacement sets, the problem was similar to the one observed in the Robo-Erectus Jr in the previous experiments due to the faulty robot. In both simulation and real robot experiments, the robots managed to correct their wrong roles after few seconds. For the no replacement set, all the robots managed to switch to a higher priority role, except for one trial where the actual robot maintained its role, but this was because the robot's program entered into a deadlock and has to be reset.

Set	Wrong Formation	Wrong Role
Simulation (Replacement)	7%	7%
Robo-Erectus Jr (Replacement)	15%	23%
Simulation (No Replacement)	0%	0%
Robo-Erectus Jr (No Replacement)	0%	7%

Table 5.2 Percentage of wrong formation and wrong role selection for possible changes with faulty robots.

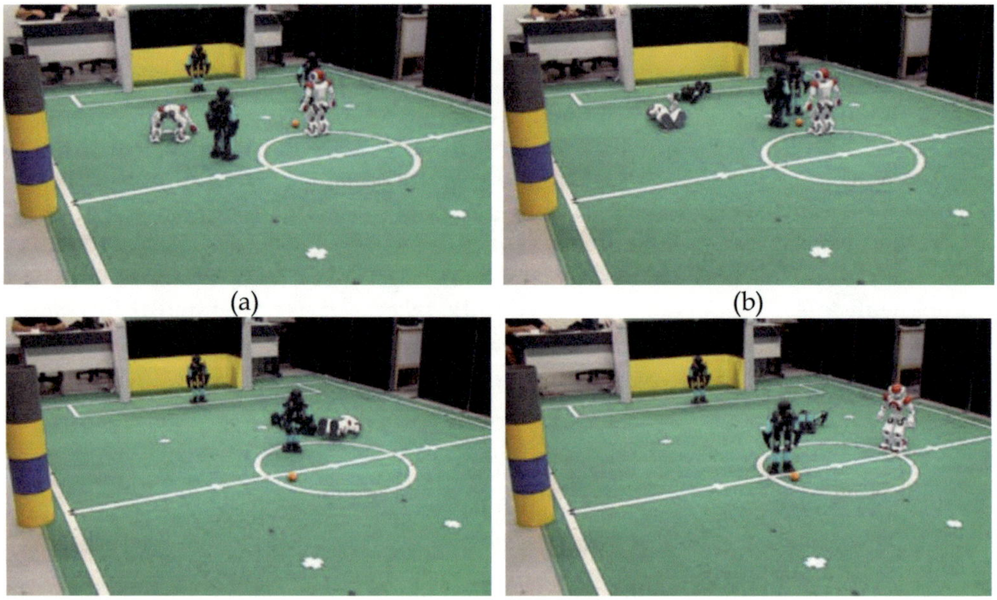

(a) (b)

(c) (d)

Fig. 5.1 Sequence of the game with formation At Least Draw with formation one Goalie and two defenders.

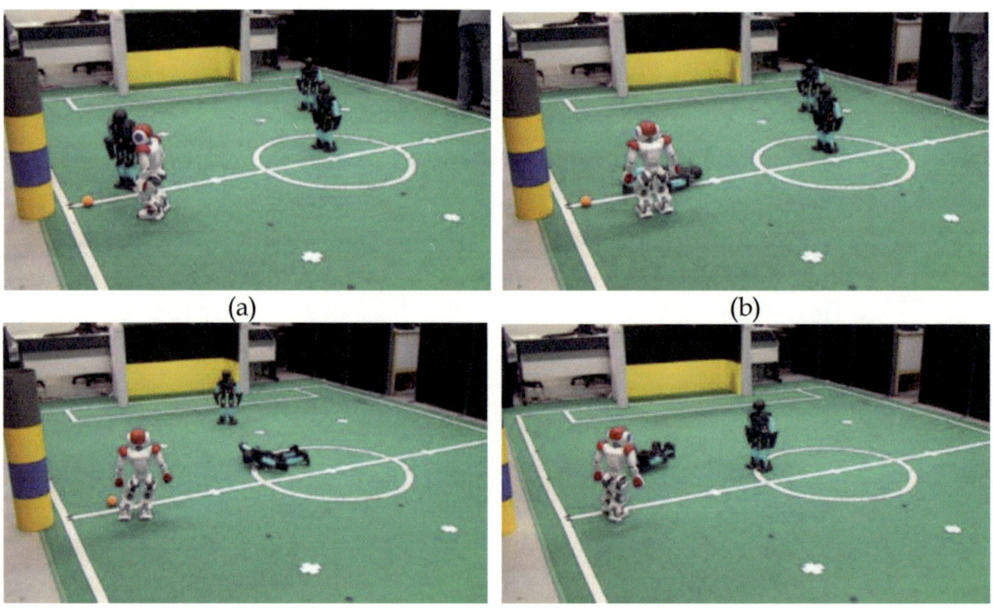

(a) (b)

(c) (d)

Fig. 5.2 Sequence of a game with a strategy Must Win with a formation three strikers. Experiment with one faulty robot.

In this set of experiments, the Robo-Erectus Jr played against a team of Nao robots with a fixed formation of goalie, defender and striker. Figure 5.1 shows a sequence of the game with Nao robots with the strategy of At Least Draw, with a formation goalie, defender, defender. In Figure 5.2 the strategy employed is Must Win with an All Out formation. In this game one robot was taken out to simulate a faulty robot. Robot in Figure 5.2(b) is taken out as can be seen in Figure 5.2(c). In Figure 5.2(c) both robots try to approach to the ball.

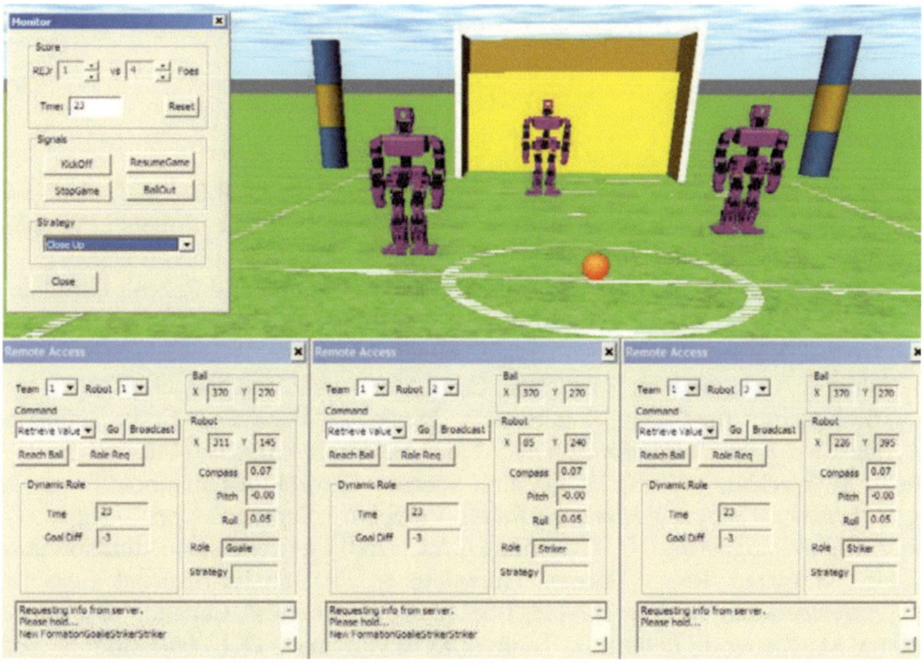

Fig. 5.3 Screenshot from the Virtual-RE with a Close-up Strategy and a formation Goalie and two strikers. This is during the first half of extra time.

The set of experiments with the simulator Virtual-RE were conducted in a similar way as the real Robo-Erectus Jr, i.e. three versus three games with different strategies (see Figure 5.3). Due to the flexibility that the simulator offered, we were able to set, besides the normal game, scenarios where the robots were placed in certain positions, with specific roles, score, and time; to test situations like approaching to the ball, or self-positioning. For the self-positioning behaviour 85% of trials was successful. This is because we placed the striker inside the goal keeper area or in one of the corners, and the robot should go back to its half of the field, in the way back the robot should avoid other robots, but sometimes it collides with them, falling and not reaching the position on time.

6. Conclusion

Humanoid robot soccer is getting popular recently, and the advances in the technology have made it possible to see exciting displays from the humanoid robots. Nevertheless, most of the teams have achieved a position where the robots are able to show several skills. It is challenging to have a team of robots displaying team behaviour without interfering with each other and supporting their objectives. The work presented here uses a dynamic role assignation but is based on the strategy that the team has for the game. Besides, other factors like remaining time and goal difference are used to determine the new formation of the team. The roles of the players underlie team behaviour, while the strategy defines the objective of the team. Experimental results support our proposal, showing that the percentage of wrong role selection among the robots is low less than 20%. Results also prove that this wrong role selection is soon corrected by the robots after negotiation. Future work is to deal with situations of negotiation when having more players, and to define more team behaviours with this framework.

7. References

Acosta Calderon, C.A.; Zhou, C.; Yue, P.K.; Wong, M. & Mohan, R.E. (2007). A Distributed Embedded Control Architecture for Humanoid Soccer Robots. *Proceedings of Advances in Climbing and Walking Robots*, pp. 487-496, Singapore, July 2007.

Acosta Calderon, C.A.; Mohan, R.E. & Zhou, C. (2007). A Humanoid Robotic Simulator with Application to RoboCup, *Proceedings of IEEE Latin American Robotic Simposium*, Mexico, November 2007.

Acosta Calderon, C.A.; Mohan, R.E.; Zhou, C.; Hu, L.; Yue, P.K. & Hu, H. (2008) A Modular Architecture for Soccer Robots with Distributed Behaviour Control, *International Journal of Humanoid Robotics*, Vol. 5, No. 3, September, pp. 397–416.

Behnke, S.; & Stueckler, J. (2008). Hierarchical Reactive Control for Humanoid Soccer Robots. *International Journal of Humanoid Robots*, Vol 5, No 3, September, pp. 375-396.

Coelho, A.L.V.; Weingaertner, D. & Gomide, F.A.C. (2001). Evolving Coordination Strategies in Simulated Robot Soccer. *Proceeding of the 5th International Conference on Autonomous Agents*, pp. 147-148, 1-58113-326-X, Montreal, Canada, May-June 2001.

Friedmann, M.; Kiener, J.; Petters, S.; Thomas, D. & von Stryk, O. (2006). Modular Software Architecture for Teams of Cooperating, Heterogeneous Robots. *Proceedings of IEEE International Conference on Robotics and Biomimetics*, pp 24-29, Kunming, China, December 2006.

Kitano, H. & Asada, H. (2000). The RoboCup Humanoid Challenge As The Millennium Challenge for Advanced Robotics. *Advanced Robotics*, Vol. 13, No. 8, pp723–736.

Phillips, M. & Veloso, M. (2009). Robust Support Role in Coordinated Two-Robot Soccer Attack, In: *RoboCup 2008, LNAI 5399*, Iocchi, L, Matsubara, H., Weitzenfeld, A. & Zhou, C., pp. 235-246, Springer-Verlag, 3-642-02920-5, Germany.

Risler, M. & von Stryk, O. (2008). Formal Behavior Specification of Multi-robot Systems Using Hierarchical State Machines in XABSL. *Proceedings of AAMAS08-Workshop on Formal Models and Methods for Multi-Robot Systems*, Estoril, Portugal, May 2008.

Work, H.; Chown, E.; Hermans, T.; Butterfield, J. & McGranaghan, M. (2009). Player Positioning in the Four-Legged League, In: *RoboCup 2008, LNAI 5399*, Iocchi, L, Matsubara, H., Weitzenfeld, A. & Zhou, C., pp. 391-402, Springer-Verlag, 3-642-02920-5, Germany.

Zickler, S. & Veloso, M. (2009). Playing Creative Soccer: Randomized Behavioral Kinodynamic Flanning of Robot Tactics, In: *RoboCup 2008, LNAI 5399*, Iocchi, L, Matsubara, H., Weitzenfeld, A. & Zhou, C., pp. 414-425, Springer-Verlag, 3-642-02920-5, Germany.

Zhou, C. & Yue, P.K. (2004). Robo-Erectus: A Low Cost Autonomous Humanoid Soccer Robot. *Advanced Robotics*, Vol. 18, No. 7, pp. 717–720.

7

Evolving Fuzzy Rules for Goal-Scoring Behaviour in Robot Soccer

Jeff Riley
RMIT University
Australia

1. Introduction

If a soccer player is able to learn behaviours it should exhibit in response to stimuli, it may adapt to unpredictable, dynamic environments. Even though the overall objective a player is expected to achieve is able to be described, it is not always possible to precisely describe the behaviours a player should exhibit in achieving that objective. If a function can be described to evaluate the results of the player's behaviour against the desired outcome, that function can be used by some reinforcement learning algorithm to evolve the behaviours necessary to achieve the desired objective.

Fuzzy Sets (Zadeh 1965; Kandel 1986; Klir and Folger 1988; Kruse, Gebhardt et al. 1994) are powerful tools for the representation of uncertain and vague data. Fuzzy inference systems make use of this by applying approximate reasoning techniques to make decisions based on such uncertain, vague data. However, a Fuzzy Inference System (FIS) on its own is not usually self-adaptive and not able to modify its underlying rulebase to adapt to changing circumstances.

There has not been a great deal of success in the automatic generation of robot soccer players, and in fact hand-coded players, or players with hand-coded skills, generally outplay automatically generated players. Genetic algorithms (Holland 1975) are adaptive heuristic search algorithms premised on the evolutionary ideas of natural selection. By combining the adaptive learning capabilities of the genetic algorithm (GA) with the approximate reasoning capabilities of the fuzzy inference system, a hybrid system is produced, and the expectation is that this hybrid system will be capable of learning the behaviours a player needs to exhibit in order to achieve a defined objective - in this case developing goal-scoring behaviour. While the combination of genetic algorithms and fuzzy inference systems have been studied in other areas, they have not generally been studied in an environment as complex, uncertain and dynamic as the robot soccer environment.

2. Related Work

2.1 RoboCup

The Robot World Cup Initiative, RoboCup, has become an international research and education initiative, providing a standard platform and benchmark problem for research in

the fields of artificial intelligence and robotics. It provides a realistic research environment by using a soccer game as a platform for a wide range of research problems including autonomous agent design, multi-agent collaboration, real-time reasoning, reactive behaviour and intelligent robot control (Kitano, Asada et al. 1997a; Kitano, Asada et al. 1997b; Kitano, Tambe et al. 1997).
RoboCup currently consists of three major domains: RoboCupSoccer, RoboCupRescue and RoboCupJunior. The RoboCupSoccer domain includes a simulation league and is the environment used for the RoboCup part of this work.

2.1.1 The RoboCupSoccer Simulation League

The RoboCupSoccer Simulation League provides a simulated but realistic soccer environment which obviates the need for robot hardware and its associated difficulties, allowing researchers to focus on issues such as autonomous agent design, learning, planning, real-time multi-agent reasoning, teamwork and collaboration. The RoboCupSoccer simulator has been in continual development since its inception in 1995, and allows researchers to study many aspects of machine learning techniques and multi-agent systems in a complex, dynamic domain. The RoboCupSoccer environment is described in detail in (Noda 1995; Noda, Matsubara et al. 1998; Noda and Stone 2001).

2.2 The SimpleSoccer Environment

The RoboCupSoccer simulation league is an important and useful tool for multi-agent and machine learning research which provides a distributed, multi-agent environment in which agents have an incomplete and uncertain world view. The RoboCupSoccer state-space is extremely large, and the agent perception and action cycles in the RoboCupSoccer environment are asynchronous, sometimes resulting in long and unpredictable delays in the completion of actions in response to some stimuli. The large state-space, the inherent delays, and the uncertain and incomplete world view of the agents can increase the learning cycle of some machine learning techniques onerously.
There is a large body of work in the area of the application of machine learning techniques to the challenges of RoboCupSoccer (e.g. (Luke 1998a; Luke 1998b; Luke, Hohn et al. 1998; Ciesielski and Wilson 1999; Stone and Veloso 1999; Uchibe 1999; Ciesielski and Lai 2001; Riedmiller, Merke et al. 2001; Stone and Sutton 2001; Ciesielski, Mawhinney et al. 2002; Lima, Custódio et al. 2005; Riedmiller, Gabel et al. 2005)), but because the RoboCupSoccer environment is so large, complex and unpredictable, the extent to which such techniques can meet these challenges is not certain. The SimpleSoccer environment was designed and developed to address the problem of the complexity of the RoboCupSoccer environment inhibiting further research, and is described in detail in (Riley 2003) and (Riley 2007).

2.3 Machine Learning, Evolutionary Algorithms and Simulated Robot Soccer

Machine learning and evolutionary algorithms in various forms have been applied to the problem of robot soccer. Some representative examples, with emphasis on evolutionary algorithms, are described briefly in the following paragraphs.
Two early attempts to learn competent soccer playing skills from scratch via genetic programming are described in (Luke, Hohn et al. 1998) and (Andre and Teller 1999). Both set out to create complete, cooperative soccer playing teams, but neither achieved that

objective. The objective in (Luke, Hohn et al. 1998) was scaled back to attempt to evolve co-operative behaviour over hand-coded low-level behaviours. The players in (Andre and Teller 1999) developed some successful individual behaviours with the use of a sophisticated composite fitness function, but the objective of collaborative team behaviour was not realised.

In (Luke 1998b) a team of soccer players with a rich set of innate soccer-playing skills was developed, using genetic programming and co-evolution, that worked through sub-optimal behaviour described as "kiddie-soccer" (where all players chase the ball) to reasonable goal-scoring and defensive behaviour.

A layered learning technique was introduced in (Stone 1998) and (Stone and Veloso 2000), the essential principle of which is to provide the algorithm with a bottom-up hierarchical decomposition of a large task into smaller sub-tasks, or layers, and have the algorithm learn each sub-task separately and feed the output of one learned sub-task into the next layer.

The layered learning technique is based upon the following four principles (Stone and Veloso 2000):

- A mapping directly from inputs to outputs is not tractably learnable.
- A bottom-up, hierarchical task decomposition is given.
- Machine learning exploits data to train and/or adapt. Learning occurs separately at each level.
- The output of learning in one layer feeds into the next.

Stone and Veloso used the layered learning technique to produce good results when training robot soccer players for RoboCupSoccer (Stone and Veloso 2000).

Keepaway Soccer (Stone, Sutton et al. 2001) is a sub-domain of robot soccer in which the objective is not to score goals but to gain and maintain possession of the ball. There are two teams in keepaway soccer: the *keepers* and the *takers*. The task of the keepers is to maintain possession of the ball, while the objective of the takers is to take the ball away from the keepers. The keepaway soccer field is generally smaller than the RoboCupSoccer field, and no goal areas are required. The keepaway soccer teams are usually smaller than a full team in RoboCupSoccer, and are often numerically unbalanced (e.g. 3 vs 2 Keepaway Soccer (Kuhlmann and Stone 2004)).

Gustafson (Gustafson 2000) and Gustafson and Hsu (Gustafson and Hsu 2001; Hsu, Harmon et al. 2004) applied genetic programming and the layered learning technique of Stone and Veloso to keepaway soccer. For this method the problem was decomposed into smaller sub-problems, and genetic programming applied to the sub-problems sequentially - the population in the last generation of a sub-problem was used as the initial population of the next sub-problem. The results presented by Gustafson and Hsu indicate that for the problem studied layered learning in genetic programming outperformed the standard genetic programming method.

Asada et al. (Asada, Noda et al. 1996) describe the *Learning from Easy Missions (LEM)* method, in which a reinforcement learning technique (*Q-learning, (Watkins 1989)*) is used to teach a robot soccer player to kick a ball through a goal. The reinforcement learning technique implemented requires the robot soccer player to be capable of discriminating a finite set of distinct world states and also be capable of taking one of a finite set of actions. The robot's world is then modelled as a Markov process, making stochastic transitions based on the current state and action taken. A significant problem with this method is that for a real robot in the real world, or the simulation of a real robot in the real world, the state and action spaces are continuous spaces that are not adequately represented by finite sets. Asada et al. overcome this by constructing a set of sub-states into which the representation of the robot's world is divided, and similarly a set of sub-actions into which the robot's full range of actions is divided. This is roughly analogous to the fuzzy sets for input variables and actions implemented for this work.

The LEM method involves using human input to modify the starting state of the soccer player, beginning with easy states and progressing over time to more difficult states. In this way the robot soccer player learns easier sub-tasks allowing it to use those learned sub-tasks to develop more complex behaviour enabling it to score goals in more difficult situations. Asada et al. concede that the LEM method has limitations, particularly with respect to constructing the state space for the robot soccer player. Asada et al. also point out that the method suffers from a lack of historical information that would allow the soccer player to define context, particularly in the situation where the player is between the ball and the goal: with only current situation context the player does not know how to move to a position to shoot the ball into the goal (or even that it should). Some methods suggested by Asada et al. to overcome this problem are to use task decomposition (i.e. find ball, position ball between player and goal, move forward, etc.), or to place reference objects on the field (corner posts, field lines, etc.) to give the player some context. It is also interesting to note that after noticing that the player performed poorly whenever it lost sight of the ball, Asada et al. introduced several extra states to assist the player in that situation: the *ball-lost-into-right* and *ball-lost-into-left* states, and similarly for losing sight of the goal, *goal-lost-into right* and *goal-lost-into-left* states. These states, particularly the *ball-lost-into-right* and *ball-lost-into-left* states are analogous to the default hunt actions implemented as part of the work described in this chapter, and another indication of the need for human expertise to be injected to adequately solve the problem.

Di Pietro et al. (Di Pietro, While et al. 2002) reported some success using a genetic algorithm to train 3 keepers against 2 takers for keepaway soccer in the RoboCup soccer simulator. Players were endowed with a set of high-level skills, and the focus was on learning strategies for keepers in possession of the ball.

Three different approaches to create RoboCup players using genetic programming are described in (Ciesielski, Mawhinney et al. 2002) – the approaches differing in the level of innate skill the players have. In the initial experiment described, the players were given no innate skills beyond the actions provided by the RoboCupSoccer server. The third experiment was a variation of the first experiment. Ciesielski et al. reported that the players from the first and third experiments – players with no innate skills - performed poorly. In the second experiment described, players were given some innate higher-level hand-coded skills such as the ability to kick the ball toward the goal, or to pass to the closest teammate. The players from the second experiment – players with some innate hand-coded skills – performed a little more adequately than the other experiments described. Ciesielski et al. concluded that the robot soccer problem is a very difficult problem for evolutionary algorithms and that a significant amount of work is still needed for the development of higher-level functions and appropriate fitness measures.

Using keepaway soccer as a machine learning testbed, Whiteson and Stone (Whiteson and Stone 2003) used neuro-evolution to train keepers in the Teambots domain (Balch 2005). In that work the players were able to learn several conceptually different tasks from basic skills to higher-level reasoning using "concurrent layered learning" - a method in which predefined tasks are learned incrementally with the use of a composite fitness function. The player uses a hand-coded decision tree to make decisions, with the leaves of the tree being the learned skills.

Whiteson et al. (Whiteson, Kohl et al. 2003; Whiteson, Kohl et al. 2005) study three different methods for learning the sub-tasks of a decomposed task in order to examine the impact of injecting human expert knowledge into the algorithm with respect to the trade-off between:

- making an otherwise unlearnable task learnable
- the expert knowledge constraining the hypothesis space
- the effort required to inject the human knowledge.

Coevolution, layered learning, and concurrent layered learning are applied to two versions of keepaway soccer that differ in the difficulty of learning. Whiteson et al. conclude that given a suitable task decomposition an evolutionary-based algorithm (in this case neuroevolution) can master difficult tasks. They also conclude, somewhat unsurprisingly, that the appropriate level of human expert knowledge injected and therefore the level of constraint depends critically on the difficulty of the problem.

Castillo et al. (Castillo, Lurgi et al. 2003) modified an existing RoboCupSoccer team - the 11Monkeys team (Kinoshita and Yamamoto 2000) - replacing its offensive hand-coded, state dependent rules with an XCS genetic classifier system. Each rule was translated into a genetic classifier, and then each classifier evolved in real time. Castillo et al. reported that their XCS classifier system outperformed the original 11Monkeys team, though did not perform quite so well against other, more recently developed, teams.

In (Nakashima, Takatani et al. 2004) Nakashima et al. describe a method for learning certain strategies in the RoboCupSoccer environment, and report some limited success. The method uses an evolutionary algorithm similar to evolution strategies, and implements mutation as the only evolutionary operator. The player uses the learned strategies to decide which of several hand-coded actions will be taken. The strategies learned are applicable only when the player is in possession of the ball.

Bajurnow and Ciesielski used the SimpleSoccer environment to examine genetic programming and layered learning for the robot soccer problem (Bajurnow and Ciesielski 2004). Bajurnow and Ciesielski concluded that layered learning is able to evolve goal-scoring behaviour comparable to standard genetic programs more reliably and in a shorter time, but the quality of solutions found by layered learning did not exceed those found using standard genetic programming. Furthermore, Bajurnow and Ciesielski claim that layered learning in this fashion requires a *"large amount of domain specific knowledge and programmer effort to engineer an appropriate layer and the effort required is not justified for a problem of this scale."* (Bajurnow and Ciesielski 2004), p.7.

Other examples of research in this or related areas can be found in, for example, (Luke and Spector 1996) where breeding and co-ordination strategies were studied for evolving teams in a simple predator/prey environment; (Stone and Sutton 2001; Kuhlmann and Stone 2004; Stone, Sutton et al. 2005) where reinforcement learning was used to train players in the keepaway soccer environment; (Lazarus and Hu 2003) in which genetic programming was used in a specific training environment to evolve goal-keeping behaviour for RoboCupSoccer; (Aronsson 2003) where genetic programming was used to develop a team of players for RoboCupSoccer; (Hsu, Harmon et al. 2004) in which the incremental reuse of intermediate solutions for genetic programming in the keepaway soccer environment is studied.

3. The Player

3.1 Player Architecture

The traditional decomposition for an intelligent control system is to break processing into a chain of information processing modules proceeding from sensing to action (Fig. **1**).

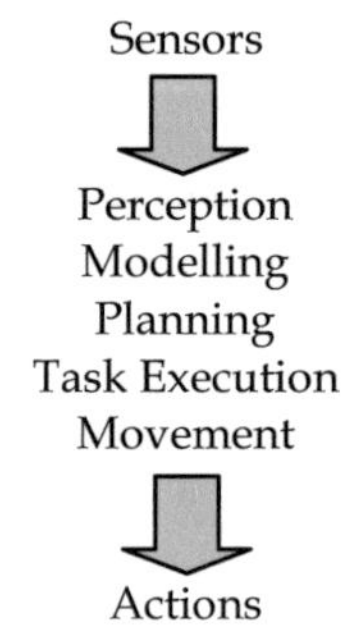

Fig. 1. Traditional Control Architecture

The control architecture implemented for this work is similar to the subsumption architecture described in (Brooks 1985). This architecture implements a layering process where simple task achieving behaviours are added as required. Each layer is behaviour producing in its own right, although it may rely on the presence and operation of other layers. For example, in Fig. 2 the *Movement* layer does not explicitly need to avoid obstacles: the *Avoid Objects* layer will take care of that. This approach creates players with reactive architectures and with no central locus of control (Brooks 1991).

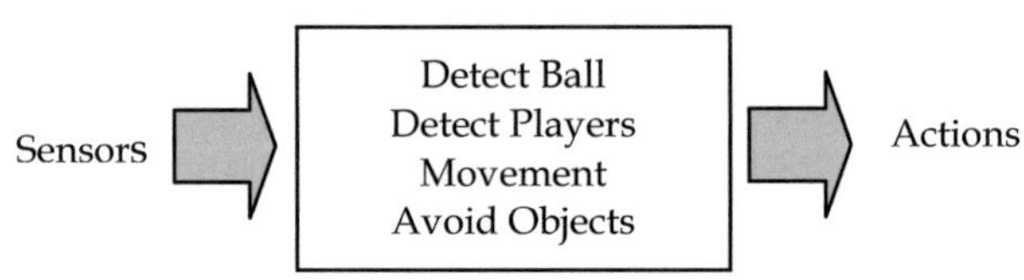

Fig. 2. Soccer Player Layered Architecture

For the work presented here, the behaviour producing layers are implemented as fuzzy if-then rules and governed by a fuzzy inference system comprised of the fuzzy rulebase, definitions of the membership functions of the fuzzy sets operated on by the rules in the rulebase, and a reasoning mechanism to perform the inference procedure. The fuzzy inference system is embedded in the player architecture, where it receives input from the soccer server and generates output necessary for the player to act Fig. **3**.

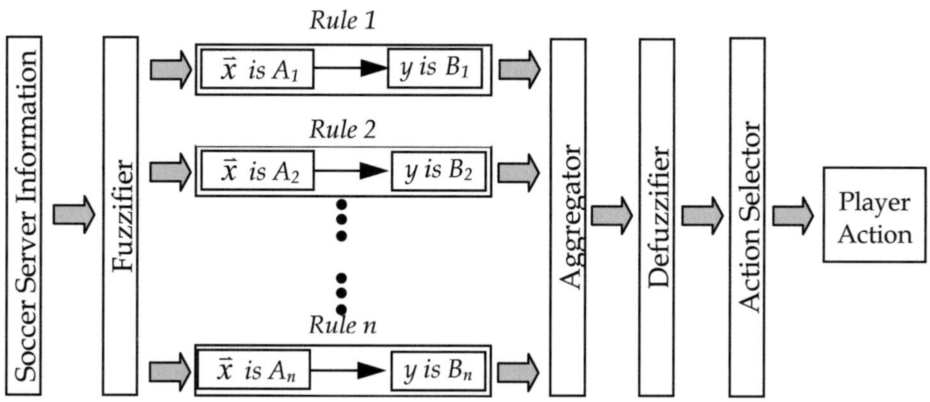

Fig. 3. Player Architecture Detail

3.1.1 Soccer Server Information

The application by the inferencing mechanism of the fuzzy rulebase to external stimuli provided by the soccer server results in one or more fuzzy rules being executed and some resultant action being taken by the client. The external stimuli used as input to the fuzzy inference system are a subset of the visual information supplied by the soccer server: only sufficient information to situate the player and locate the ball is used. The environments studied in this work differ slightly with regard to the information supplied to the player:

- In the RoboCupSoccer environment the soccer server delivers regular *sense, visual* and *aural* messages to the players. The player implemented in this work uses only the *object name, distance* and *direction* information from the visual messages in order to determine its own position on the field and that of the ball. The player ignores any aural messages, and uses the information in the sense messages only to synchronise communication with the RoboCupSoccer server. Since the information supplied by the RoboCupSoccer server is not guaranteed to be complete or certain, the player uses its relative distance and direction from all fixed objects in its field of vision to estimate its position on the field. The player is then able to use the estimate of its position to estimate the direction and distance to the known, fixed location of its goal. The player is only aware of the location of the ball if it is in its field of vision, and only to the extent that the RoboCupSoccer server reports the relative direction and distance to the ball.

- In the SimpleSoccer environment the soccer server delivers only regular visual messages to the players: there are no aural or sense equivalents. Information supplied by the SimpleSoccer server is complete, in so far as the objects actually with the player's field of vision are concerned, and certain. Players in the SimpleSoccer environment are aware at all times of their exact location on the field, but are only aware of the location of the ball and the goal if they are in the player's field of vision. The SimpleSoccer server provides the *object name, distance* and *direction* information for objects in a player's field of vision. The only state information kept by a player in the SimpleSoccer environment is the co-ordinates of its location and the direction in which it is facing.

3.1.2 Fuzzification

Input variables for the fuzzy rules are fuzzy interpretations of the visual stimuli supplied to the player by the soccer server: the information supplied by the soccer server is fuzzified to represent the degree of membership of one of three fuzzy sets: *direction*, *distance* and *power*; and then given as input to the fuzzy inference system. Output variables are the fuzzy actions to be taken by the player. The universe of discourse of both input and output variables are covered by fuzzy sets (direction, distance and power), the parameters of which are predefined and fixed. Each input is fuzzified to have a degree of membership in the fuzzy sets appropriate to the input variable.

Both the RoboCupSoccer and the SimpleSoccer servers provide crisp values for the information they deliver to the players. These crisp values must be transformed into linguistic terms in order to be used as input to the fuzzy inference system. This is the fuzzification step: the process of transforming crisp values into degrees of membership for linguistic terms of fuzzy sets. The membership functions shown in Fig. 4 on are used to associate crisp values with a degree of membership for linguistic terms. The parameters for these fuzzy sets were not learned by the evolutionary process, but were fixed empirically. The initial values were set having regard to RoboCupSoccer parameters and variables, and fine-tuned after minimal experimentation in the RoboCupSoccer environment.

3.1.3 Implication and Aggregation

The core section of the fuzzy inference system is the part which combines the facts obtained from the fuzzification with the rule base and conducts the fuzzy reasoning process: this is where the fuzzy inferencing is performed. The FIS model used in this work is a Mamdani FIS (Mamdani and Assilian 1975). The method implemented to apply the result of the antecedent evaluation to the membership function of the consequent is the correlation minimum, or *clipping* method, where the consequent membership function is truncated at the level of the antecedent truth. The aggregation method used is the min/max aggregation method as described in (Mamdani and Assilian 1975). These methods were chosen because they are computationally less complex than other methods and generate an aggregated output surface that is relatively easy to defuzzify.

3.1.4 Defuzzification

The defuzzification method used is the *mean of maximum* method, also employed by Mamdani's fuzzy logic controllers. This technique takes the output distribution and finds its mean of maxima in order to compute a single crisp number. This is calculated as follows:

$$z = \sum_{i=1}^{n} \frac{z_i}{n}$$

where z is the mean of maximum, z_i is the point at which the membership function is maximum, and n is the number of times the output distribution reaches the maximum level.

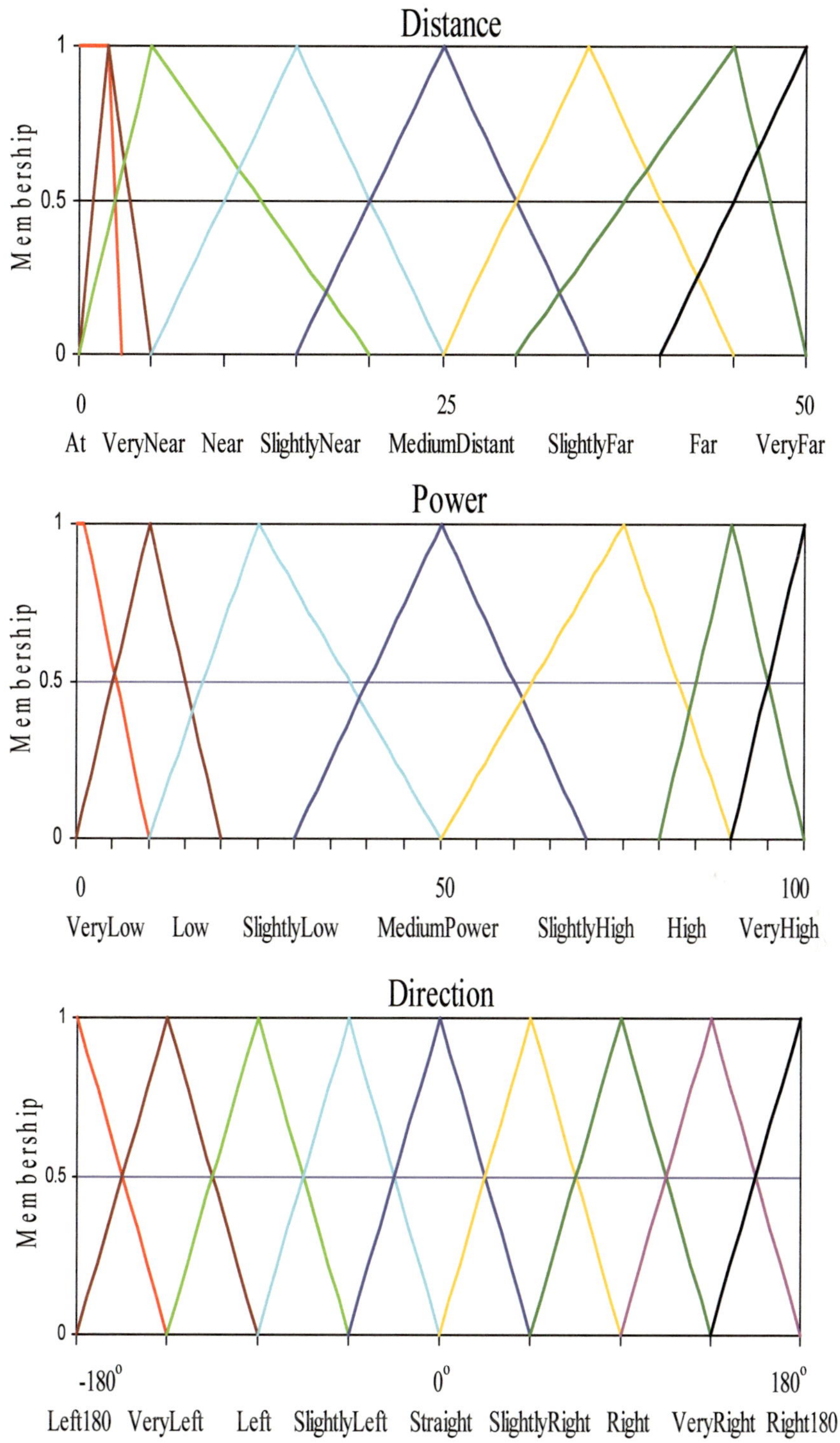

Fig. 4. Distance, Power and Direction Fuzzy Sets

An example outcome of this computation is shown in Fig. **5**. This method of defuzzification was chosen because it is computationally less complex than other methods yet produces satisfactory results.

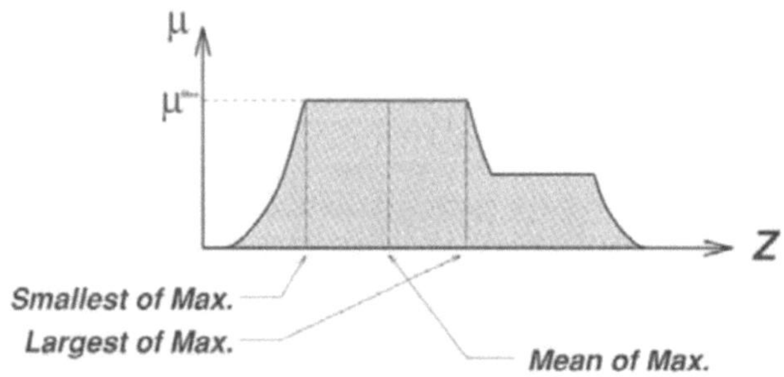

Fig. 5. Mean of Maximum defuzzification method
(Adapted from (Jang, Sun et al. 1997))

3.1.5 Player Actions

A player will perform an action based on its skillset and in response to external stimuli; the specific response being determined in part by the fuzzy inference system. The action commands provided to the players by the RoboCupSoccer and SimpleSoccer simulation environments are described in (Noda 1995) and (Riley 2007) respectively. For the experiments conducted for this chapter the SimpleSoccer simulator was, where appropriate, configured for RoboCupSoccer emulation mode.

3.1.6 Action Selection

The output of the fuzzy inference system is a number of *(action, value)* pairs, corresponding to the number of fuzzy rules with unique consequents. The *(action, value)* pairs define the action to be taken by the player, and the degree to which the action is to be taken. For example:

(KickTowardGoal, power)

(RunTowardBall, power)

(Turn, direction)

where *power* and *direction* are crisp values representing the defuzzified fuzzy set membership of the action to be taken.

Only one action is performed by the player in response to stimuli provided by the soccer server. Since several rules with different actions may fire, the action with the greatest level of support, as indicated by the value for truth of the antecedent, is selected.

3.2 Player Learning

This work investigates the use of an evolutionary technique in the form of a messy-coded genetic algorithm to efficiently construct the rulebase for a fuzzy inference system to solve a particular optimisation problem: goal-scoring behaviour for a robot soccer player. The flexibility provided by the messy-coded genetic algorithm is exploited in the definition and format of the genes on the chromosome, thus reducing the complexity of the rule encoding from the traditional genetic algorithm. With this method the individual player behaviours are defined by sets of fuzzy if-then rules evolved by a messy-coded genetic algorithm. Learning is achieved through testing and evaluation of the fuzzy rulebase generated by the genetic algorithm. The fitness function used to determine the fitness of an individual rulebase takes into account the performance of the player based upon the number of goals scored, or attempts made to move toward goal-scoring, during a game.

The genetic algorithm implemented in this work is a messy-coded genetic algorithm implemented using the Pittsburgh approach: each individual in the population is a complete ruleset.

4. Representation of the Chromosome

For these experiments, a chromosome is represented as a variable length vector of genes, and rule clauses are coded on the chromosome as genes. The encoding scheme implemented exploits the capability of messy-coded genetic algorithms to encode information of variable structure and length. It should be noted that while the encoding scheme implemented is a messy encoding, the algorithm implemented is the classic genetic algorithm: there are no primordial or juxtapositional phases implemented.

The basic element of the coding of the fuzzy rules is a tuple representing, in the case of a rule premise, a fuzzy clause and connector; and in the case of a rule consequent just the fuzzy consequent. The rule consequent gene is specially coded to distinguish it from premise genes, allowing multiple rules, or a ruleset, to be encoded onto a single chromosome.

For single-player trials, the only objects of interest to the player are the ball and the player's goal, and what is of interest is where those objects are in relation to the player. A premise is of the form:

(Object, Qualifier, {Distance | Direction}, Connector)

and is constructed from the following range of values:

Object:	{ BALL, GOAL }
Qualifier:	{ IS, IS NOT }
Distance:	{ AT, VERYNEAR, NEAR, SLIGHTLYNEAR, MEDIUMDISTANT, SLIGHTLYFAR, FAR, VERYFAR }
Direction:	{ LEFT180, VERYLEFT, LEFT, SLIGHTLYLEFT, STRAIGHT, SLIGHTLYRIGHT, RIGHT, VERYRIGHT, RIGHT180 }
Connector:	{ AND, OR }

Each rule consequent specifies and qualifies the action to be taken by the player as a consequent of that rule firing thus contributing to the set of *(action, value)* pairs output by the fuzzy inference system. A consequent is of the form:

(Action, {Direction | Null}, {Power | Null})

and is constructed from the following range of values (depending upon the skillset with which the player is endowed):

Action:	{ TURN, DASH, KICK, RUNTOWARDGOAL, RUNTOWARDBALL, GOTOBALL, KICKTOWARDGOAL, DRIBBLETOWARDGOAL, DRIBBLE, DONOTHING }
Direction:	{ LEFT180, VERYLEFT, LEFT, SLIGHTLYLEFT, STRAIGHT, SLIGHTLYRIGHT, RIGHT, VERYRIGHT, RIGHT180 }
Power:	{ VERYLOW, LOW, SLIGHTLYLOW, MEDIUMPOWER, SLIGHTLYHIGH, HIGH, VERYHIGH }

Fuzzy rules developed by the genetic algorithm are of the form:

if Ball is Near and Goal is Near then KickTowardGoal Low
if Ball is Far or Ball is SlightlyLeft then RunTowardBall High

In the example chromosome fragment shown in Fig. 6 the shaded clause has been specially coded to signify that it is a consequent gene, and the fragment decodes to the following rule:

if Ball is Left and Ball is At or Goal is not Far then Dribble Low

In this case the clause connector *OR* in the clause immediately prior to the consequent clause is not required, so ignored.

(Ball, is Left, And)	(Ball, is At, Or)	(Goal, is not Far, Or)	(Dribble, Null, Low)

Fig. 6. Messy-coded Genetic Algorithm Example Chromosome Fragment

Chromosomes are not fixed length: the length of each chromosome in the population varies with the length of individual rules and the number of rules on the chromosome. The number of clauses in a rule and the number of rules in a ruleset is only limited by the maximum size of a chromosome. The minimum size of a rule is two clauses (one premise and one consequent), and the minimum number of rules in a ruleset is one. Since the cut, splice and mutation operators implemented guarantee no out-of-bounds data in the resultant chromosomes, a rule is only considered invalid if it contains no premises. A complete ruleset is considered invalid only if it contains no valid rules. Some advantages of using a messy encoding in this case are:

- a ruleset is not limited to a fixed size
- a ruleset can be overspecified (i.e. clauses may be duplicated)
- a ruleset can be underspecified (i.e. not all genes are required to be represented)
- clauses may be arranged in any way

An example complete chromosome and corresponding rules are shown in Fig. 7 (with appropriate abbreviations).

(B,N,O)	(B,nF,A)	(G,N,A)	(RB,n,L)	(B,A,A)	(G,vN,O)	(KG,n,M)	(B,L,A)	(T,L,n)

Premise	Consequent

Rule 1: if Ball is Near or Ball is not Far and Goal is Near then RunTowardBall Low
Rule 2: if Ball is At and Goal is VeryNear then KickTowardGoal MediumPower
Rule 3: if Ball is Left then Turn Left

Fig. 7. Chromosome and corresponding rules

In contrast to classic genetic algorithms which use a fixed size chromosome and require *"don't care"* values in order to generalise, no explicit *don't care* values are, or need be, implemented for any attributes in this method. Since messy-coded genetic algorithms encode information of variable structure and length, not all attributes, particularly premise variables, need be present in any rule or indeed in the entire ruleset. A feature of the messy-coded genetic algorithm is that the format implies *don't care* values for all attributes since any premise may be omitted from any or all rules, so generalisation is an implicit feature of this method.

For the messy-coded genetic algorithm implemented in this work the selection operator is implemented in the same manner as for classic genetic algorithms. Roulette wheel selection was used in the RoboCupSoccer trials and the initial SimpleSoccer trials. Tests were conducted to compare several selection methods, and elitist selection was used in the remainder of the SimpleSoccer trials. Crossover is implemented by the *cut* and *splice* operators, and mutation is implemented as a single-allele mutation scheme.

5. Experimental Evaluation

A series of experiments was performed in both the RoboCupSoccer and the SimpleSoccer simulation environments in order to test the viability of the fuzzy logic-based controller for the control of the player and the genetic algorithm to evolve the fuzzy ruleset. The following sections describe the trials performed, the parameter settings for each of the trials and other fundamental properties necessary for conducting the experiments.

An initial set of 20 trials was performed in the RoboCupSoccer environment in order to examine whether a genetic algorithm can be used to evolve a set of fuzzy rules to govern the behaviour of a simulated robot soccer player which produces consistent goal-scoring behaviour. This addresses part of the research question examined by this chapter.

Because the RoboCupSoccer environment is a very complex real-time simulation environment, it was found to be prohibitively expensive with regard to the time taken for the fitness evaluations for the evolutionary search. To overcome this problem the SimpleSoccer environment was developed so as to reduce the time taken for the trials. Following the RoboCupSoccer trials, a set of similar trials was performed in the SimpleSoccer environment to verify that the method performs similarly in the new environment.

Trials were conducted in the SimpleSoccer environment where the parameters controlling the operation of the genetic algorithm were varied in order to determine the parameters that should be used for the messy-coded genetic algorithm in order to produce acceptable results.

5.1 Trials

For the results reported, a single trial consisted of a simulated game of soccer played with the only player on the field being the player under evaluation. The player was placed at a randomly selected position on its half of the field and oriented so that it was facing the end of the field to which it was kicking. For the RoboCupSoccer trials the ball was placed at the centre of the field, and for the SimpleSoccer trials the ball was placed at a randomly selected position along the centre line of the field.

5.2 Fitness Evaluation

The objective of the fitness function for the genetic algorithm is to reward the fitter individuals with a higher probability of producing offspring, with the expectation that combining the fittest individuals of one generation will produce even fitter individuals in later generations. All fitness functions implemented in this work indicate better fitness as a lower number, so representing the optimisation of fitness as a minimisation problem.

5.2.1 RoboCupSoccer Fitness Function

Since the objective of this work was to produce goal-scoring behaviour, the first fitness function implemented rewarded individuals for goal-scoring behaviour only, and was implemented as:

$$f = \begin{cases} 1.0 & , goals = 0 \\ \dfrac{1.0}{2.0 \times goals} & , goals > 0 \end{cases}$$

where *goals* is the number of goals scored by the player during the trial.

Equation 1 RoboCupSoccer Simple Goals-only Fitness Function

In early trials in the RoboCupSoccer environment the initial population of randomly generated individuals demonstrated no goal-scoring behaviour, so the fitness of each individual was the same across the entire population. This lack of variation in the fitness of the population resulted in the selection of individuals for reproduction being reduced to random choice. To overcome this problem a composite fitness function was implemented which effectively decomposes the difficult problem of evolving goal-scoring behaviour essentially from scratch - actually from the base level of skill and knowledge implicit in the primitives supplied by the environment – into two less difficult problems:

- evolve ball-kicking behaviour, and
- evolve goal-scoring behaviour from the now increased base level of skill and knowledge

In the RoboCupSoccer trials, individuals were rewarded for, in order of importance:

- the number of goals scored in a game
- the number of times the ball was kicked during a game

This combination was chosen to reward players primarily for goals scored, while players that do not score goals are rewarded for the number of times the ball is kicked on the assumption that a player which actually kicks the ball is more likely to produce offspring capable of scoring goals. The actual fitness function implemented in the RoboCupSoccer trials was:

$$f = \begin{cases} \begin{cases} 1.0 & , kicks = 0 \\ 1.0 - \dfrac{kicks}{2.0 \times ticks} & , kicks > 0 \end{cases} & , goals = 0 \\ \dfrac{1.0}{2.0 \times goals} & , goals > 0 \end{cases}$$

where

goals = the number of goals scored by the player during the trial
kicks = the number of times the player kicked the ball during the trial
ticks = the number of RoboCupSoccer server time steps of the trial

Equation 2 RoboCupSoccer Composite Fitness Function

5.2.2 SimpleSoccer Fitness Function

A similar composite fitness function was used in the trials in the SimpleSoccer environment, where individuals were rewarded for, in order of importance:

- the number of goals scored in a game
- minimising the distance of the ball from the goal

This combination was chosen to reward players primarily for goals scored, while players that do not score goals are rewarded on the basis of how close they are able to move the ball to the goal on the assumption that a player which kicks the ball close to the goal is more likely to produce offspring capable of scoring goals. This decomposes the original problem of evolving goal-scoring behaviour into the two less difficult problems:

- evolve ball-kicking behaviour that minimises the distance between the ball and goal
- evolve goal-scoring behaviour from the now increased base level of skill and knowledge

The actual fitness function implemented in the SimpleSoccer trials was:

$$f = \begin{cases} \begin{cases} 1.0 & , kicks = 0 \\ 0.5 + \dfrac{dist}{2.0 \times fieldLen} & , kicks > 0 \end{cases} & , goals = 0 \\ \dfrac{1.0}{2.0 \times goals} & , goals > 0 \end{cases}$$

where

goals = the number of goals scored by the player during the trial
kicks = the number of times the player kicked the ball during the trial
dist = the minimum distance of the ball to the goal during the trial
fieldLen = the length of the field

Equation 3 SimpleSoccer Composite Fitness Function

The difference between the composite fitness function implemented in the RoboCupSoccer environment and the composite fitness function implemented in the SimpleSoccer environment is just an evolution of thinking - rewarding a player for kicking the ball often when no goal is kicked could reward a player that kicks the ball very often in the wrong direction more than a player that kicks the ball fewer times but in the right direction. The SimpleSoccer implementation of the composite fitness function rewards players more for kicking the ball closer to the goal irrespective of the number of times the ball was kicked. This is considered a better approach to encourage behaviour that leads to scoring goals.

5.2.3 Fitness Values

To facilitate the interpretation of fitness graphs and fitness values presented throughout this chapter, following is an explanation of the fitness values generated by the fitness functions used in this work. All fitness functions implemented in this work generate a real number R, where $0.0 < R \leq 1.0$, $R = 1.0$ indicates no ball movement and $R \approx 0.0$ indicates very good performance - smaller fitness values indicate better performance.
For ball movement in the RoboCupSoccer environment where a composite fitness function is implemented, fitness values are calculated in the range $x \leq R \leq y$, where $x = 0.5$ and $y = 1.0$. For ball movement in the SimpleSoccer environment where a composite fitness function is implemented, fitness values are calculated in the range $x \leq R \leq y$, where $x \approx 0.5$ and $y \approx 0.77$. Where a simple goals-only fitness function is implemented, ball movement alone is not rewarded: if no goals are scored the fitness function assigns $R = 1.0$. In both environments all fitness functions assign discrete values for goal-scoring, depending upon the number of goals scored. Table 1 summarises the fitness values returned by the various fitness functions.

		Simple Goals-only Fitness Function	RoboCupSoccer Composite Fitness Function	SimpleSoccer Composite Fitness Function
No Goals Scored	No Ball Movement	1.0000	1.0000	1.0000
	Ball Movement	*n/a*	[0.5, 1.0]	[~0.5, ~0.77]
Goals Scored	1	0.5000	0.5000	0.5000
	2	0.2500	0.2500	0.2500
	3	0.1667	0.1667	0.1667
	4	0.1250	0.1250	0.1250
	5	0.1000	0.1000	0.1000
	6	0.0833	0.0833	0.0833
	7	0.0714	0.0714	0.0714
	8	0.0625	0.0625	0.0625
	9	0.0556	0.0556	0.0556
	10	0.0500	0.0500	0.0500
	...	...	...	...
	n	$0.5/n$	$0.5/n$	$0.5/n$

Table 1. Fitness Assignment Summary

Parameter	Value
Maximum Chromosome Length	64 genes
Population Size	200
Maximum Generations	25
Selection Method	Roulette Wheel
Crossover Method	Single Point
Crossover Probability	0.8
Mutation Rate	10%
Mutation Probability	0.35

Table 2. Genetic Algorithm Control Parameters

In initial trials in the RoboCup environment players were evaluated over five separate games and then assigned the average fitness value of those games. Since each game in the Robocup environment is played in real time, this was a very time consuming method. The results of experiments where the player's fitness was calculated as the average of five games were compared with results where the player's fitness was assigned after a single game and were found to be almost indistinguishable. Due to the considerable time savings gained by assigning fitness after a single game, this is the method used throughout this work. Since players evolved using the average fitness method are exposed to different starting conditions they may be more robust than those evolved using single-game fitness, but the effect is extremely small considering the number of different starting positions players could be evaluated against and the fact that the starting positions of the player and ball really only affect the first kick of the ball.

5.3 Control Parameters

The genetic algorithm parameters common to all 20 initial trials in both the RoboCupSoccer and SimpleSoccer environments are shown in Table 2.

A game was terminated when:

- the target fitness of 0.05 was reached
- the ball was kicked out of play (RoboCupSoccer only)
- the elapsed time expired:
  - 120 seconds real time for RoboCupSoccer
  - 1000 ticks of simulator time for SimpleSoccer
- A period of no player movement or action expired
  - 10 seconds real time for RoboCupSoccer
  - 100 ticks of simulator time for SimpleSoccer

The target fitness of 0.05 reflects a score of 10 goals in the allotted playing time. This figure was chosen to allow the player a realistic amount of time to develop useful strategies yet terminate the search upon finding an acceptably good individual.

Two methods of terminating the evolutionary search were implemented. The first stops the search when a specified maximum number of generations have occurred; the second stops the search when the best fitness in the current population becomes less than the specified target fitness. Both methods were active, with the first to be encountered terminating the search. Early stopping did not occur in any of the experiments reported in this chapter.

6. Results

The following sections describe the results for the experiments performed for both the RoboCupSoccer and the SimpleSoccer environments. Discussion and analysis of the results is also presented.

6.1 RoboCupSoccer Initial Trial Results

Fig. **8** shows the average fitness of the population after each generation for each of the 20 trials for the RoboCupSoccer environment, showing that the performance of the population does improve steadily and, in some of the trials, plateaus towards a fitness of 0.5, or goal-scoring behaviour. Fig. **9** shows the best individual fitness from the population after each generation for each of 20 trials for the RoboCupSoccer environment, showing that good individuals are found after very few generations, in contrast to the gradual improvement in average fitness shown in Fig. **8**.

Fig. **10** is another visualisation of the progressive learning of the population from generation to generation, showing that not only do more players learn to kick goals over time, they learn to kick more goals more quickly. The histogram shows, for the initial 20 RoboCupSoccer trials, the average percentage of the population which scored 0, 1, 2, 3, 4 or more than 4 goals for each generation. The maximum number of goals scored by any individual was 3.

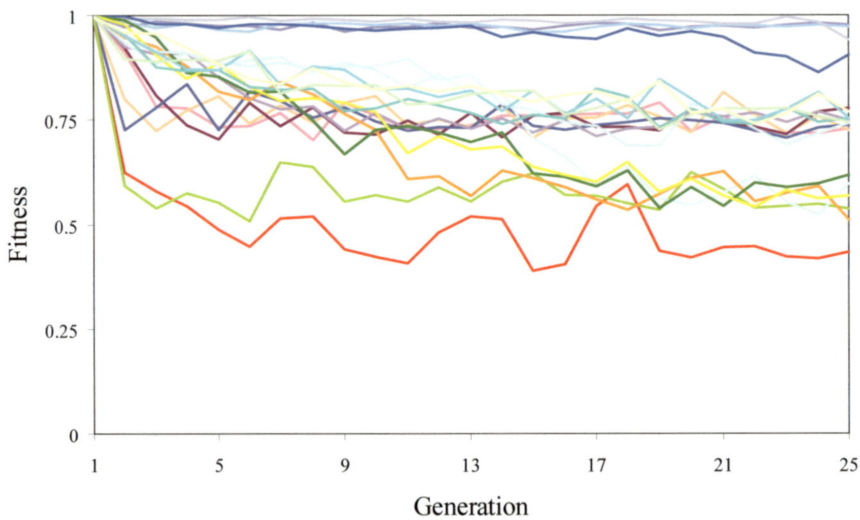

Fig. 8. RoboCupSoccer: Average Fitness - Initial 20 Trials

6.2 SimpleSoccer Initial Trial Results

Fig. **11** shows the average fitness of the population after each generation for each of the 20 trials for the SimpleSoccer environment, and as for the RoboCupSoccer trials, this graph shows that the performance of the population does improve steadily and plateaus, but unlike the RoboCupSoccer trials the average performance of the population does not approach a fitness of 0.5, or goal-scoring behaviour. Fig. **11** also shows that the average fitness curves for the SimpleSoccer trials are more tightly clustered than those of the

RoboCupSoccer trials, and plateau towards a fitness of around 0.75 which, in the SimpleSoccer environment indicates ball-kicking behaviour rather than goal-scoring behaviour.

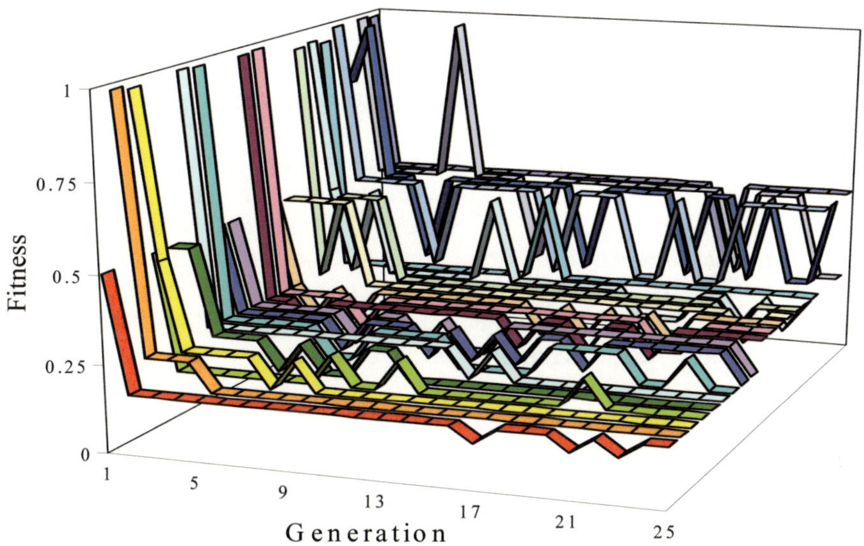

Fig. 9. RoboCupSoccer: Best Fitness - Initial 20 Trials

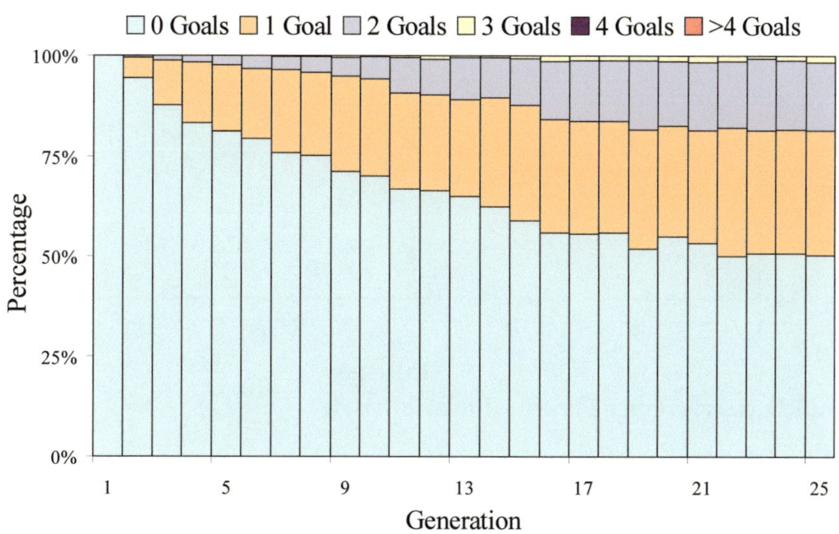

Fig. 10. RoboCupSoccer: Frequency of Individuals Scoring Goals

Fig. **12** shows the best individual fitness from the population after each generation for each of 20 trials for the SimpleSoccer environment and, as for the RoboCupSoccer trials, this graph shows that good individuals are found after very few generations. It is evident from a comparison of Fig. **9** and Fig. **12** that while good individuals are found quickly in both environments, the algorithm seems to be more stable in the RoboCupSoccer environment. The data shows that once a good individual is found in the RoboCupSoccer environment, good individuals are then more consistently found in future generations than in the SimpleSoccer environment.

Fig. **13** shows, for the initial 20 SimpleSoccer trials, the average percentage of the population which scored 0, 1, 2, 3, 4 or more than 4 goals for each generation. The maximum number of goals scored by an individual was 10. The contrast with the equivalent graph for the RoboCupSoccer environment (Fig. **10**) is striking since, although some individuals in the SimpleSoccer environment scored more goals than any individual in the RoboCupSoccer environment, the average goal-scoring behaviour of the population was less developed in the SimpleSoccer environment. This inconsistency is likely to be an indication that the combination of parameters used for the SimpleSoccer environment causes the genetic algorithm to converge more quickly than in the RoboCupSoccer environment, and a possible explanation for the lower average performance of the population when compared to that of the RoboCupSoccer environment as seen in Fig. **11**. Since these SimpleSoccer experiments were performed primarily as a comparison with the RoboCupSoccer experiments the genetic algorithm parameters were kept the same, but the soccer simulator implementations differ considerably and no tuning of simulator parameters to ensure similar performance was performed.

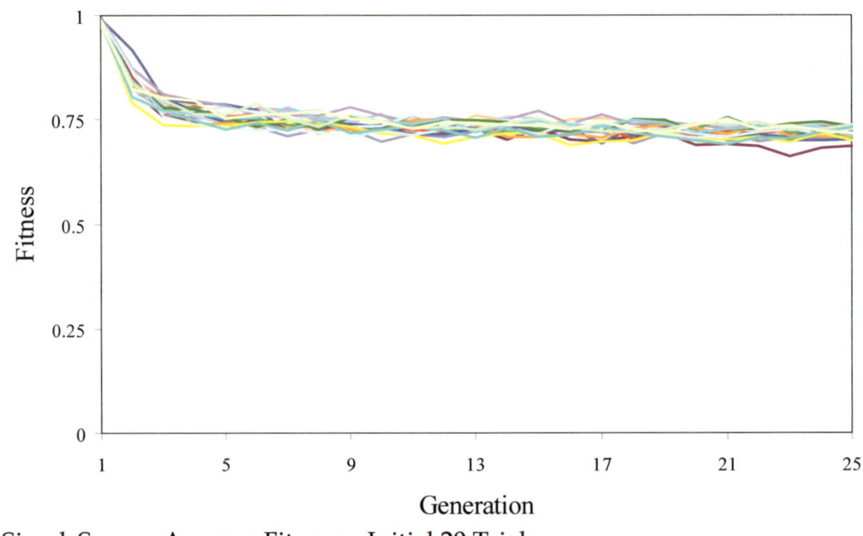

Fig. 11. SimpleSoccer: Average Fitness - Initial 20 Trials

6.3 SimpleSoccer as a Model for RoboCupSoccer

While the difference in the results of the experiments in the RoboCupSoccer and SimpleSoccer environments indicate that SimpleSoccer is not an exact model of RoboCupSoccer (as indeed it was not intended to be), there is a broad similarity in the results which is sufficient to indicate that the SimpleSoccer environment is a good simplified model of the RoboCupSoccer environment. Because SimpleSoccer is considered a reasonable model for RoboCupSoccer, and to take advantage of the significantly reduced training times provided by the SimpleSoccer environment when compared to RoboCupSoccer, all results reported in the remainder of this chapter are for experiments conducted exclusively in the SimpleSoccer environment.

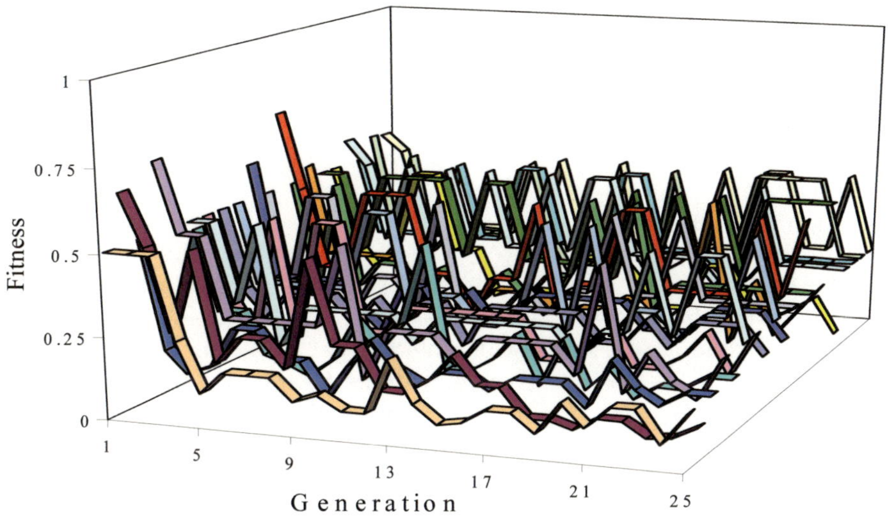

Fig. 12. SimpleSoccer: Best Fitness - Initial 20 Trials

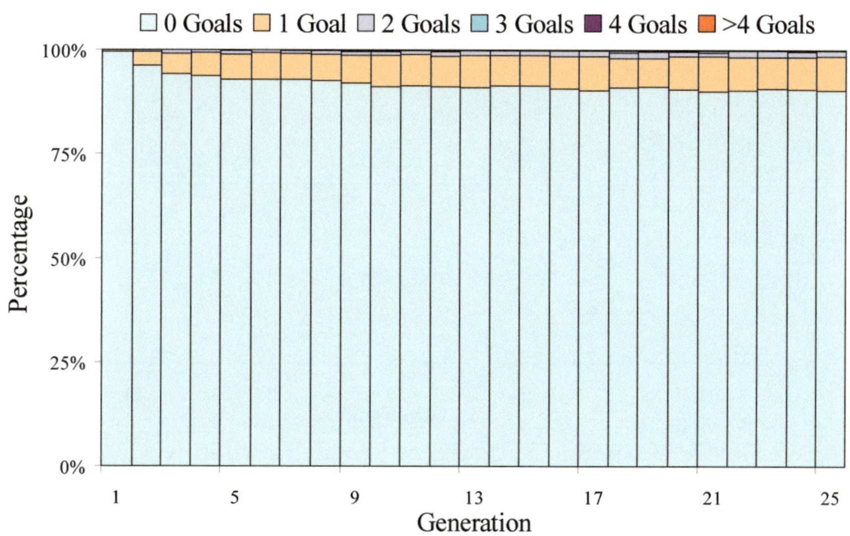

Fig. 13. SimpleSoccer: Frequency of Individuals Scoring Goals

6.4 GA Parameter Determination

Several experiments were conducted in order to determine a set of genetic algorithm parameters conducive to producing acceptable results. The following GA parameters were varied in these trials:

- Maximum Chromosome Length
- Population Size
- Selection Method
- Crossover Method
- Mutation Rate
- Maximum Generations

The values for the parameters shown in Table 2 on are used as control values, and in each of the trials presented in the following sections the value for a single GA parameter is varied. In the experiments conducted a series of 10 trials was performed for each different value of the parameter being varied, and in each case, with the exception of the experiment varying the maximum number of generations, the results presented are the averages of the 10 trials – each line on the graphs shown represents the average of the 10 trials. For the experiment varying the maximum number of generations, only 10 trials were conducted, and the results for each trial is reported individually – each line on the graph represents a single trial.

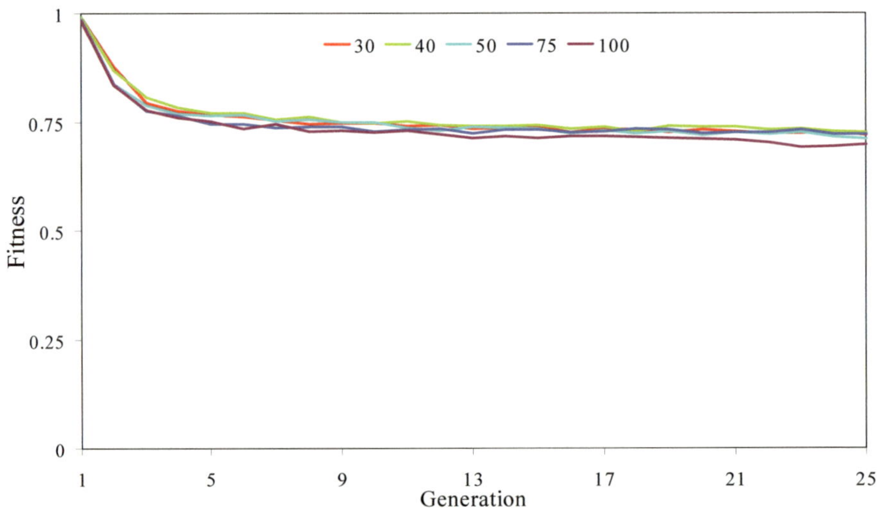

Fig. 14. Average Fitness: Maximum Chromosome Length Variation

6.4.1 Maximum Chromosome Length

While the actual length of the chromosome, measured as the number of genes on the chromosome, may vary depending upon the location of the cut point during the cut-and-splice operation of crossover, the maximum length of the chromosome is fixed throughout the evolutionary process. In order to determine if the maximum length of the chromosome is a significant factor in determining the quality of the evolutionary search, and if so what value is a good value, a series of trials was performed with different maximums for the chromosome length. Fig. **14** shows the average fitness of the population throughout the evolutionary process. It is evident from Fig. **14** that while a maximum chromosome length of 100 offers a very slight advantage, it is not significant. This is further substantiated by Fig. **15** which shows the best fitness in the population throughout the evolutionary process. The results shown indicate that while the method is not sensitive in any significant way to variations in the maximum chromosome length, a maximum chromosome length of somewhere between 50 and 100 genes, and most probably between 50 and 75 genes, produces less variation in the best fitness over the duration of the process.

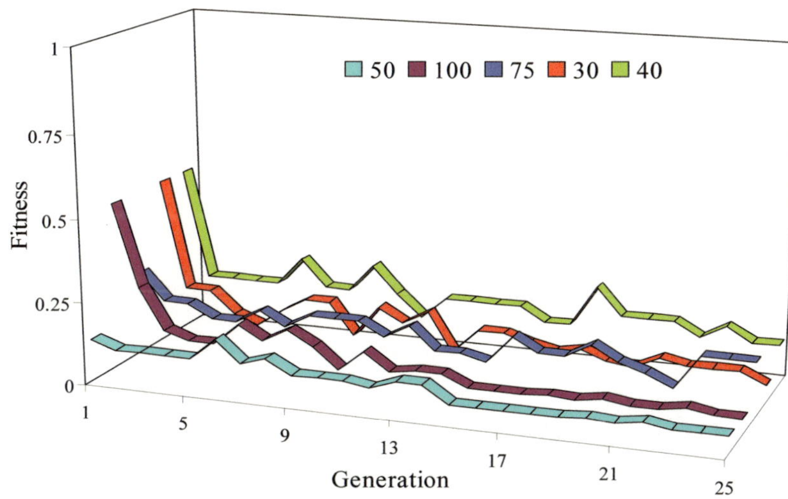

Fig. 15. Best Fitness: Maximum Chromosome Length Variation

Since the actual chromosome length may vary up to the maximum, the average chromosome length for each of the variations in maximum length was measured throughout the evolutionary process, as was the average number of valid rules per chromosome. These data are shown in Fig. **16** and Fig. **17** respectively. The chromosome lengths in the initial populations are initialised randomly between the minimum length of two genes (the smallest number of genes for a valid ruleset) and the maximum chromosome length, so the average chromosome lengths for the initial population shown in Fig. **16** are as expected. All trials show the average chromosome length rising in the initial few generations, then settling to around two-thirds of the maximum length. Given this, and since single-point crossover was used in these trials, with the cut point chosen randomly and chromosomes truncated at the maximum length after the cut-and-splice operation, the results indicate that chromosome length is unaffected by selection pressure. However, Fig. **17** shows the average number of rules per chromosome rising for all of the trials. This would indicate that there is some selection pressure for more rules per chromosome or shorter rules, but since the chromosome length is bounded, so is the number of rules per chromosome. Though outside the scope of this chapter, some further trials to investigate whether the pressure is for more rules or shorter rules, and the optimum number and/or length of rules per chromosome, would be useful work to undertake in the future.

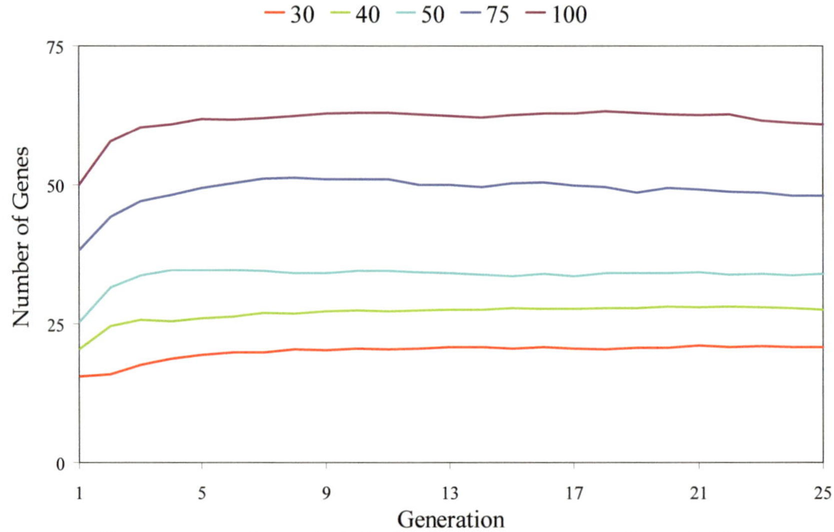

Fig. 16. Average Chromosome Length: Maximum Chromosome Length Variation

6.4.2 Population Size

Since the size of the population differs in this experiment, the number of generations was also varied according to the population size for each set of 10 trials to ensure the comparison between population sizes was for the same number of evaluations. Fig. **18** shows the average fitness of the population over 10,000 evaluations, and Fig. **19** shows the best fitness values for the same trials. It can be seen from the graphs that varying the population size has little effect on the population average fitness with only marginally better results for smaller population sizes, and a similarly small effect on individual best fitness, with larger populations producing slightly more stable results.

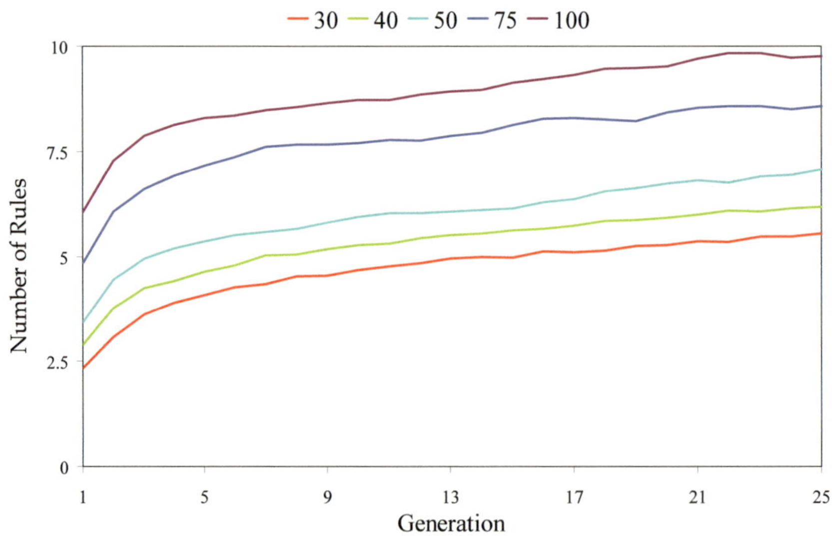

Fig. 17. Average Valid Rules per Chromosome: Maximum Chromosome Length Variation

Overall, the difference in performance between the population sizes tested is not significant, suggesting that it is the number of solutions evaluated, or the extent of the search, that is a significant factor affecting performance. This is consistent with the findings of other work in the area (Luke 2001).

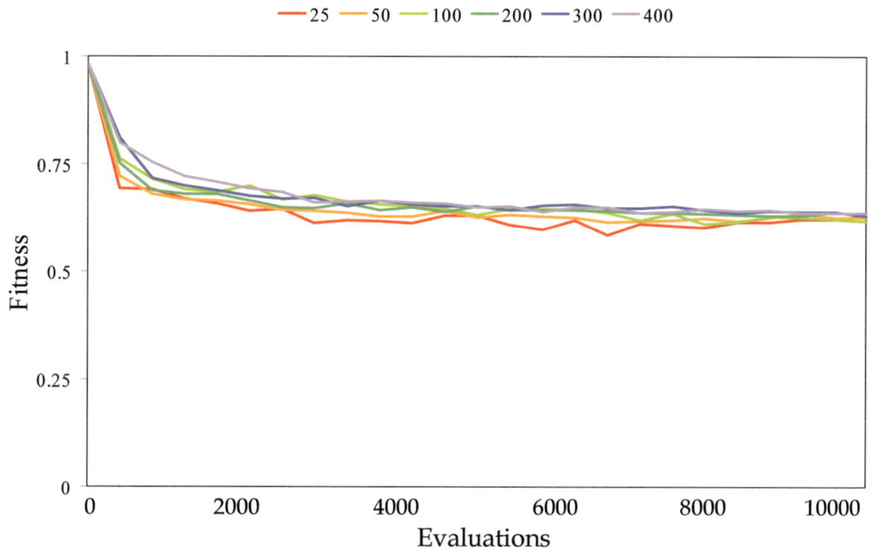

Fig. 18. Average Fitness: Population Size Variation

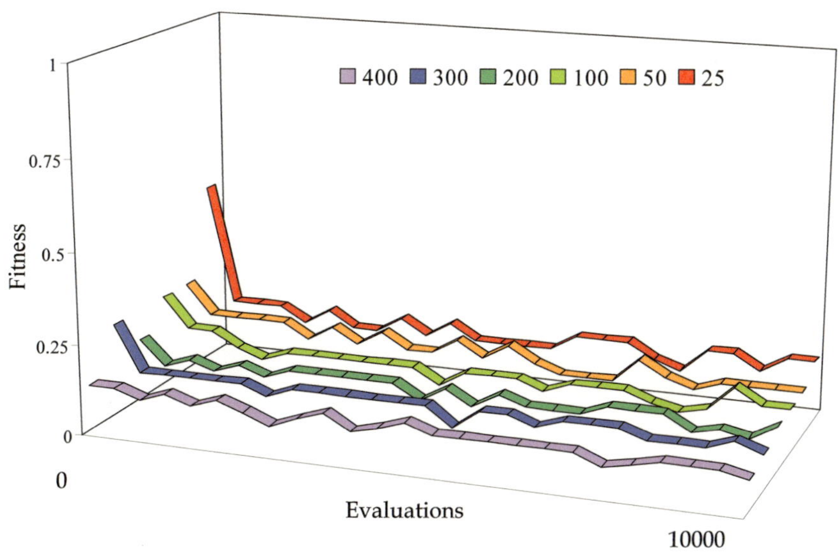

Fig. 19. Best Fitness: Population Size Variation

6.4.3 Selection Methods

Trials were conducted to compare the performance of three methods of selection: roulette wheel selection, tournament selection with tournament size = 2 and tournament selector = 0.75, and elitist selection with 5% retention. The population average fitness for the 10 trials conducted for each method is shown in Fig. **20**, and shows clearly that in terms of the population average fitness the method of selection is not a significant determinant. Fig**.** 21 shows the best fitness curves for these trials and shows that the elitist method produces similar results to the tournament method, with the roulette wheel method producing slightly less stable results. The good performance of the elitist method is probably due to the stochastic nature of the environment. Since the placement of the ball and the player is random, a player evaluated twice would likely be assigned different fitness values for each evaluation. The elitist method works well in this type of environment, allowing the better solutions to be evaluated several times thus allowing the reliability of the estimate of fitness to increase over time.

6.4.4 Crossover Methods

Since the chromosomes involved in crossover may be of different lengths, crossover methods that assume equal length chromosomes are not defined. The performance of two methods of crossover was compared: one-point and two-point. Fig. **22** shows the population average fitness over the duration of the evolutionary process, and Fig. **23** shows the best fitness values for the same period. It can be seen from this data that there is no meaningful difference in performance between the two methods, either with respect to the population average fitness or the best fitness achieved. While two-point crossover is more disruptive than one-point crossover, it is not clear from this data if a much more disruptive crossover method, such as uniform crossover, would significantly affect the performance of the method. It is likely that the messy-coding of the genetic algorithm and the rules-based nature of the representation causes the method to be somewhat less sensitive to disruptive crossover.

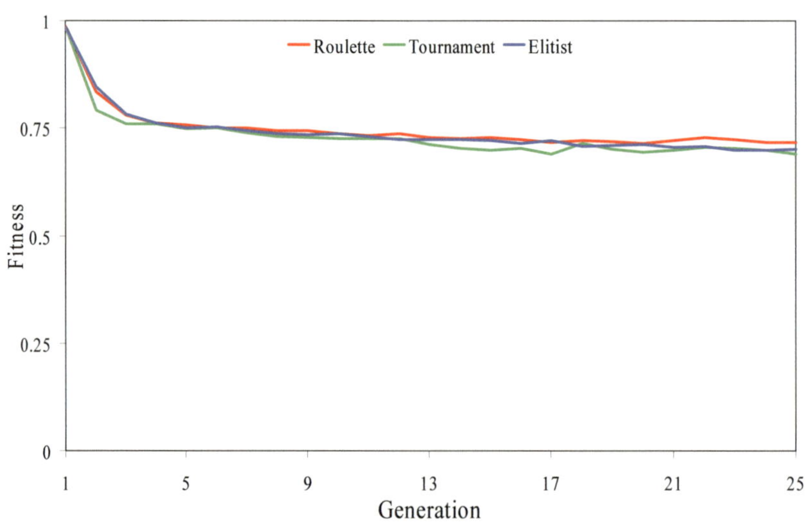

Fig. 20. Average Fitness: Selection Method Variation

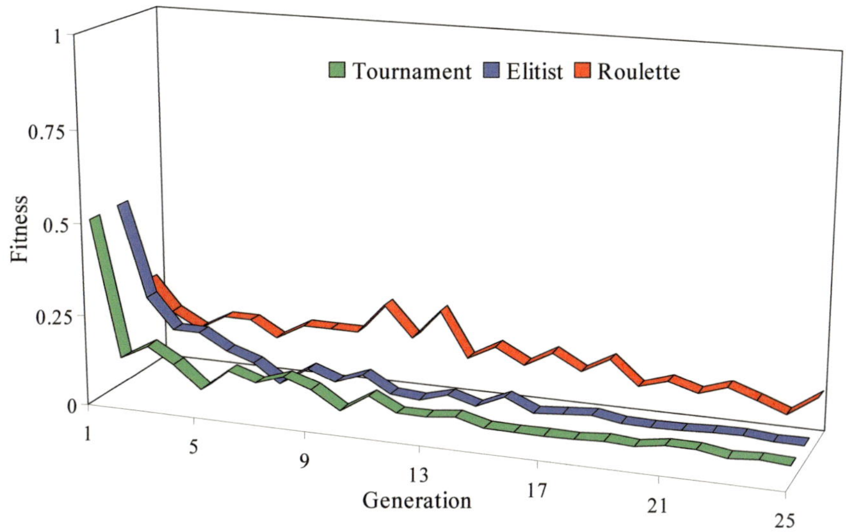

Fig. 21. Best Fitness: Selection Method Variation

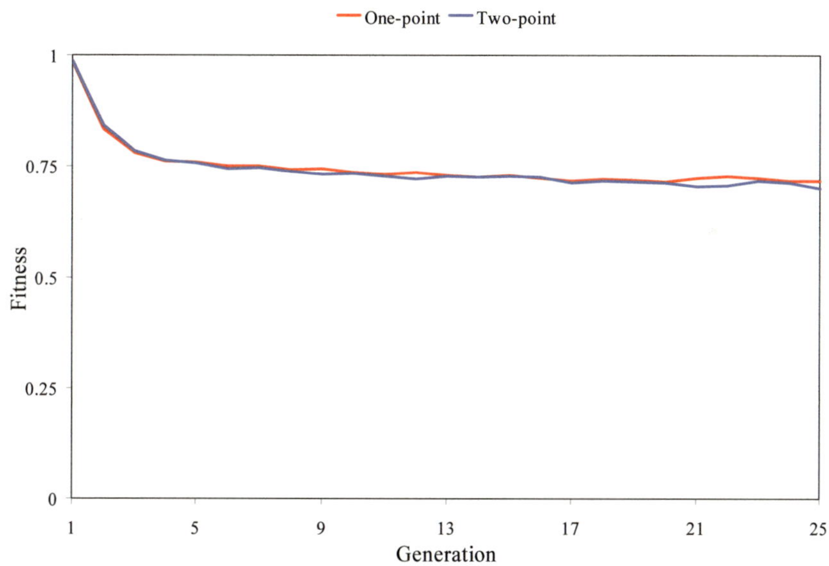

Fig. 22. Average Fitness: Crossover Method Variation

6.4.4 Mutation Rate

To determine the effect of the rate of mutation on the evolutionary process, 10 trials for each of several mutation rates were performed, and the averages of those trials presented. Fig. **24** shows the population average fitness for each mutation rate tested, and Fig. **25** the best fitness for those mutation rates throughout the evolutionary search. While varying the mutation rate has only a marginal effect on the population average fitness and, for the most part the individual best fitness, a mutation rate of 15% does seem to improve the population average fitness slightly, and the individual best fitness more markedly. This suggests that a mutation rate of 15% is the best balance between maintaining sufficient diversity in the population to help drive the evolutionary process while minimising the disruption to the good building blocks being created throughout the process.

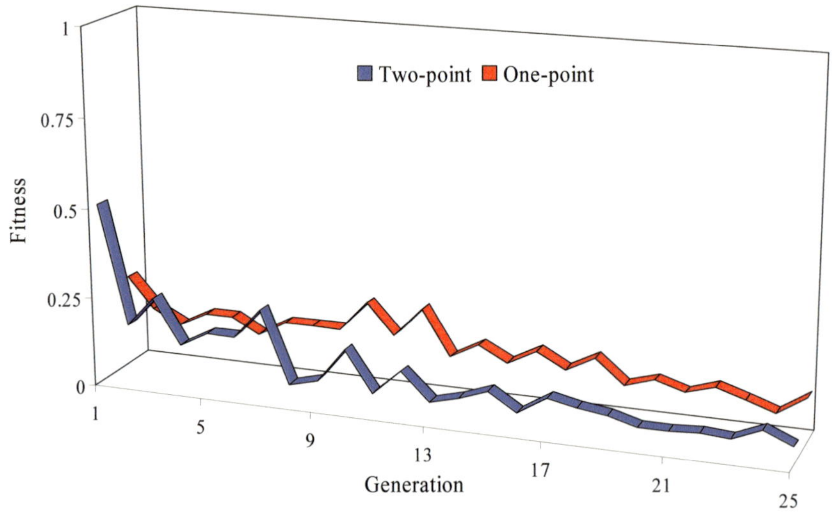

Fig. 23. Best Fitness: Crossover Method Variation

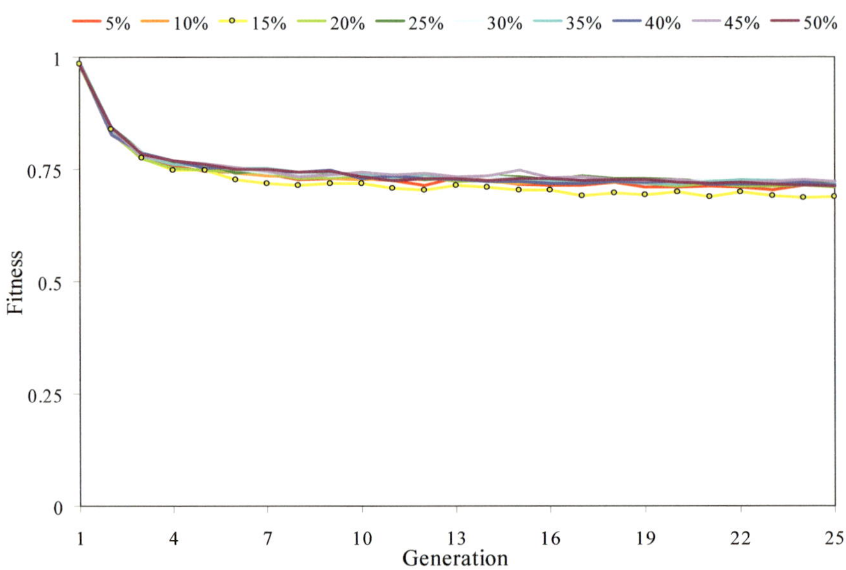

Fig. 24. Average Fitness: Mutation Rate Variation

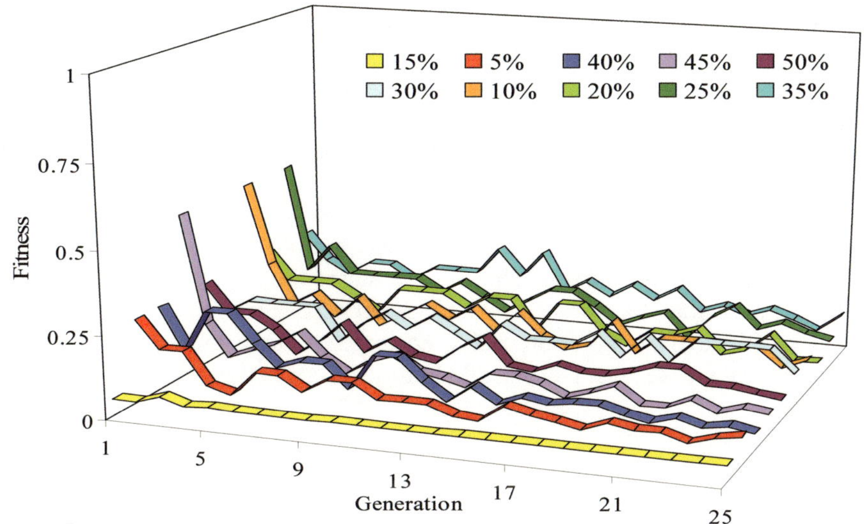

Fig. 25. Best Fitness: Mutation Rate Variation

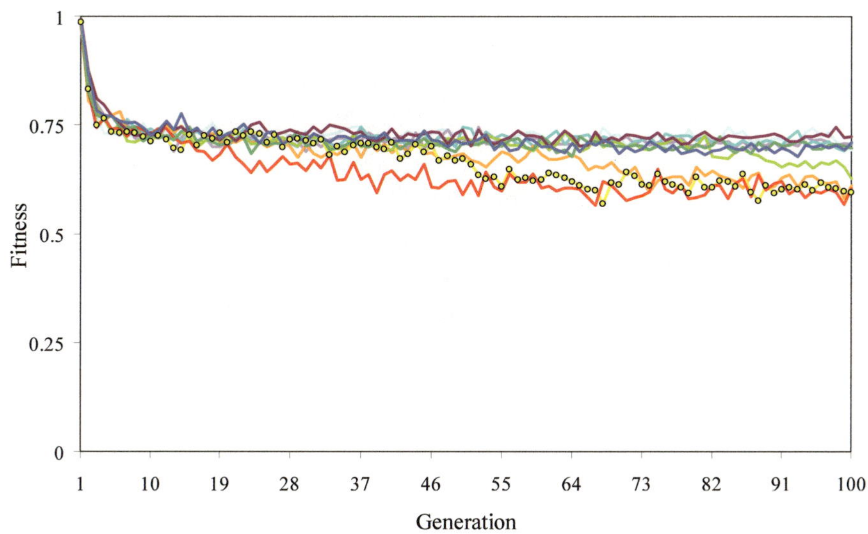

Fig. 26. Average Fitness: 100 Generations

6.4.4 Maximum Generations

In order to determine the effect of allowing the evolutionary process to continue for an extended time, a series of 10 trials was conducted with each trial continuing the evolutionary process for 100 generations. Two graphs of the results of these trials are presented. Fig. **26** shows the average fitness of the population for each of the 10 trials, and it can be seen that for more than half the trials the average fitness does not improve significantly after the tenth generation. Fig. **27** shows the best individual fitness from the population after each generation for each of the trials, and presents a similar scenario to that of the average fitness values: the best fitness does not improve significantly in most of the trials after the first few generations, though for a small proportion of the trials some significantly fitter individuals are evolved. These graphs suggest that although for some instances continuing to allow the population to evolve for an extended period can produce an improved population average, and that in those instances the best performing individuals from the population are consistently better, there is no real advantage in extending the evolutionary process. In almost every case an individual from the first 10 to 15 generations achieved the equal best fitness seen over the 100 generations, so given that the objective is to find good goal-scoring behaviour there would seem to be no real advantage in extended evolution of the population. This is a similar to the result reported in (Luke 2001), where Luke suggests that for some problems genetic programming encounters a critical point beyond which further evolution yields no improvement. Luke further suggests that performing many shorter experiments is likely to produce better results than a single very long experiment.

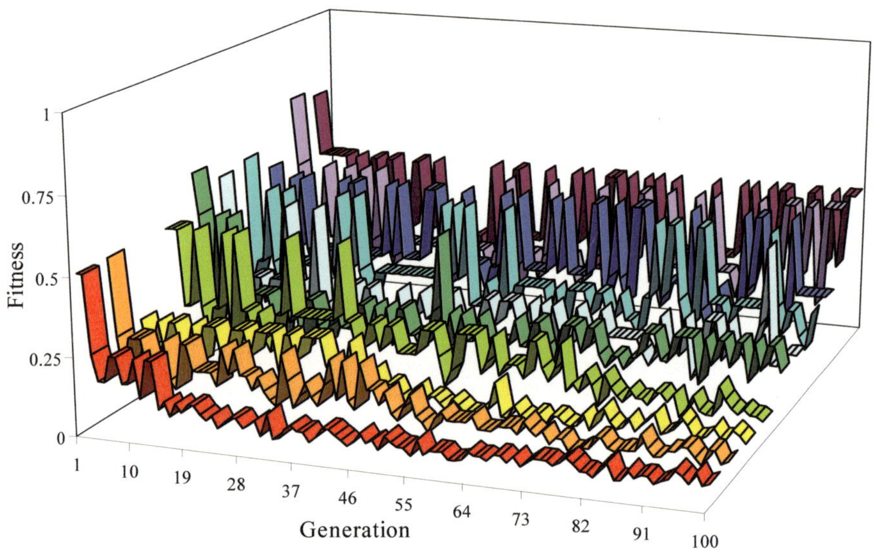

Fig. 27. Best Fitness: 100 Generations

6.4.5 Gene Pools

The trials reported in the previous section provide an opportunity to study, in broad terms, the genetic makeup of the population being evolved. For each of the 100 generations, Fig. **28** shows the average number of genes per chromosome, premises per chromosome, rules per chromosome, and valid rules per chromosome, with standard deviations for each. This graph shows that average chromosome length does not grow uncontrollably, and in fact plateaus at about 2/3 the maximum possible length. The average number of rules per chromosome, and hence the average number of consequents per chromosome, grows steadily throughout the evolutionary run. This agrees with the data presented earlier for the maximum chromosome length variation trials. It is interesting to note that the number of rules per chromosome is still increasing after 100 generations. Since the minimum number of genes per rule is 2 the number of rules per chromosome is bounded by half the chromosome length, and although the graph shows the number of rules approaching the upper bound, it has not reached that figure after 100 generations. Though outside the scope of this chapter, some more experimentation to observe the effect of reaching the upper bound would be useful work to undertake in the future.

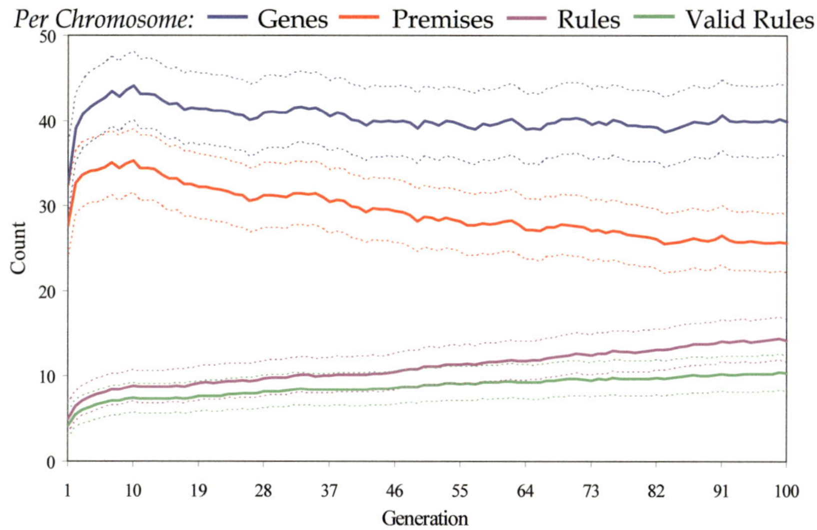

Fig. 28. Population Composition Mean and Standard Deviation: 100 Generations

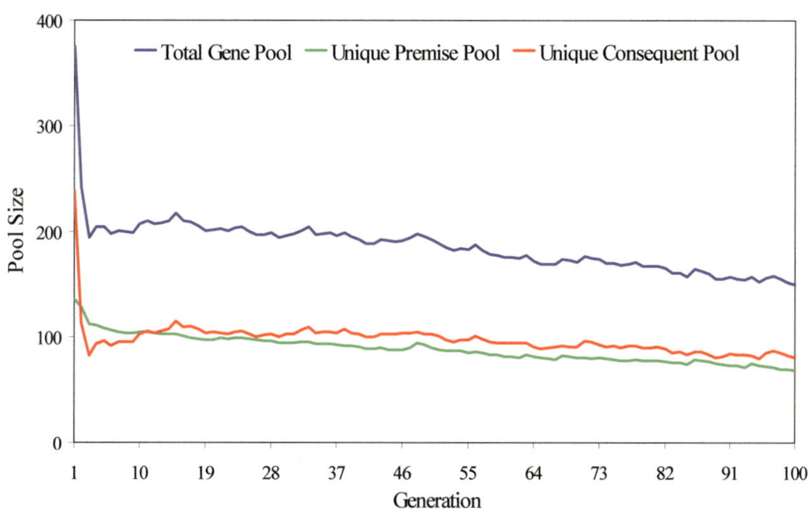

Fig. 29. Gene Pools: 100 Generations

The population gene pool sizes throughout the evolutionary process are show in Figure 29. The graph shows raw numbers of unique premise genes, unique consequent genes, and the total gene pool available to each generation of individuals. It is evident from the graph that the pool of unique premises falls slowly, but steadily, from the first generation, while the pool of unique consequent genes stays reasonably constant for close to 40 generations after an initial decrease. This is not unexpected, and is an indication that some selection pressure is evident, but that the number of rules remains fairly constant.

6.5 Performance Comparison – GA vs Random Search

In this section, in order to gauge the relative difficulty of the problem, the results obtained using the messy-coded GA search are compared to results obtained from random search. The messy-coded GA results shown are the average of the 10 trials conducted for the "maximum generations" experiments described earlier. The random search technique was simply the random generation and evaluation of a "population" of 500 individuals repeated 40 times – to equal the number of evaluations completed for the messy-coded GA trials. The "population" average fitness is shown in Fig. **30**, and the best individual fitness at intervals of 500 evaluations is shown in Fig. 31. The average fitness curves are included only to illustrate that the genetic algorithm is able to consistently improve the quality of the population for the duration of the evolutionary process: random search would not be expected to perform in the same manner. The best fitness curves (Fig. 31) show that although random search is able to find individuals that exhibit goal-scoring behaviour (i.e. fitness ≤ 0.5), evolutionary search finds better individuals and finds them more consistently, indicating that evolutionary search is not only a more successful search technique than random search, it is more robust.

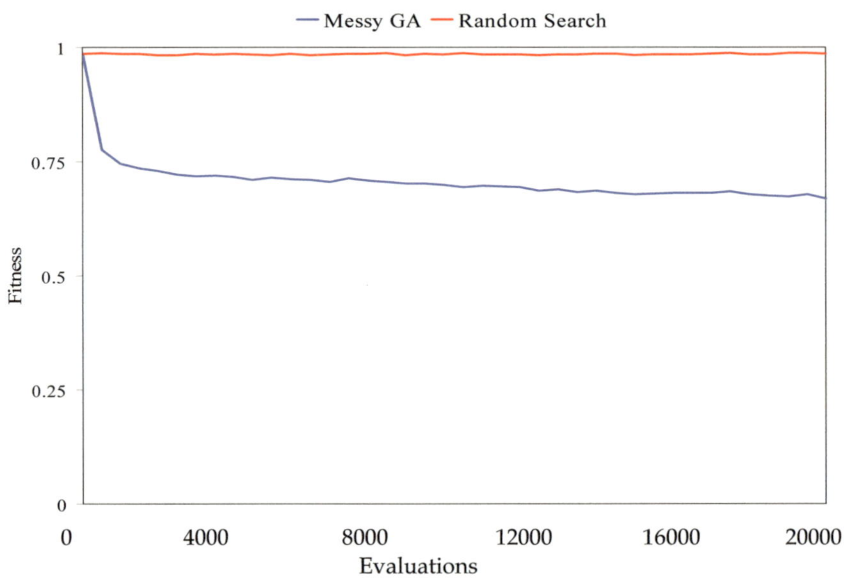

Fig. 30. Average Fitness: Random Search Comparison

As noted, random search does successfully find individuals with reasonably good goal-scoring skills. This result is a little surprising at first glance, but on closer inspection of the problem an explanation does present itself – the problem of finding goal-scoring behaviour *in the solution space defined*, whether by evolutionary search or random search, is not as difficult as it first seems, and this is because the solution space defined is not the one envisaged when first considering the problem. The solution space defined is one populated by players with mid-level, hand-coded skills available to them, as well as a "smart" default hunt action, which is a much richer solution space than the one usually envisaged when considering the problem of evolving goal-scoring behaviour "from scratch". As evidenced by the results of the random search shown here, the density of "reasonably good" solutions in the solution space is sufficiently high that random search will occasionally, and with some consistency, find one of those "reasonably good" solutions.

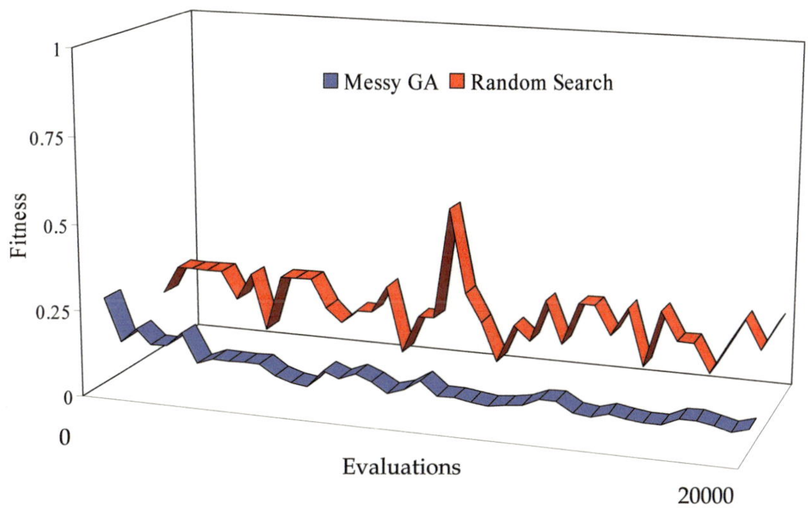

Fig. 31. Best Fitness: Random Search Comparison

7. Summary and Discussion

The work presented in this chapter has provided an implementation and empirical analysis of a fuzzy logic-based robot soccer player and the messy-coded genetic algorithm training algorithm. Several trials were performed to test the capacity of the method to produce goal-scoring behaviour. The results of the trials performed indicate that the player defined by the evolved fuzzy rules of the controller is capable of displaying consistent goal-scoring behaviour. This outcome indicates that for the problem of developing goal-scoring behaviour in a simulated robot soccer environment, when the initial population is endowed with a set of mid-level hand-coded skills, taking advantage of the flexible representation afforded by the messy-coded genetic algorithm and combining that with a fuzzy logic-based controller enables a fast and efficient search technique to be constructed.

Several experiments were performed to vary the genetic algorithm parameters being studied. The results of those tests indicate that within the range of the values tested, most parameters have little effect on the performance of the search. The *Maximum Chromosome Length* and *Selection Method* parameters had a marginal influence over the efficacy of the search, and although better performance was sometimes achieved after a long period of evolution, the *Maximum Generations* parameter is not considered to have a large effect on the performance of the algorithm after an upper bound of about 15 generations.

8. References

Andre, D. and A. Teller (1999). Evolving Team Darwin United. M. Asada and H. Kitano eds, RoboCup-98: Robot Soccer World Cup II. Lecture Notes in Computer Science, Springer-Verlag.

Aronsson, J. (2003). Genetic Programming of Multi-agent System in the RoboCup Domain. Masters Thesis, Department of Computer Science. Lund, Sweden, Lund Institute of Technology.

Asada, M., S. Noda, et al. (1996). "Purposive Behavior Acquisition for a Real Robot by Vision-Based Reinforcement Learning." Machine Learning **23**(2-3): 279-203.

Bajurnow, A. and V. Ciesielski (2004). Layered Learning for Evolving Goal Scoring Behaviour in Soccer Players. G. Greenwood, ed., Proceedings of the 2004 Congress on Evolutionary Computation, Vol. 2, p. 1828-1835, IEEE.

Balch, T. (2005). Teambots Domain, http://www.teambots.org.

Brooks, R. (1985). Robust Layered Control System for a Mobile Robot. A.I. Memo 864, Massachusetts Institute of Technology, Artificial Intelligence Laboratory.

Brooks, R. (1991). "Intelligence Without Representation." Artificial Intelligence **47**: 139-159.

Castillo, C., M. Lurgi, et al. (2003). Chimps: An Evolutionary Reinforcement Learning Approach for Soccer Agents. Proceedings of the 2003 IEEE International Conference on Systems, Man and Cybernetics, Vol. 1, p. 60-65.

Ciesielski, V. and S. Y. Lai (2001). Developing a Dribble-and-Score Behaviour for Robot Soccer using Neuro Evolution. Proceedings of the Fifth Australia-Japan Joint Workshop on Intelligent and Evolutionary Systems, p. 70-78, Dunedin, New Zealand.

Ciesielski, V., D. Mawhinney, et al. (2002). Genetic Programming for Robot Soccer. Proceedings of the RoboCup 2001 Symposium. Lecture Notes in Artificial Intelligence, p. 319-324.

Ciesielski, V. and P. Wilson (1999). Developing a Team of Soccer Playing Robots by Genetic Programming. Proceedings of the Third Australia-Japan Joint Workshop on Intelligent and Evolutionary Systems, p. 101-108, Canberra, Australia.

Di Pietro, A., L. While, et al. (2002). Learning in RohoCup Keepaway Using Evolutionary Algorithms. Langdon et al., eds, Proceedings of the Genetic and Evolutionary Computation Conference, p. 1065-1072, New York, NY, Morgan Kaufmann.

Gustafson, S. M. (2000). Layered Learning in Genetic Programming for a Cooperative Robot Soccer Problem. Masters Thesis, Department of Computing and Information Science, College of Engineering. Manhattan, KS, Kansas State University.

Gustafson, S. M. and W. H. Hsu (2001). Layered Learning in Genetic Programming for a Co-operative Robot Soccer Problem. Proceedings of the Fourth European Conference on Genetic Programming, Lake Como, Italy, Springer.

Holland, J. (1975). Adaptation in Natural and Artificial Systems. Ann Arbor, The University of Michigan Press.

Hsu, W. H., S. J. Harmon, et al. (2004). Empirical Comparison of Incremental Reuse Strategies in Genetic Programming for Keep-Away Soccer. Late Breaking Papers of the 2004 Genetic and Evolutionary Computation Conference, Seattle WA.

Jang, J.-S., C.-T. Sun, et al. (1997). Neuro-Fuzzy and Soft Computing, Prentice Hall.

Kandel, A. (1986). Fuzzy Mathematical Techniques with Applications, Addison-Wesley, Reading MA.

Kinoshita, S. and Y. Yamamoto (2000). 11Monkeys Description. Veloso et al., eds, Proceedings of Robocup-99: Robot Soccer World Cup III. Lecture Notes In Computer Science, Vol. 1856, p. 550-553, Springer-Verlag, London.

Kitano, H., M. Asada, et al. (1997a). RoboCup: The Robot World Cup Initiative. Proceedings of the First International Conference on Autonomous Agents, p. 340-347, Marina Del Rey, CA.

Kitano, H., M. Asada, et al. (1997b). RoboCup: A Challenge Problem for AI. AI Magazine, 18(1): p.73-85. **18**.

Kitano, H., M. Tambe, et al. (1997). The RoboCup Synthetic Agent Challenge 97. Proceedings of the Fifteenth International Joint Conference on Artificial Intelligence, p. 24-29, Nagoya, Japan.

Klir, G. J. and T. A. Folger (1988). Fuzzy Sets, Uncertainty and Information, Prentice Hall.

Kruse, R., J. Gebhardt, et al. (1994). Foundations of Fuzzy Systems, Wiley.

Kuhlmann, G. and P. Stone (2004). Progress in Learning 3 vs. 2 Keepaway. D. Polani et al., eds, RoboCup-2003: Robot Soccer World Cup VII, Springer Verlag, Berlin.

Lazarus, C. and H. Hu (2003). Evolving Goalkeeper Behaviour for Simulated Soccer Competition. Proceedings of the Third IASTED International Conference on Artificial Intelligence and Applications, Benalmádena, Spain.

Lima, P., L. Custódio, et al. (2005). RoboCup 2004 Competitions and Symposium: A Small Kick for Robots, a Giant Score for Science. AI Magazine 6(2). **6**.

Luke, S. (1998a). Evolving SoccerBots: A Retrospective. Proceedings of the Twelfth Annual Conference of the Japanese Society for Artificial Intelligence, Tokyo, Japan.

Luke, S. (1998b). Genetic Programming Produced Competitive Soccer Softbot Teams for RoboCup97. J. Koza et al., eds, Proceedings of the Third Annual Genetic Programming Conference, p. 204-222, Morgan Kaufmann, San Francisco.

Luke, S. (2001). When Short Runs Beat Long Runs. Proceedings of the 2001 Genetic and Evolutionary Computation Conference, p. 74-80, San Francisco CA.

Luke, S., C. Hohn, et al. (1998). Co-evolving Soccer Softbot Team Coordination with Genetic Programming. H. Kitano, ed., RoboCup-97: Robot Soccer World Cup I. Lecture Notes in Artificial Intelligence, p. 398-411, Springer-Verlag, Berlin.

Luke, S. and L. Spector (1996). Evolving Teamwork and Coordination with Genetic Programming. J.R. Koza et al., eds, Proceedings of the First Annual Conference on Genetic Programming, p. 150-156, Cambridge MA, The MIT Press.

Mamdani, E. and S. Assilian (1975). "An Experiment in Linguistic Synthesis with a Fuzzy Logic Controller." International Journal of Man-Machine Studies **7**(1): 1-13.

Nakashima, T., M. Takatani, et al. (2004). An Evolutionary Approach for Strategy Learning in RoboCup Soccer. Proceedings of the 2004 IEEE International Conference on Systems, Man and Cybernetics, Vol. 2, p. 2023-2028.

Noda, I. (1995). Soccer Server: A Simulator of Robocup. Proceedings of AI Symposium '95. Japanese Society for Artificial Intelligence, pp. 29-34.

Noda, I., H. Matsubara, et al. (1998). "Soccer Server: A Tool for Research on Multiagent Systems." Applied Artificial Intelligence **12**: 233-250.

Noda, I. and P. Stone (2001). The RoboCup Soccer Server and CMUnited: Implemented Infrastructure for MAS research. T. Wagner and O. Rana, eds, International Workshop on Infrastructure for Multi-Agent Systems (Agents 2000). Lecture Notes in Computer Science, p. 94-101, Barcelona, Spain.

Riedmiller, M., T. Gabel, et al. (2005). Brainstormers 2D - Team Description 2005. Team Description Papers, Proceedings of RoboCup 2005 (CD) (to appear).
Riedmiller, M., A. Merke, et al. (2001). Karlsruhe Brainstormers - a Reinforcement Learning Approach to Robotic Soccer. P. Stone, T. Balch and G. Kraetszchmar, eds, RoboCup-2000: Robot Soccer World Cup IV. Lecture Notes in Artificial Intelligence., Springer Verlag, Berlin.
Riley, J. (2003). The SimpleSoccer Machine Learning Environment. S.-B. Cho, H. X. Nguen and Y. Shan, eds, Proceedings of the First Asia-Pacific Workshop on Genetic Programming, p. 24-30, Canberra, Australia.
Riley, J. (2007). Learning to Play Soccer with the SimpleSoccer Robot Soccer Simulator. Robotic Soccer. P. Lima. Vienna, I-Tech Education and Publishing,: 281-306.
Stone, P. (1998). Layered Learning in Multiagent Systems. PhD Thesis, Computer Science Department, Technical Report CMU-CS98-187, Carnegie Mellon University.
Stone, P. and R. Sutton (2001). Scaling Reinforcement Learning Toward RoboCup Soccer. Proceedings of the Eighteenth International Conference on Machine Learning, Williamstown MA.
Stone, P., R. S. Sutton, et al. (2005). "Reinforcement Learning for RoboCup-Soccer Keepaway." Adaptive Behavior **13**(3): 165-188.
Stone, P., R. S. Sutton, et al. (2001). Reinforcement Learning for 3 vs. 2 Keepaway. Robocup 2000: Robot Soccer World Cup IV. P. Stone, T.R. Balch, and G.K. Kraetzschmar, eds. Lecture Notes In Computer Science, vol. 2019, p. 249-258, Springer-Verlag, London.
Stone, P. and M. Veloso (1999). Team-partitioned, Opaque-transition Reinforcement Learning. Proceedings of the Third International Conference on Autonomous Agents, Seattle WA.
Stone, P. and M. M. Veloso (2000). Layered Learning. Proceedings of the Eleventh European Conference on Machine Learning, p. 369-381, Springer, Berlin.
Uchibe, E. (1999). Cooperative Behavior Acquisition by Learning and Evolution in a Multi-Agent Environment for Mobile Robots. PhD Thesis, Osaka University.
Watkins, C. (1989). Learning from Delayed Rewards. PhD Thesis, King's College, University of Cambridge.
Whiteson, S., N. Kohl, et al. (2003). Evolving Keepaway Soccer Players through Task Decomposition. E. Cantu-Paz et al., eds, Genetic and Evolutionary Computation - GECCO-2003, volume 2723 of Lecture Notes in Computer Science, p. 356-368, Chicago IL, Spinger-Verlag.
Whiteson, S., N. Kohl, et al. (2005). "Evolving Keepaway Soccer Players through Task Decomposition." Machine Learning **59**: 5-30.
Whiteson, S. and P. Stone (2003). Concurrent Layered Learning. Proceedings of the Second International Joint Conference on Autonomous Agents and Multiagent Systems, p. 193-200.
Zadeh, L. (1965). "Fuzzy Sets." Journal of Information and Control **Vol. 8**.

8

FIRA Mirosot Robot Soccer System Using Fuzzy Logic Algorithms

*Elmer A. Maravillas, PhD and **Elmer P. Dadios, PhD
* *Cebu Institute Of Technology (CIT), N. Bacalso Avenue, 6000 Cebu City, Philippines*
***De La Salle University-Manila, Taft Avenue, 1000 Manila, Philippines*

1. Introduction

In November of 1996, the first Micro-Robot World Cup Soccer Tournament (MIROSOT) was held in Korea participated in by several countries. Three robots per team play soccer on a 130 cm x 150 cm soccer field. There were more than 25 research papers submitted on the proceedings and technical sessions dealing with various aspects of robot soccer system development and game control strategies (1996 Micro-Robot World Cup Soccer Tournament Proceedings, November 9-12, 1996, KAIST, Taejon, KOREA). From then on several robot soccer tournaments were held in many places of the world.

A robot soccer system is a multi-agent intelligent control system composed of two or more robots, vision system, communication equipment, and a personal computer. Each robot in the system has its own movement mechanism and therefore can move independently as well as cooperate with other robots.

Robot soccer game is an ideal venue for finding solutions to the many challenging problems facing the multi-agent robotic system. These problems include coordination between robots, motion planning of robots, recognition of objects visually, obstacle avoidance, and so on [24]. The robots must be taught different behaviors so that they will be able to react appropriately in any given situation. These behaviors shall be put into action in coordination with other robots so that a specific game plan shall be accomplished. It is therefore imperative that better soccer players (robots) are made to follow a deliberate plan of action from a module that provides very swift decisions especially in critical situations, which in a real-time game can mean a difference of many goals [25].

Since 1996 there are already lots of developments in multi-agent robotic system as a result of the MIROSOT undertaking. The behaviors of robots were improved dramatically as major breakthroughs in the control system are piling one after another.

This chapter will focus on the development of intelligent behaviors for the Mirosot wheeled robots to generate human interest in the framework of entertainment using fuzzy logic algorithms. Algorithms for scorer, goalie, obstacle detection and avoidance will be developed into the robot soccer system. In addition, a game strategy is also implemented in real-time using fuzzy logic algorithms.

2. Fuzzy Logic FIRA Robot Soccer Game On-line Scorer

Since 1996 when the first FIRA robot soccer world cup tournament was held in South Korea, scoring for robot soccer games were always done by a human scorer. Because of the human's susceptibility to biases and assumptions many researchers have proven that the reliability of most decisions made by humans is greatly undermined. Therefore faulty decision-making tends to be present in almost all systems managed by humans employing only instincts as their tool. In several competitions such as horse racing, dog racing, track and field, swimming, etc., video camera is used on the finish line to determine the true winner of hotly contested competition.

The existing FIRA robot soccer system uses a camera also but as a component of its vision system hardware for locating the positions of the robots and the ball. The information gathered through this camera is fed to the computer for use in the game strategy adopted by the competing teams. The decision-making, as far as scoring is concerned, is still made by a human game official.

This section presents an artificial scoring system for FIRA robot soccer games given the Cartesian coordinates of the ball. It aims to resolve very contentious scoring situations that a human scorer will most likely fail to recognize. The system incorporates cheering sounds and physical coordinated/synchronized movements of robots after a score is made. Fuzzy logic system is a formal methodology for representing, manipulating, and implementing a human's heuristic knowledge about how to make decisions [1,2,3]. Fuzzy logic scorer for FIRA robot soccer game is an artificial decision maker that operates in real-time. Figure 1 shows the block diagram of the fuzzy logic scorer. The system has four main components: (1) the "rule-base" holds the knowledge in the form of rules, of how best to score a FIRA robot soccer game; (2) the inference mechanism evaluates which control rules are relevant at the current time and then decides what the input to the scoring process should be; (3) the fuzzification interface simply modifies the inputs so that they can be interpreted and compared to the rules in the rule base; and (4) the defuzzification interface converts the conclusions reached by the inference mechanism into inputs to the scoring process. The fuzzy scorer takes the Cartesian coordinates (x,y) of the ball and output a **score (1)** or **noscore(0)** depending on the *points* computed by the defuzzification interface which ranges from 0 to 100. The scoring process receives the value of *points* and gives output according to the following:

$$\textbf{output} = \begin{cases} \textbf{1 if } \textit{points} >= \textbf{85} \\ \textbf{0 if } \textit{points} < \textbf{85} \end{cases}$$

Figure 2 shows the Cartesian coordinates of FIRA robot soccer system's soccer field. The shaded portions of the field indicate the goal areas on both sides. The Cartesian coordinates for the LEFT goal are –10cm <= x <= 0cm, 45cm <= y <=85cm, and for the RIGHT goal are 150cm <= x <= 160cm, 45cm <= y <= 85cm.

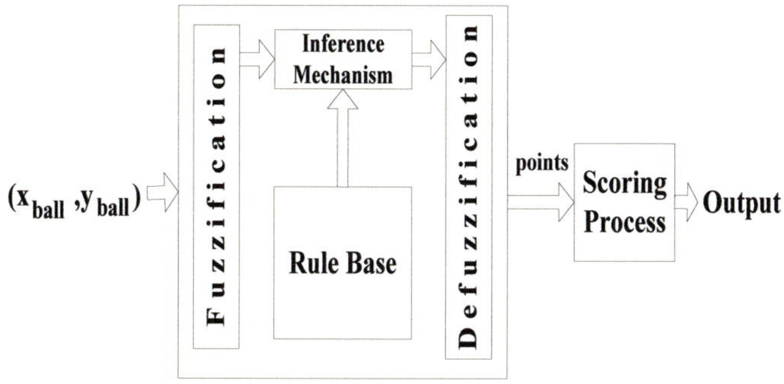

Fig. 1. Fuzzy scorer for FIRA Robot Soccer System.

When the ball's coordinates fall within these ranges, a scoring process will be done to determine whether a score could be awarded or not. The scoring process will be invoked only if the ball arrives within the specified areas of the field considered very close to the goals. Figure 3 shows some of these critical instances on the LEFT goal where the scoring process could be invoked. The same occurrences could happen on the RIGHT goal. In Figure 3, the ball's positions show that a greater portion of its body is beyond the goal line towards the goal area. Since the goal line is imaginary to the human referee, a particular team could be deprived of a score if the human referee fails to recognize the scoring situation. With fuzzy scorer, these things are not possible to happen.

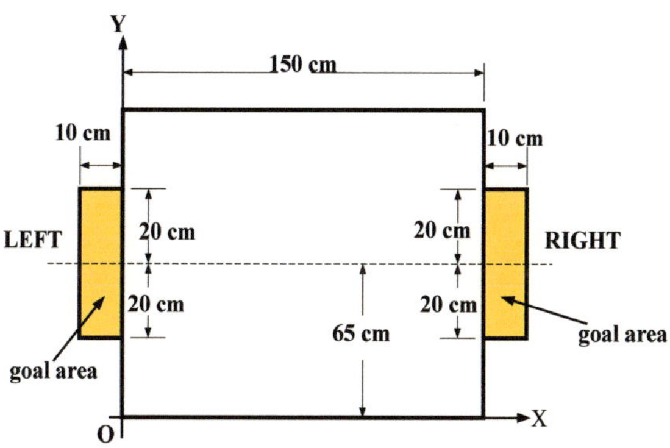

Fig. 2. Cartesian coordinates of FIRA robot soccer system's playing field.

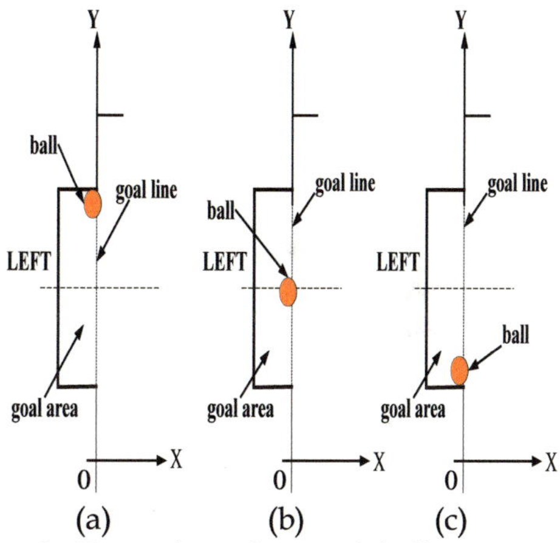

Fig. 3. Examples of critical ball positions where scoring process could be invoked.

Fuzzy sets map each element of the subsets of the "universe of discourse" to a continuous membership value (membership grade) ranging from 0 to 1. Input and output entities of a fuzzy logic system are composed or several maps (also called membership functions) with varying shapes common of which are, gaussian, triangular, and trapezoidal. Each map is assigned a linguistic term that best describes a range of values of the input and output entities. Figures 4 & 5 show the membership functions of the ball's x-coordinate. **CL** functions' maximum value is placed on the goal line (at 0 cm). Figure 6 shows the membership functions of the ball's y-coordinate. **CR's** maximum value is located at 65 cm. The maximum values of **LR** and **UR** are placed at the extremities of the goal area. Figure 7 shows the membership functions of the consequence. Defuzzification of the relevant membership functions of the consequence generates a value that will be assigned to *points* (0-100) as input to the scoring process. The value of *points* determines whether a score is to be awarded or not. It should be noted that the shapes of the membership functions are products of heuristics (trial and error).

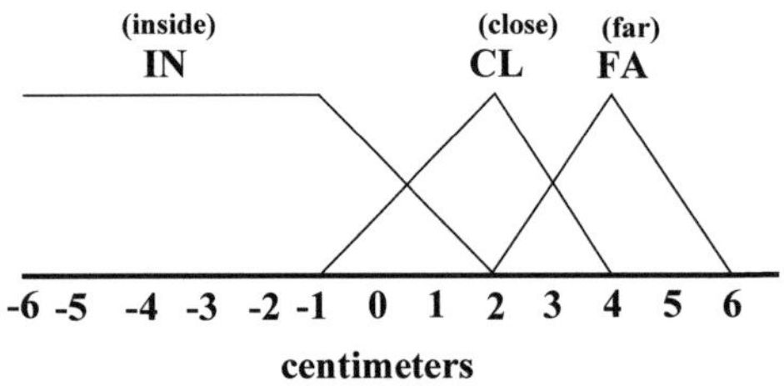

Fig. 4. Membership functions of ball's X-coordinate.

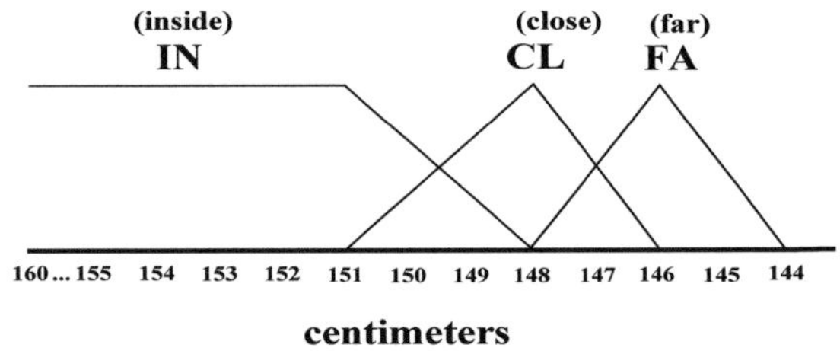

Fig. 5. Membership functions of ball's X-coordinate.

Reliable scoring decisions are made only when there is a clear outline of the set of rules provided by experts. Inputs are associated with each other from which a consequence will be assigned. The table that shows the consequences resulting from all possible associations of the input quantities is called the fuzzy associative memory (FAM) which represents the rule base of the fuzzy logic scorer. Table 1 shows the FAM of the fuzzy logic scorer.
The rules whose antecedents have membership function values greater than zero participate in the defuzzification process using their individual consequences to determine the *points* to be used in the scoring process.

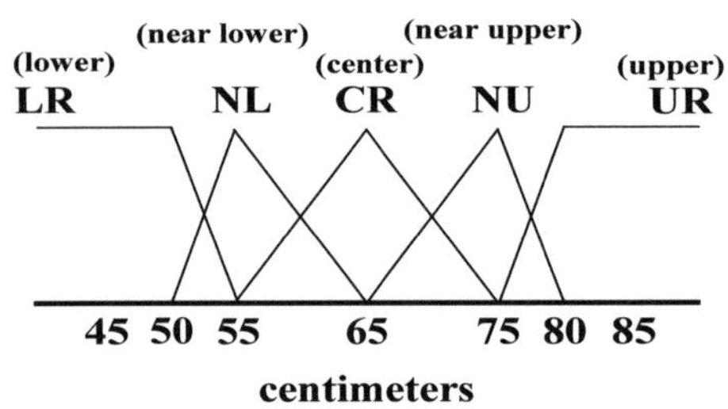

Fig. 6. Membership functions of ball's Y-coordinate.

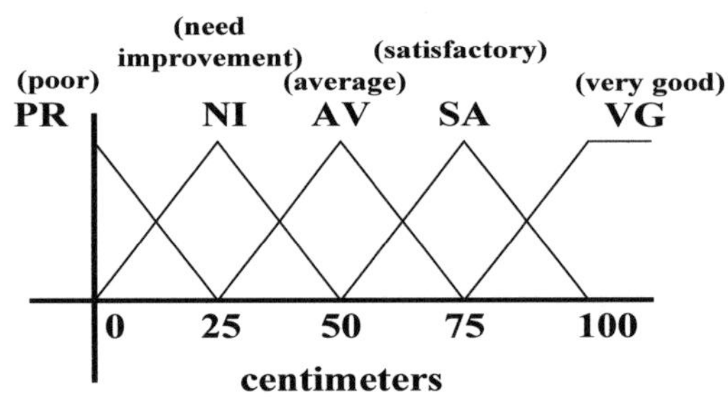

Fig. 7. Membership functions of consequence.

		Ball's y-coordinates				
Ball's x-coordinates		LR	NL	CR	NU	UR
	IN	VG	VG	VG	VG	VG
	CL	NI	AV	SA	AV	NI
	FA	PR	NI	AV	NI	PR

Table 1. Fuzzy logic scorer fuzzy associative memory (FAM).

2.1 Incorporating Sounds

Adding sounds to robot soccer games can make the game more entertaining and brings the game closer to the real world. Sounds can give feedback as an appreciation when a team scores and as criticism when a team commits foul action on the field.

Different recorded sounds can be played as the game progresses. Instances where playing of sounds is appropriate are: (1) when robots are executing field demonstrations simulating a cheering competition music can be played, (2) when a team made a goal a sound of applause can be played, (3) when a team committed foul a booing sound can be played, (4) when a team executes a defense strategy a sound shouting defense! Defense! ... may be played, and (5) when a team executes an offense strategy a sound of encouragement may be played. Sometimes a game situation requires the playing of different tunes at the same time.

Support for playing waveform audio is included in Visual C++. The Windows multimedia library *winmm.lib* and the *Wave class* library give a fast and handy way for adding sound effects to any Windows 95, Windows 98, or Windows NT application. A class can be created also to play multiple wave files at the same time. The *Wave class* is only added to any Visual C++ project, after linking the *winmm.lib*, it is now ready to add sound to the application. There are only three (3) methods required from the *Wave class* that enables an application to play sound from given medium. First, the recorded sound must be loaded into memory by the Load method. Second, the loaded sound will be played by the Play method. Third, played sound will be stopped by the Stop method [4].

Just like real soccer games, the playing of sounds and robot movements must occur simultaneously but independent of each other. When a team scores a goal, the robots of such team celebrates by moving around the robot soccer field while the sound of applause is playing. These tasks cannot be accomplished in ordinary sequential execution of components in a program. Because when the sound instructions grab the control of execution, the robots have to stop or enter in an uncontrollable situation. It is only when the sound finished playing that the robots gain control of execution. It is at this point that a separate line of execution is deemed necessary to allow playing of sounds on different occasions as the game progresses without encroaching on the roles of the robots.

The concept of multithreading is appropriate in this respect. A *thread* is a path of execution through a process' code. A preemptive scheduler inside the operating system divides CPU time among active threads so that they appear to run simultaneously. Secondary threads are ideal for performing tasks such as playing sounds while robots are doing their thing on the robot soccer field.

2.2 Robots Synchronized Movements

When the robots are playing soccer, their movements are not predetermined but depend most on the position of the ball and the game strategy being applied. However, when robots celebrate resulting from a score being made, their movements are predetermined so that proper coordination is attained thus making it more appealing to the audience. Different paths can be generated and loaded in memory at the start of the game so that they are readily available when needed. Geometric curves may be used as patterns for these coordinated movements of robots. Figure 14 shows the robots converging at the center of the playing field after a score has been made.

2.3 Experimental Results Of Scorer

Figures 8 to 10 show the performance of the fuzzy scorer on the left goal of the playing field. Three experiments were conducted on the left goal, namely, on the upper portion of the goal where the ball is made to approach the goal area with $y_{ball} > 75$, another where $y_{ball} = 65 \pm 5$, and another with $y_{ball} < 55$. The figures show the values of *points*, output of the defuzzification interface, as the ball approaches the goal line towards the goal area. Values of *points* equal or greater than 85 means that the scorer have decided it is a score. Experiments show that as the ball crosses the goal line ($x_{ball} = 0$), the fuzzy scorer gives *points* values that translate to a score.

The above experiments were also done on the right side goal of the playing field the results of which are shown on Figures 11 to 13. Consistent with the results of the above experiments, the fuzzy scorer gives *points* values that also translate to score as the ball passes thru the goal line ($x_{ball} = 150$) towards the goal area.

Figure 14 shows the coordinated movements of the robots after their attacker made a score. These actuations mimic the behavior of a human soccer player after making a score.

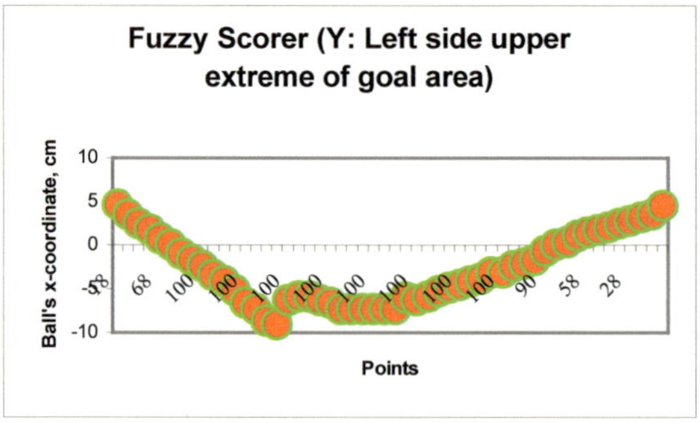

Fig. 8. Fuzzy scorer performance on the left side upper extreme portion of goal area.

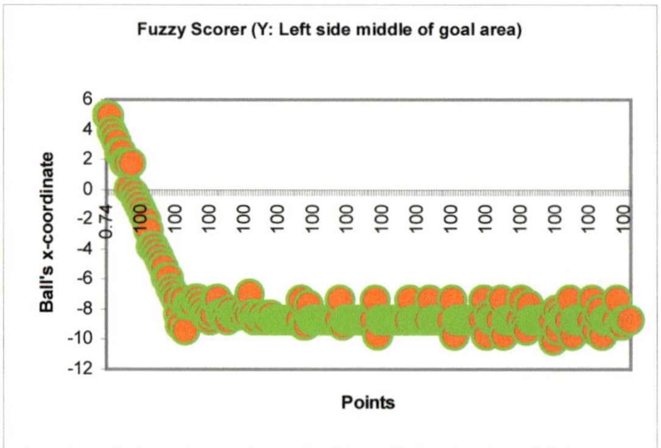

Fig. 9. Fuzzy scorer performance on the left side portion of goal area.

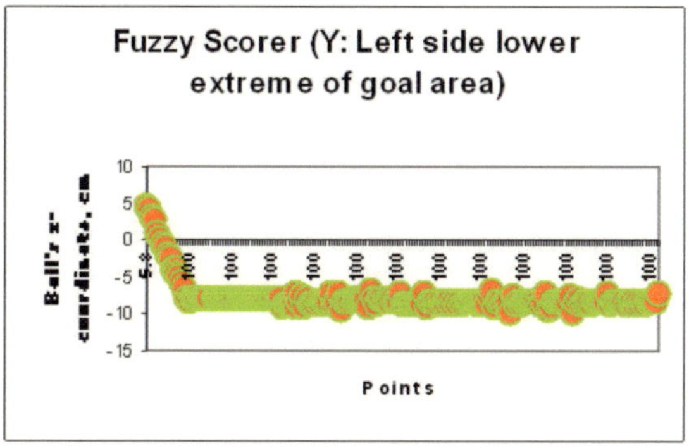

Fig. 10. Fuzzy scorer performance on the left side lower extreme portion of goal area.

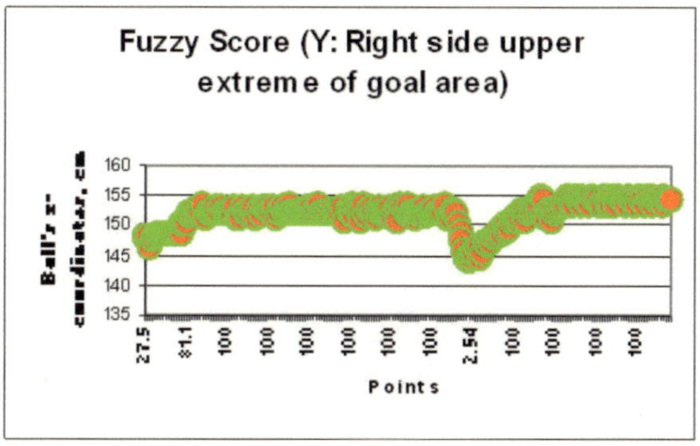

Fig. 11. Fuzzy scorer performance on the right side upper extreme portion of goal area.

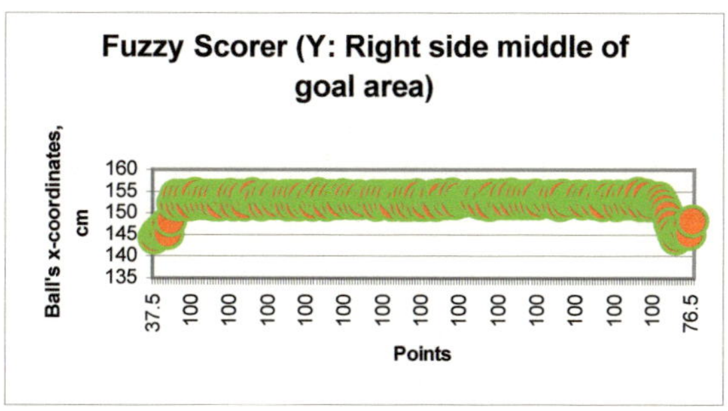

Fig. 12. Fuzzy scorer performance on the right side middle portion of goal area.

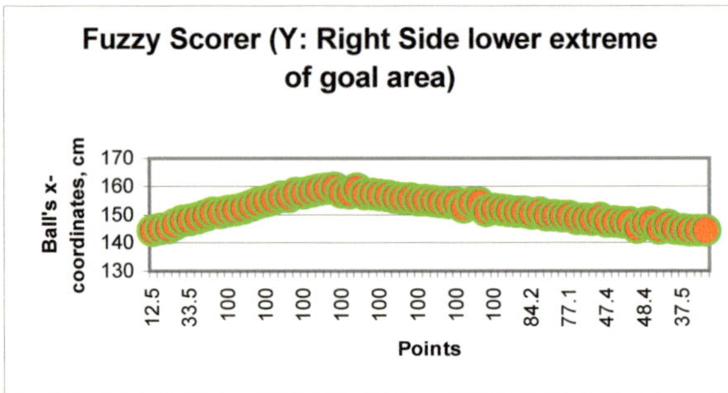

Fig. 13. Fuzzy scorer performance on the right side lower extreme portion of goal area.

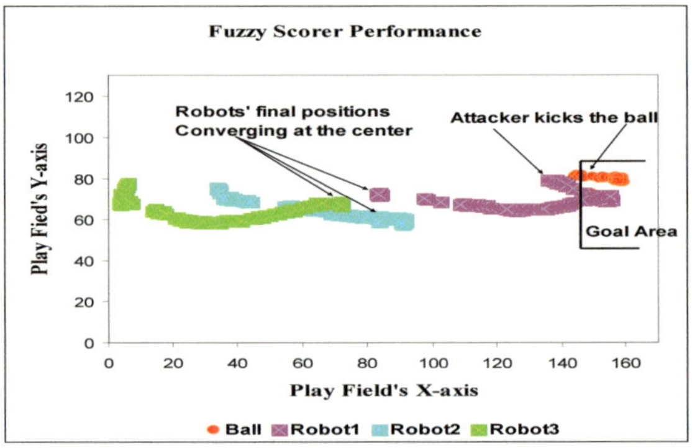

Fig. 14. Fuzzy scorer performance showing coordinated movement of robots by converging at the center after the attacker made a score.

The performance of the fuzzy scorer is greatly dependent on the shapes of the membership functions of x_{ball} and y_{ball}. Inaccurate adjustment of these shapes would result in a score even if the ball has not yet touched the goal line or no score in spite of the ball's being inside the goal area. The speed of the ball can also affect the performance of the fuzzy scorer. If the ball crosses the goal line at high speed, a situation that seldom happens in actual games, the vision system of the robot soccer will not be reliable enough in tracking the ball's position since its frame grabbing speed is constant. This will cause the fuzzy scorer to decide affirmatively a little late, that is, when the ball is already few units from the goal line towards the goal area.

The performance on synchronized movements of the robots after a score is made, depends on the ability of the robots to follow exactly the path design. The robots have difficulty in positioning themselves on the different points that composes the designated path. Actually, the robots cannot be perfectly positioned on the points where we want them to be. But the closeness of the robots to these points can be optimized.

3. Soccer Robot Goalie Strategy Using Fuzzy Logic

This section presents a control strategy for the FIRA robot soccer game goalie using fuzzy logic systems. This strategy adopts the divide and conquer concept by decomposing the goalie's task of defending the goal into four (4) categories. Consequently, each category is characterized by different goalie-ball situations and will be handled by a separate rule-base. The first fuzzy system takes the x-coordinates of the robot and the ball then outputs the category number that determines the rule-base to be used in the second fuzzy system. The second fuzzy system handles the goalie's movements. It takes the current y-coordinate of the goalie-robot and the ball then outputs the y-coordinate of the goalie's next position (destination) which is located along its line-of-action (a line with predefined x-coordinate where the goalie moves back and forth just in front of the goal it is defending). Experiment shows that this strategy is feasible, efficient, and robust.

Robot soccer games are scored by the number of times the ball actually entered the opponent goal. The entity that is tasked to prevent the ball from entering the goal is called the goalie. The goalie plays a very vital role in the game since a weak goalie means a lost game. It moves over an area just in front of the goal, called goal area, to block and clear the ball out from this area. Thus, preventing the opponent team from scoring.

Conventional goalie strategies [5,6,7,8] would use mathematics to predict the next position of the ball and subsequently generates a path to block it. But the data utilized in the mathematical process are imprecise in the sense that they are the result of the image processing procedure of the robot soccer vision system that is laden with varying degrees of imperfections aside from the fact that robot soccer is a very dynamic system. Thus, the results would have far-reaching effects on the actual performance of the goalie robot.

In [6], the goalie predicts the point of intersection between the goalline and the path of the ball and moves to it to block the ball when it threatens to enter the goal. When the ball is far from the goal area, the goalie moves to the center of the goalline. If the ball is blocked and sticks on the goalie, the goalie makes a spin move to drive the ball away from the goal area. This move is a bit dangerous because the possibility of having the ball driven to its own goal is present. Four states are assumed by the goalie in [7]. The first state is assumed when the game begins by moving the goalie to the center of the goal. The second state is assumed when the point of entry of the moving ball towards the goal is predicted and so the goalie moves to that point. The third state is assumed when the ball and the center point of the goal are connected that it is necessary for the goalie to move between this connection. The last state is assumed when the ball is in the goal area that the goalie has to kick it out. The goalie strategy described in [8] places the goalie in front of the goal until the ball arrives at the goal area. The goalie would follow the ball location in the y direction when the ball is outside the goal area. The goalie would move towards the ball as soon as the ball enters the goal area.

The goalie strategies discussed above do have some of the necessary qualities of an effective goal defender. What is lacking however is the fact that these strategies do not provide solution when the goalie is trapped and do not prevent the goalie from pushing the ball towards its own goal. This paper offers an encompassing solution to all possible problems that may prevent the goalie from performing its task of defending the goal effectively. The use of fuzzy logic systems in determining the next position of the goalie not only take care of the imprecise data to dampen its negative effects but also provide sufficient measures to counter the countless possibilities that might confront the goalie.

3.1 Description of the Goalie Function

The importance of the goalie's task in defending the goal cannot be overemphasized. Experience tells us that the most effective way to defend the goal is to have the goalie move back-and-forth along a specified line (line-of-action) just in front of the goal without ever changing its orientation on it (the goalie should either be oriented 90^0 or 270^0 with respect to the x-axis). This means that if the goalie happens to be moderately far from its line-of-action, it should return immediately to it but with careful consideration on the ball's position.

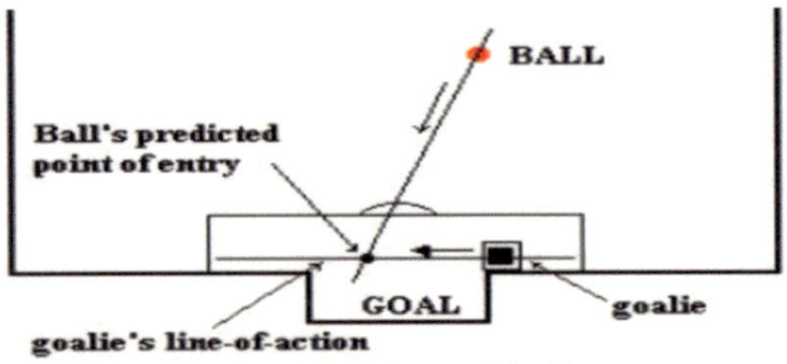

Fig. 15. Category 1 goalie and ball situation.

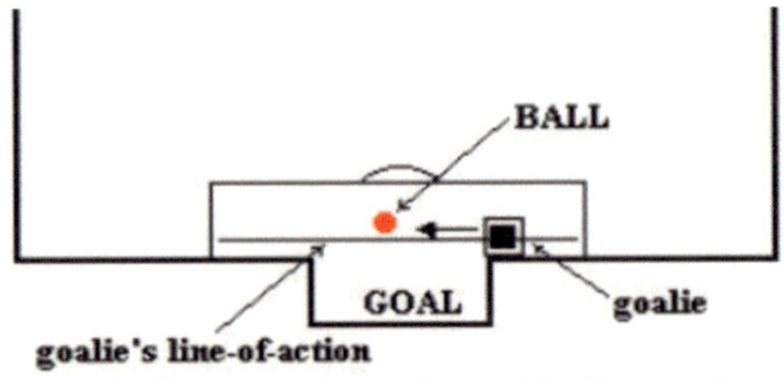

Fig. 16. Category 2 goalie and ball situation.

When the goalie is on its line-of-action as shown in Figure 15, it can block the moving ball by moving ahead to the position (along the goalie's line-of-action) where the ball is supposed to enter. This is referred to as category 1. However, when the ball is sensed to be within the goal area, as shown Figure 16, the goalie immediately kicks the ball out as it is about to cross the line-of-action and immediately moves back to prevent from being blocked by an opponent robot. This is referred to as category 2.

As shown in Figure 17, the goalie is far from its line of action so it has to be returned back to it but the point where the robot should enter must be carefully chosen so as not to accidentally push the ball into its own goal if it happens to be around. Otherwise, if the ball is not in the goal area, the goalie can go back to its line-of-action on the point nearest it. This is referred to as category 3. When the goalie is trapped in its own goal as shown in Figure 18, it has to move back to its line-of-action immediately to block or kick the ball out. This situation is referred to as category 4.

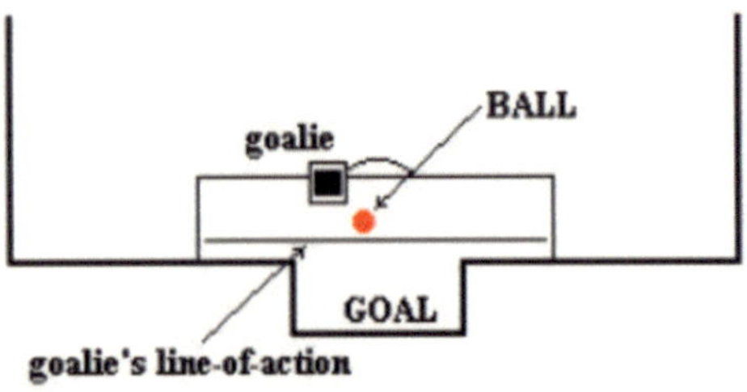

Fig. 17. Category 3 goalie and ball situation.

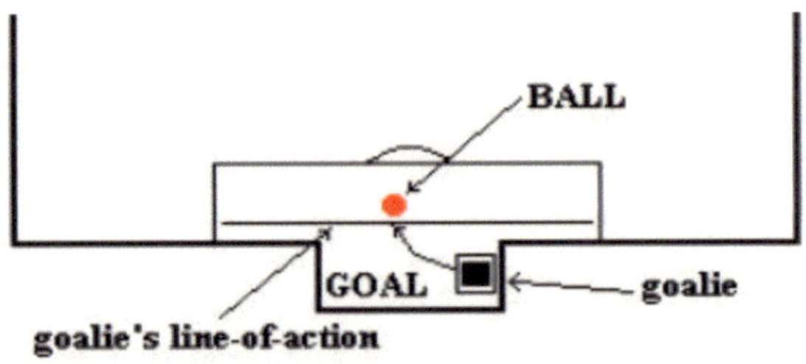

Fig. 18. Category 4 goalie and ball situation.

3.2 Predicting the Position of the Moving Ball

For category 1 goalie-ball situation, the strategy predicts the point where the ball would intersect along the line-of-action of the goalie (see Figure 15). By inspection the path of the moving ball is always a series of straight lines. Being so, it only requires two points to sense the direction of the ball at any given time and that is by knowing its current and previous positions. When the difference between the x-coordinates of the former and the latter is increasing, the goalie doesn't have to worry because the ball is going away from the home goal. Otherwise, the slope of the line connecting these two points can be computed and then the line is extended to the line-of-action of the goalie by finding their point of intersection. If the point of intersection falls in front of the home goal, the goalie has to move ahead to this point to be able to block the ball there. When the slope of the path of the moving ball is infinity, it means that the direction of the ball is vertical and the goalie would have to block the connection of the ball to any point of the goal by following the ball in the y direction.

3.3 Goalie Fuzzy Logic System

The objective of goalie fuzzy logic system is to generate the y-coordinate of the goalie's next position given its current position and the ball's. The line-of-action of the goalie is predefined therefore the point of its next position, where the y-coordinate is output of the fuzzy logic system, lies on this line. The block diagram of the fuzzy logic system used in this research is shown in Figure 19. The goalie moves back-and-forth on its line-of-action so its x-coordinate (X in Figure 19) is predefined and it is fed directly to the goalie. The determination of the y-coordinate for the goalie's next position begins by feeding the x-coordinates of the goalie and ball (Xb and Xr) to the first fuzzy system which outputs the rule-base number that will be used in the second fuzzy system. The y-coordinates of the ball and goalie (Yb and Yr) are inputs to the second fuzzy system that outputs the y-coordinate of the goalie's next position.

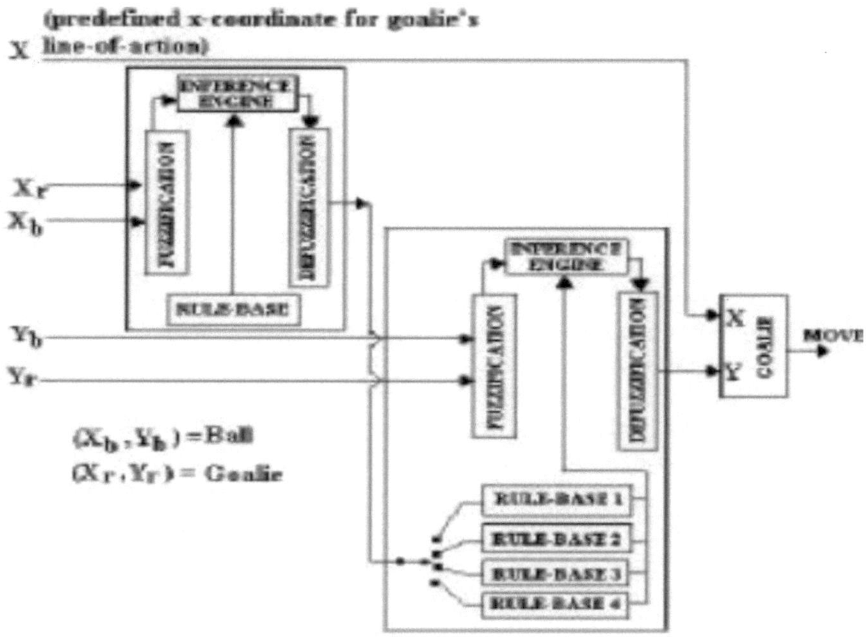

Fig. 19. Goalie's fuzzy logic system block diagram.

3.4 Membership Functions

Table 2 shows the variables used to represent the membership functions (fuzzy sets) of the goalie fuzzy logic system with their corresponding linguistic definitions to describe the system's behavior.

Figure 20 shows the membership functions for the x-coordinates of the ball and the goalie used in the first fuzzy logic system of the goalie. Please refer to Table 2 for the variables' descriptions. In Figure 6 the minimum crisp value of **SO** is pegged at 4 cm from the goal-line (at 0 cm) instead at the goal-line itself to ensure that when the goalie goes closer to the goal-line than 4 cm, the first fuzzy logic system declares it as a trap situation (category 4) and the second fuzzy logic system utilizes the rule-base 4 for the goalie's movement. Similarly, the minimum crisp value of **MO** is pegged at 7 cm to ensure that the goalie kicks the ball out when the ball is in the goal area.

X-coordinates (X_b and X_r)	Y-coordinates (Y_b and Y_r)	Category (1,2,3,or 4)
IN: Inside the goal SI: Slightly inside the goal ZO: At x=0 SO: Slightly outside the goal MO: Moderately outside the goal FO: Far outside the goal VF: Very far from the goal	DN: Down the origin y=0 MD: Moderately down the origin SD: Slightly down the origin GD: Going down the origin CR: At y-coordinate's center line GU: Going up SU: Slightly up MU: Moderately up TP: Top or at maximum value of y-coordinate	ONE: Category number 1 TWO: Category number 2 TRE: Category number 3 FOR: Category number 4

Table 2. Variables used in the membership functions.

Figure 20 shows the membership functions for the x-coordinates of the ball and the goalie used in the first fuzzy logic system of the goalie. Please refer to Table 2 for the variables' descriptions. In Figure 20 the minimum crisp value of **SO** is pegged at 4 cm from the goalline (at 0 cm) instead at the goalline itself to ensure that when the goalie goes closer to the goalline than 4 cm, the first fuzzy logic system declares it as a trap situation (category 4) and the second fuzzy logic system utilizes the rule-base 4 for the goalie's movement. Similarly, the minimum crisp value of **MO** is pegged at 7 cm to ensure that the goalie kicks the ball out when the ball is in the goal area. Figure 21 shows the membership functions of the y-coordinates used in the second fuzzy logic system of the goalie. The distance from **MD** to **MU** is the length of the goal area and the distance from **SD** to **SU** is the length of the goal. **CR** is the location of the horizontal centerline of the robot soccer playing field.

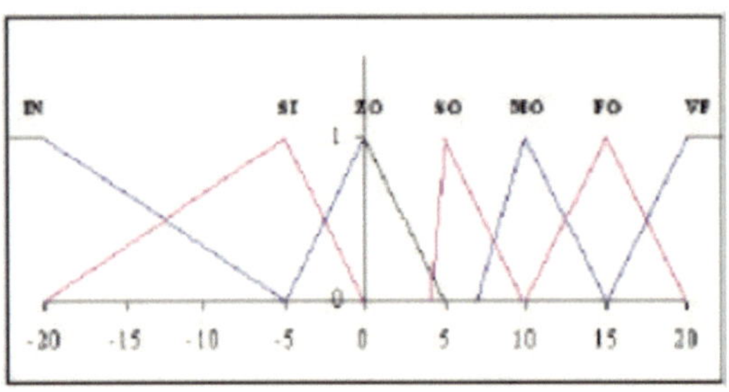

Fig. 20. Membership functions of x- coordinates.

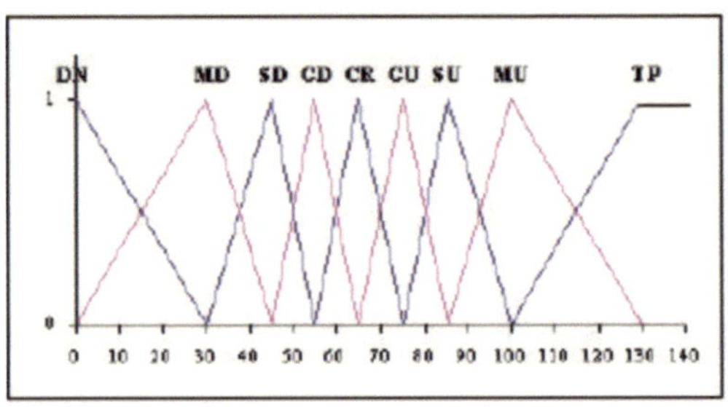

Fig. 21. Membership functions of y-coordinates.

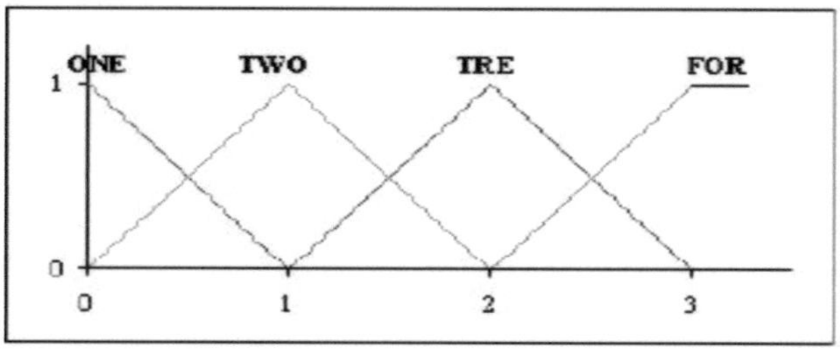

Fig. 22. Membership functions category.

3.5 Fuzzy Associative Memory (FAM) For The Goalie

Tables 3, 4, 5, 6, and 7 show the FAMs (fuzzy associative memory or rule-base) of the first and second fuzzy logic systems of the goalie. They contain the sets of rules provided by experts in associating the antecedents in order to generate the consequence. Logical implications are utilized in formulating these rules like as follows:

(1) IF **Yr** is SD(slightly down) *AND* **Yb** is GU(going up) *THEN* **Y** is GU(going up) or **SD Ù GU** or **GU**

(2) IF **Yr** is GU(going up) *AND* **Yb** is TP(top) *THEN* **Y** is SU(slightly up) or **GU Ù TP** or **SU** and so on…

The rules whose antecedents have membership function values greater than zero participate in the defuzzification process using their individual consequences. Looking at Table 5, there are cells that are empty. This is because Table 5 is a rule-base used by the second fuzzy system for category 4 which is a trap situation for the goalie. That is, the goalie is inside the goal and is being restraint by the goal's boundary walls. This means that the goalie's **Yr** cannot go beyond **SD** and **SU.**

Goalie's y-coordinates, Y_r \ Ball's y-coordinates, Y_b	DN	MD	SD	GD	CR	GU	SU	MU	TP
DN	SD	SD	SD	GD	CR	GU	SU	SU	SU
MD	SD	SD	SD	GD	CR	GU	SU	SU	SU
SD	SD	SD	SD	GD	CR	GU	SU	SU	SU
GD	SD	SD	SD	GD	CR	GU	SU	SU	SU
CR	SD	SD	SD	GD	CR	GU	SU	SU	SU
GU	SD	SD	SD	GD	CR	GU	SU	SU	SU
SU	SD	SD	SD	GD	CR	GU	SU	SU	SU
MU	SD	SD	SD	GD	CR	GU	SU	SU	SU
TP	SD	SD	SD	GD	CR	GU	SU	SU	SU

Table 3. Fuzzy Associative Memory (FAM) category 1.

Goalie's y-coordinates, Y_r \ Ball's y-coordinates, Y_b	DN	MD	SD	GD	CR	GU	SU	MU	TP
DN	SD	SU	SU	SU	SU	SU	SU	SU	SU
MD	SD	SD	SU	SU	SU	SU	SU	SU	SU
SD	SD	SD	SD	SU	SU	SU	SU	SU	SU
GD	SD	SD	SD	GD	SU	SU	SU	SU	SU
CR	SD	SD	SD	SD	CR	SU	SU	SU	SU
GU	SD	SD	SD	SD	SD	GU	SU	SU	SU
SU	SD	SD	SD	SD	SD	SD	SU	SU	SU
MU	SD	SD	SD	SD	SD	SD	SD	SU	SU
TP	SD	SD	SD	SD	SD	SD	SD	SD	SU

Table 4. Fuzzy Associative Memory (FAM) for for category 2 of the second fuzzy system.

Goalie's y-coordinates, Y_r \ Ball's y-coordinates, Y_b	DN	MD	SD	GD	CR	GU	SU	MU	TP
DN	SD	SD	MD	SD	GD	CR	GU	SU	SU
MD	SD	SD	MD	SD	GD	CR	GU	SU	SU
SD	SD	SD	GD	SD	GD	CR	GU	SU	SU
GD	SD	SD	GD	CR	GD	CR	GU	SU	SU
CR	SD	SD	GD	CR	GD	CR	GU	SU	SU
GU	SD	SD	GD	CR	GU	CR	GU	SU	SU
SU	SD	SD	GD	CR	GU	SU	GU	SU	SU
MU	SD	SD	GD	CR	GU	SU	GU	SU	SU
TP	SD	SD	GD	CR	GU	SU	MU	SU	SU

Table 5. Fuzzy Associative Memory (FAM) category 3 of the second fuzzy system.

Goalie's y-coordinates, Y_r \ Ball's y-coordinates, Y_b	DN	MD	SD	GD	CR	GU	SU	MU	TP
DN	-	-	-	-	-	-	-	-	-
MD	-	-	-	-	-	-	-	-	-
SD	GD	GD	GD	GD	CR	GU	SU	SU	SU
GD	SD	SD	SD	GD	CR	GU	SU	SU	SU
CR	SD	SD	SD	GD	CR	GU	SU	SU	SU
GU	SD	SD	SD	GD	CR	GU	SU	SU	SU
SU	SD	SD	SD	GD	CR	GU	SU	SU	SU
MU	-	-	-	-	-	-	-	-	-
TP	-	-	-	-	-	-	-	-	-

Table 6. Fuzzy Associative Memory (FAM) for for category 4 of the second fuzzy system.

3.6 Experimental Results Of Goalie Strategy

To test the performance of the goalie using the above strategies, a series of real-time experiments were run. Tests were conducted for each of the goalie-ball situation categories. Figures 23, 24, 25, 26, 27, and 28 show the results of such test runs.

Goalie's X-coordinate, X_r \ Ball's X-coordinate, X_b	IN	SI	ZE	SO	MO	FO	VF
IN	1	1	1	1	1	1	1
SI	1	1	1	1	1	1	1
ZE	1	1	1	1	1	1	1
SO	3	3	3	3	4	4	4
MO	2	2	2	2	4	4	4
FO	2	2	2	2	2	4	4
VF	2	2	2	2	2	2	2

Table 7. Fuzzy Associative Memory (FAM) first fuzzy system.

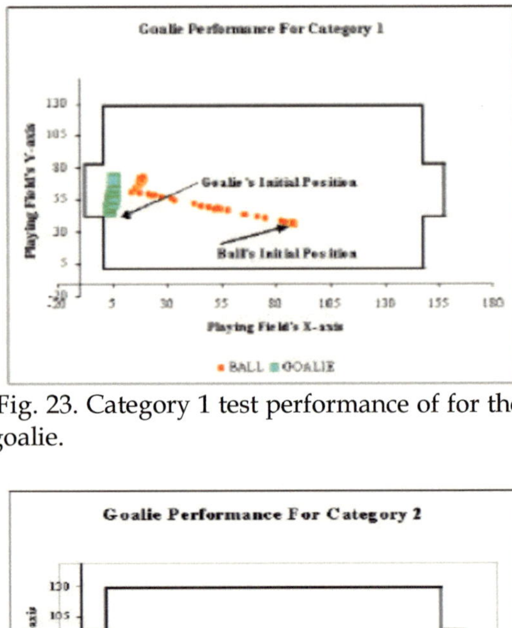

Fig. 23. Category 1 test performance of for the goalie.

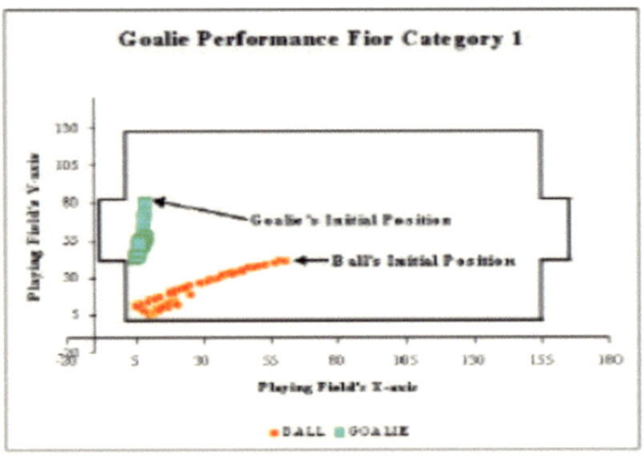

Fig. 24. Category 1 test performance of goalie.

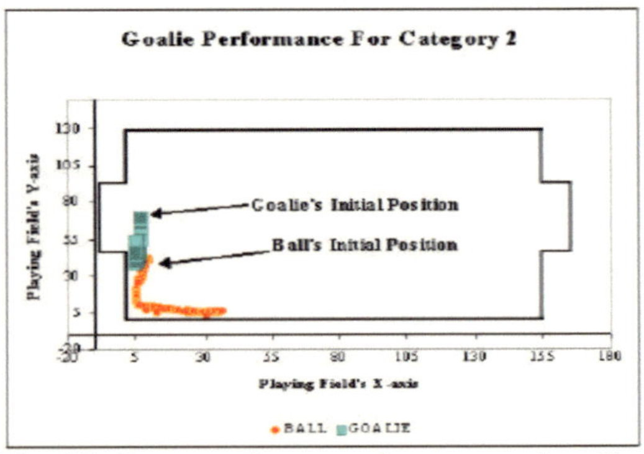

Fig. 25. Category 2 test performance of goalie.

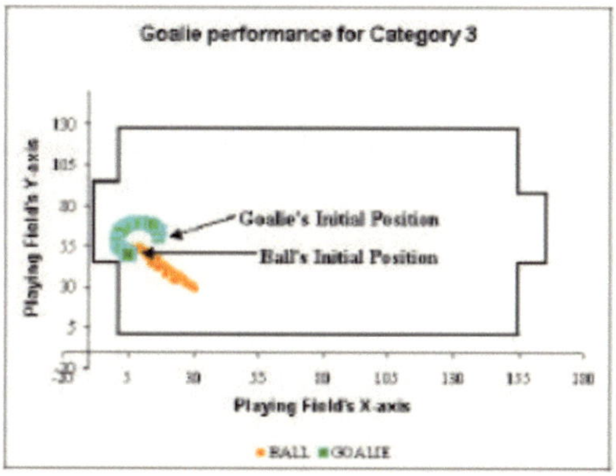

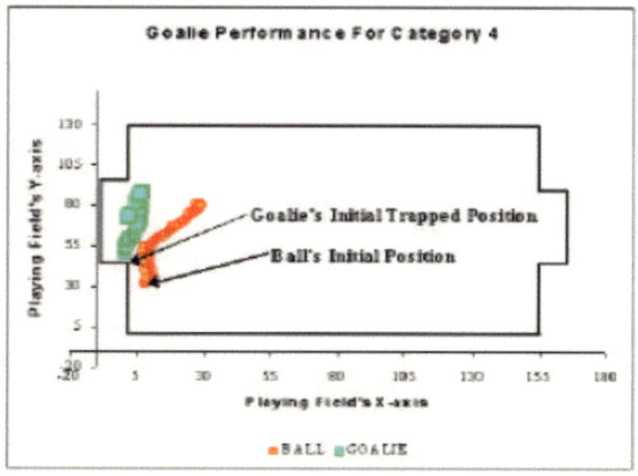

Fig. 26. Category 3 test performance of goalie. Fig. 27. Category 4 test performance of goalie.

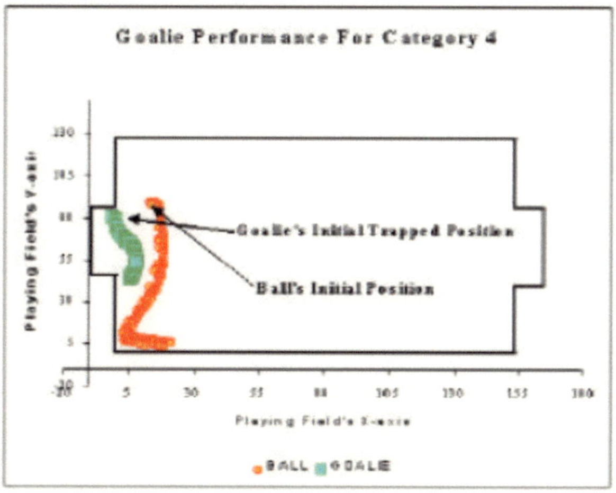

Fig. 28. Category 4 test performance of goalie.

Figure 23 shows the result of the test made on the goalie with a situation that falls under category 1. The ball was initially placed moderately far from the goal and was pushed towards the center of the goal. The goalie which was initially positioned at the lower part of the goal reacted by moving ahead towards the predicted point of intersection between the ball's path and the goalie's line-of-action and successfully blocked the ball there.

The situation in Figure 24 still falls under category 1 except that the ball was not moving towards the goal but to the lower portion of the playing field. The goalie initially on the upper part of the goal reacted by moving towards the lower part of the goal and positioned itself there to close the connecting line between the goal and the ball.

Figure 25 is a situation where the goalie is on its line-of-action and sensed the ball was entering the goal area. This situation falls under category 2. The goalie's reaction was to kick the ball out from the goal area and moved back to the lower portion of the goal again waiting for the ball's next approach.

Figure 26 shows a situation where the goalie is out from its line-of-action with the ball threatening to enter the goal. This situation falls under category 3. The goalie cannot go

directly to the ball because it would drive the ball to its own goal. Instead, the goalie made a circling move to successfully block the ball from the inside of the goal.
Figures 27 and 28 are situations where the goalie in both cases was initially trapped inside the goal. These are situations that fall under category 4. In both cases the goalie remained on its initial positions until the ball approached the goal then it followed the ball in the y direction until the ball threatens no more.

4. Obstacle Avoidance

Obstacle avoidance uses a fuzzy logic system that has the angle and distance of the robot with respect to the obstacle as its inputs and generates the velocities of the robot's left and right motors that will effectively avoid any perceived obstacle.
As shown in Figure 29, obstacle avoidance starts by detecting obstacles (walls, teammates, and opponents) within a 30 cm circle whose center is 50 cm ahead of the robot along its path towards a given destination. The detected obstacle that is nearest to the robot will be considered as the foremost obstacle and will be the one to be avoided as soon as the robot draws nearer to it by a distance lower than 35 cm. When this happens the input entities to the fuzzy logic system will be computed and the appropriate velocities for the left and right motors are generated to effectively avoid the perceived obstacle.

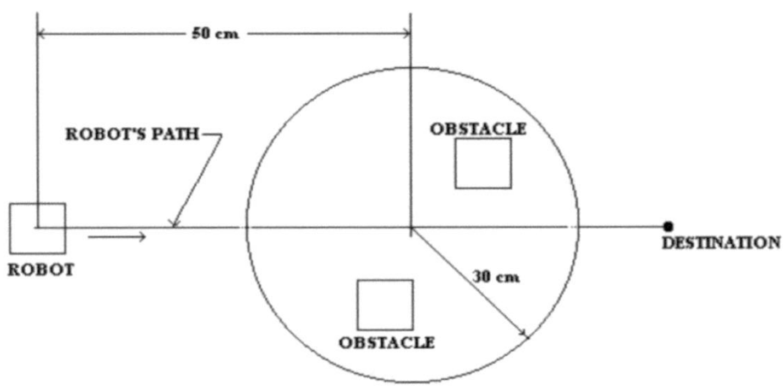

Fig. 29. Detecting obstacles.

Input Distance	Input Angle	Motor Speed
ZE: Zero	NL : Negatively Large	VL: Very Low
VN: Very Near from the obstacle	NM: Negatively Medium	ML: Moderately Low
NE: Near the obstacle	NS: Negatively Small	SL: Slightly Low
SF: Slightly far from the obstacle	ZE: Zero	LO: Low
FA: Far from the obstacle	PS: Positively Small	SH: Slightly High
VF: Very far from the obstacle	PM: Positively Medium	MH: Moderately High
	PL: Positively Large	VH: Very High

Table 8. Variables used in the membership functions of fuzzy logic obstacle avoidance.

The consequences of the fuzzy rules can be designed such that as the robot becomes nearer to the obstacle, the motors' speeds drive the robot away from it but towards the designated destination. Figures 30 to 32 show the membership functions of the inputs and out entities of the fuzzy logic system used for obstacle avoidance. Table 8 shows the description of the variables used. Tables 9 and 10 show the fuzzy associative memory (FAM) for the left and right motors.

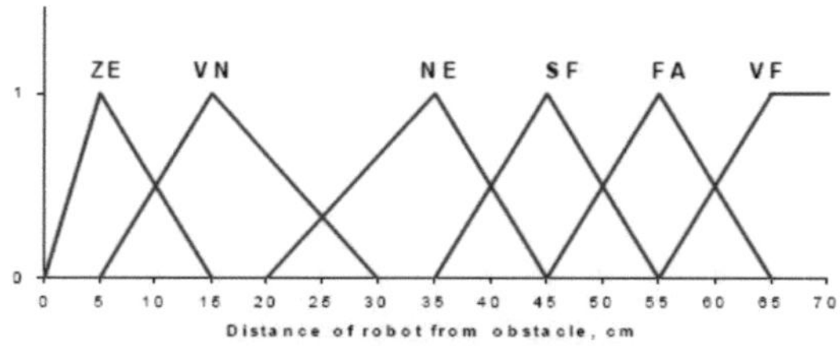

Fig. 30. Membership functions of input distance for obstacle avoidance.

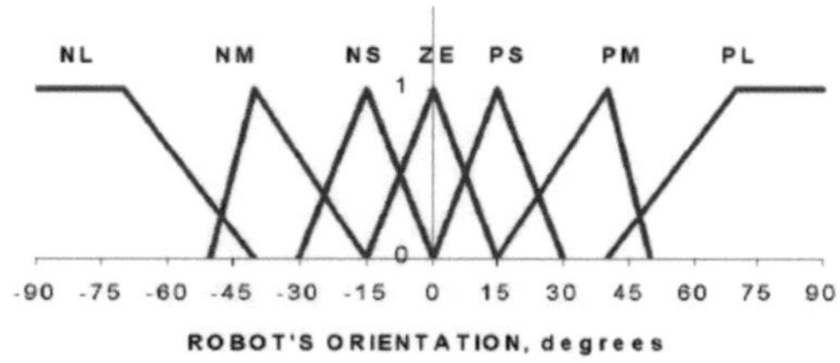

Fig. 31. Membership functions of input angle for obstacle avoidance.

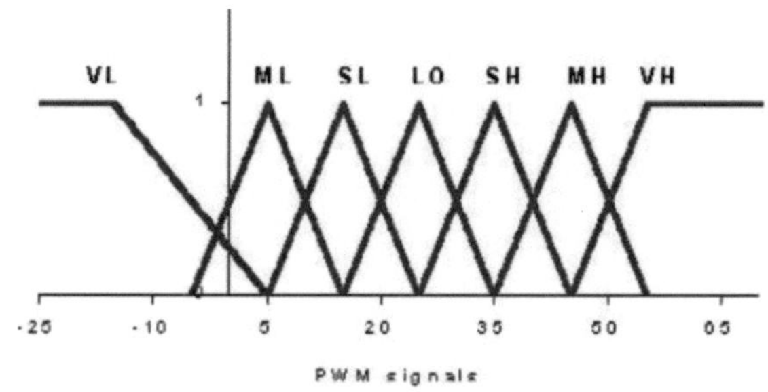

Fig. 32. Membership functions of motor speeds for obstacle avoidance.

	NL	NM	NS	Z	PS	PM	PL
ZE	SL	SL	SL	SL	VL	VL	VL
VN	SL	SL	SL	SH	ML	VL	VL
NE	VH	VH	MH	VH	SH	LO	LO
SF	VH	VH	MH	VH	SH	LO	LO
FA	VH	MH	MH	VH	SH	SH	LO
VF	VH	MH	MH	VH	SH	SH	LO

Table 9. Left Motor FAM for obstacle avoidance.

	NL	NM	NS	Z	PS	PM	PL
ZE	VL	VL	VL	SL	SL	SL	SL
VN	ML	ML	ML	SH	SL	SL	SL
NE	LO	LO	SH	VH	MH	VH	VH
SF	LO	LO	SH	VH	MH	VH	VH
FA	LO	SH	SH	VH	MH	MH	VH
VF	LO	SH	SH	VH	MH	MH	VH

Table 10. Right Motor FAM for obstacle avoidance.

4.1 Obstacle Avoidance Experimental Results

Figure 33 and Table 11 show a home robot prevents collision with other robots by effectively avoiding them before hitting the ball towards the goal.

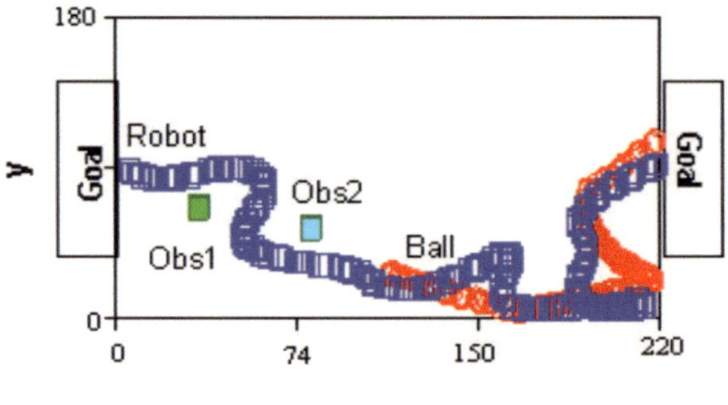

Fig. 33. An autonomous robot obstacle avoidance performance.

	Trials					
	1	2	3	4	5	Total
Avoids	2	0	2	2	2	8
Goal scores	2	2	2	2	2	10
Total points						18
Success	0.2	0.1	0.2	0.2	0.2	90%
Failure	0	0.1	0	0	0	10%

Table 11. Performance evaluation of the robot for obstacle avoidance.

5. Game Strategy

This section presents a hierarchical control for the cooperative behavior of autonomous mobile robots using fuzzy logic systems with 5 vs. 5 Mirosot as the platform for its investigation. The strategy begins by identifying the state of the game, offense or defense, using a fuzzy logic main-task-identifier. During offense a fuzzy control system is used by the ball handler to decide whether to shoot, pass, or just dribble the ball. To ensure success for these tasks they are preceded by lane checking activity to assign a lane free of obstacles. Experimental results show that the proposed strategy is effective and robust.

Multi-agent systems have attracted a lot of interests from roboticists today because of its numerous advantages over single agent systems [10,11,19,20]. One of the most famous activities of a multi-agent system is the MIROSOT, which is a soccer tournament played by autonomous robots. In multi-agent system, the agents can be made to work cooperatively. However, in [10] it was revealed that even two robots given with simple interaction tasks display complicated behaviors. Problems such as this arise because of (1) lack of knowledge and information like any other new technology, (2) sometimes conflicting goals and interactions among agents, (3) requirement of an effective learning and cooperative capabilities [19]. Faced with this predicament, researchers came up with different solutions some of which are focused on the following:(1) the generation of efficient algorithm for coordinating paths of multiple robots [9,10,13,14], (2) the use of special sensor-based navigational tools and onboard processing power [12,21], (3) stigmergic or evolutionary approach of generating cooperative behaviors [11,20], (4) development of a multi-agent learning systems for cooperation [19], etc.

Path planning and path following are the most basic behaviors of a multi-agent system. The realization of the system's goal depends primarily on the efficient navigational performance of the agents. In [9] the motions of multiple robots are coordinated by plotting their paths into a coordinate diagram and searching for a coordination configuration that is free of collision sub paths. This method requires a lot of computing time and has the possibility of ending with no solutions or deadlocks and does not suit well in a competitive environment where time is very critical. Iterative transportation technique [10] provides path planning activity by cooperative exploration of the environment by multiple robots before laying paths consisting of 1- and 2-lane types then a strategy is devised for controlling the flow of robots on said paths. This method allows the reuse of previous paths and focuses on the non-dynamic and non-hostile obstacles that make it less essential in the very competitive environment.

The main problem that this research primarily wants to address is that of cooperative behavior design problem. As stated in [16], it is a problem of investigating "how given a group of robots, an environment, and a task, cooperative behavior should arise."

In Mirosot, the atomistic approach in which no relationships between robots are assumed in rushing to the ball will not work because the robots may run into their teammates, or foul their opponents. If one is assigned to attack the ball then its teammates may not do so. This establishes relationship between robots. While robots are not attacking, they will move to specified objective positions or remain still [23,13]. If they are on collision course with other robots then implement obstacle avoidance. It is equally important to know the positions of all objects (robots and ball) in the playing field including their structure. As this will enable the team to extract information from it where a sound move can be decided upon [22].

We propose a hierarchical or top down line of control using intelligent techniques, such as fuzzy logic systems, to solve this problem. The global goals of the multi-agent system are clearly defined beforehand and imposed upon the agents by adopting centralized control [21] using global parameters from the environment dynamics [11] and fuzzy logic system to determine the desired control actions. The problem of conflicting goals and interactions among agents will be addressed.

5.1 Hierarchical Multi-agent Cooperative Control Structure

The global goal is divided into two main tasks, namely *offense-* and *defense-*states. Each of these tasks is further subdivided into subtasks that are individually executed by the agents. Each agent is given functions to be able to accomplish these subtasks and does implicit communication and coordination with other agents in the multi-agent system. The fact that no two- or more agents assume the same subtask shows some kind of communication between agents though its communicative character is not being codified or not-manifest (it is called "implicit communication"). An exploration of the environment combines the actions of all agents at the highest level of the hierarchy. This simplifies the communication between agents hence cooperation is maximized. This point is especially important since scoring a goal is also the individual goal of each agent.

Figure 34 shows the hierarchical multi-agent cooperation strategy that we propose for middle-league robot-soccer. This strategy allows the agents to master the skills of coordinated/joint execution of the main goal. Each agent concentrates in executing the individual tasks at the subtasks level. Centralized control is achieved by means of global information available to the agents to constantly guide them so their individual actions (S_1,

S_2, ..) are always in conjunction with the team's main goal. Thus, confusion resulting from conflicting goals will be avoided.

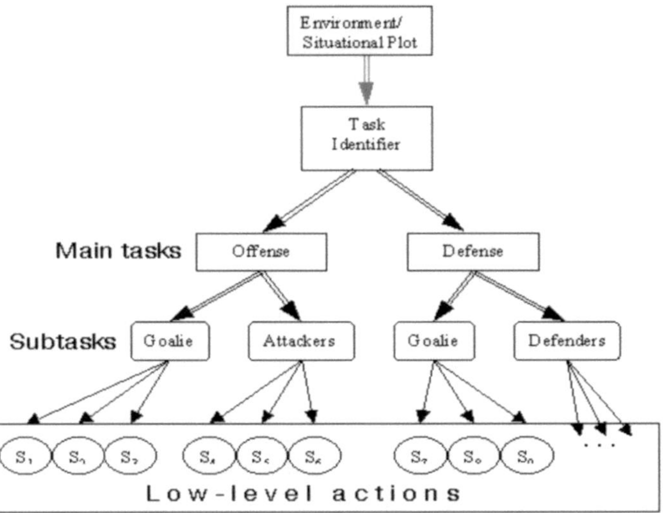

Fig. 34. Hierarchical Multi-agent Cooperation Strategy.

5.2 Roles Assignment and Robot Coordination

Individual positions of robots and ball are made available by the vision system of the robot soccer. Given these data, information that is of primary importance for the performance of the roles to be assigned to the robots is extracted at the highest level of the hierarchy. For example the predicted coordinates of the ball, which robot is nearest to the ball, the degree of opening on the opponent's goal, etc., are just few of the necessary information that play pivotal role in the outcome of the game. As soon as they are available they're immediately broadcasted to all robots thus executions of the low level actions such as blocking and intercepting the ball are straightforward and faster.

5.3 Main Task Identification

When the team is on offense, the objective is to score. While on defense, the objective is to control the ball. In this strategy, specific subtasks are in line for each of these main tasks. Given sufficient information of the environment the identification of the main task to be pursued by the team is done by means of a *main task identifier* using fuzzy logic as shown in Figure 35. The main task identifier takes the robot's orientation with respect to the ball, the calculated distance of the ball from the opponent goal, and the difference of the nearest distances between the home robots and the opponents from the ball. The fuzzy logic system output is assigned to *switchval* whose value ranges from 0 to 10. The *switchval* value will be evaluated to determine the desired state of the game using the following rules:

$$\textbf{Game State} = \begin{cases} \textbf{Offense if } 6 <= \textbf{switchval} < 11 \\ \textbf{Defense if } 0 <= \textbf{switchval} < 6 \end{cases}$$

Figs. 36, 37, 38, & 39 show the membership functions of the inputs to the main task identifier. The FAM (fuzzy associative memory) of the main task identifier will be composed of seventy-five (75) entries. It contains the set of rules provided by experts in associating the input entities in order to generate the consequence. Logical implications are utilized in formulating these rules like as follows:

(1) IF **ND** is LEW(leading widely) *AND* **D** is VN(very near) *AND* **O** is VG(very good) *THEN* **SV** is **HO**(high offense) or **LEW ∧ VN ∧ VG → HO**

(2) IF **ND** is LES(leading slightly) *AND* **D** is VF(very far) *AND* **O** is GD(good) *THEN* **SV** is MO(moderate offense) or **LES ∧ VF ∧ GD → MO** and so on ...

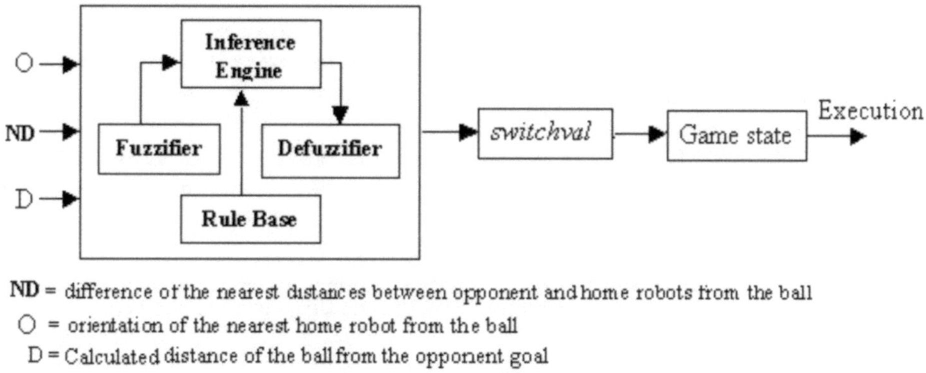

Fig. 35. Fuzzy logic main task identifier.

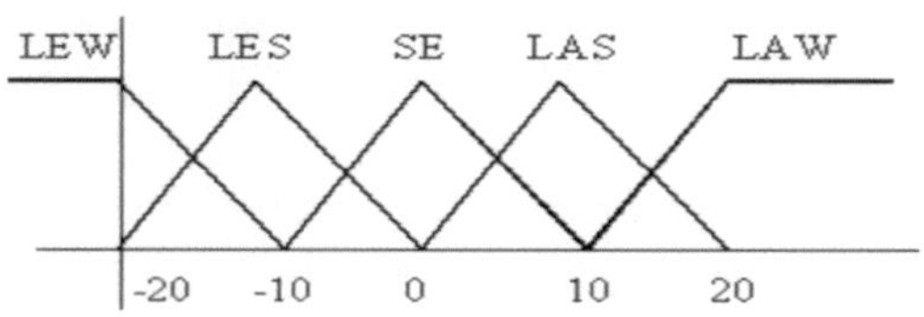

Fig. 36. Membership Functions of the difference of the nearest distances between the opponent and home robots from the ball, ND: LEW(leading widely), LES(leading slightly), SE(same), LAS(lagging slightly), LAW(lagging widely).

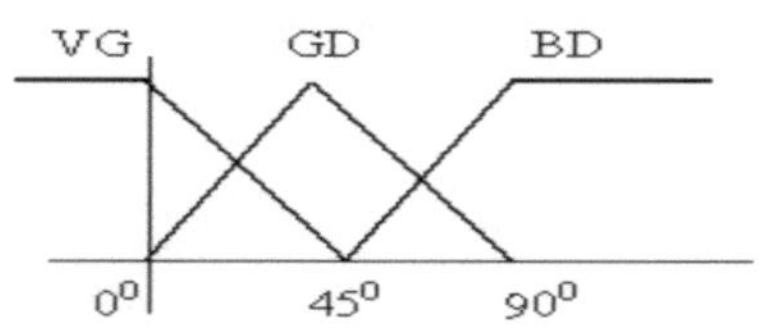

Fig. 37. Membership functions of home robot orientation with the ball, O: VG(very good), GD(good), BD(bad).

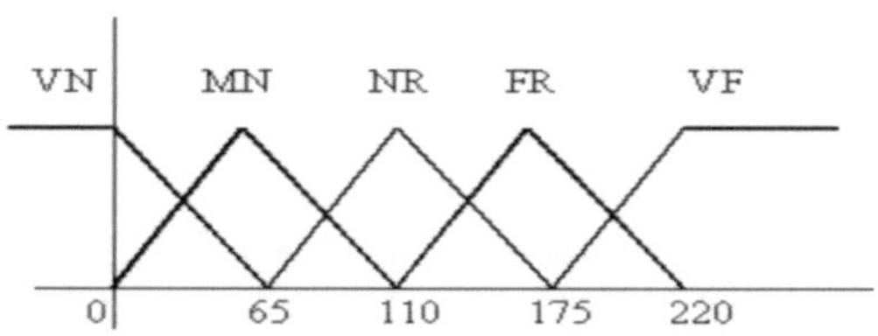

Fig. 38. Membership functions of calculated distance of ball from opponent goal, D: VN(very ear), MN(moderately near), MD(medium), MF(moderately far), VF(very far).

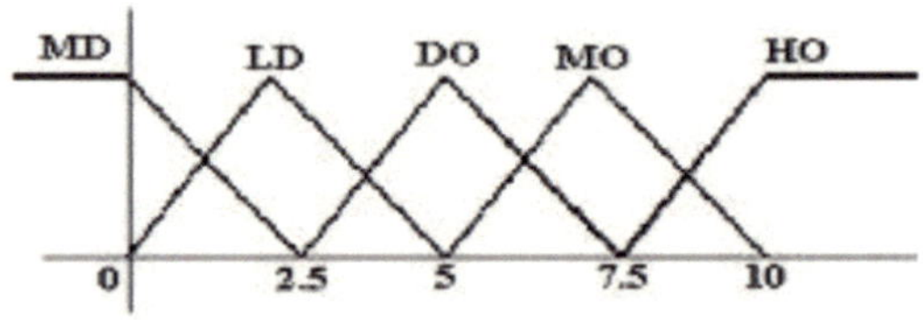

Fig. 39. Membership functions of *switchval*: MD(more defense), LD(light defense), DO(defense-offfense), MO(moderately offense), HO(high offense).

5.4 Offense Strategy

Figure 40 shows the different robot formations with respect to the different locations of the ball. The ball handler's fuzzy logic arbiter, as shown in Figure 41, takes the x- and y-coordinates of the ball, and the statuses of P_1, P_2, P_3 (free or blocked) as inputs then a decision, whether to shoot, pass, or dribble, will be made. The arbiter's fuzzy system outputs a crisp value that is assigned to the *priority* variable. The value stored in the *priority* variable is then evaluated to determine the desired action like as follows:

$$\textbf{Action} = \begin{cases} \textbf{shoot} & \textbf{if } 100 >= \textbf{priority} > 68 \\ \textbf{pass} & \textbf{if } 68 >= \textbf{priority} > 34 \\ \textbf{dribble} & \textbf{if } 34 >= \textbf{priority} > 0 \end{cases}$$

When the ball handler dribbles the ball, it is done in such a way that the ball will be brought away from the opponent robots or nearer to the opponent goal. Figs. 42, 43, 44, & 45 show the membership functions of x- and y-coordinates of the ball, of the shooting and passing lanes statuses, and of the priority variable. The FAM (fuzzy associative memory) of the ball handler's arbiter will be composed of one-hundred eighty (180) entries. It contains the set of rules provided by experts in associating the input entities in order to generate the consequence. Logical implications are utilized in formulating these rules like as follows:

(1) IF $\mathbf{X_{ball}}$ is VN(very near) *AND* $\mathbf{Y_{ball}}$ is EL(extreme low) *AND* $\mathbf{S_{pass}}$ is BK(blocked) *AND* $\mathbf{S_{shoot}}$ is CR(clear) *THEN* **Priority** is **MH**(moderately high); (VN$\wedge$EL$\wedge$BK$\wedge$CR $\rightarrow$ MH)
(2) IF $\mathbf{X_{ball}}$ is FR(far) *AND* $\mathbf{Y_{ball}}$ is NL(near low) *AND* $\mathbf{S_{pass}}$ is CR(clear) *AND* $\mathbf{S_{shoot}}$ is BK(blocked) *THEN* **Priority** is **ML**(moderately low) or FR $\wedge$ NL $\wedge$ CR $\wedge$ BK $\rightarrow$ ML and so on…

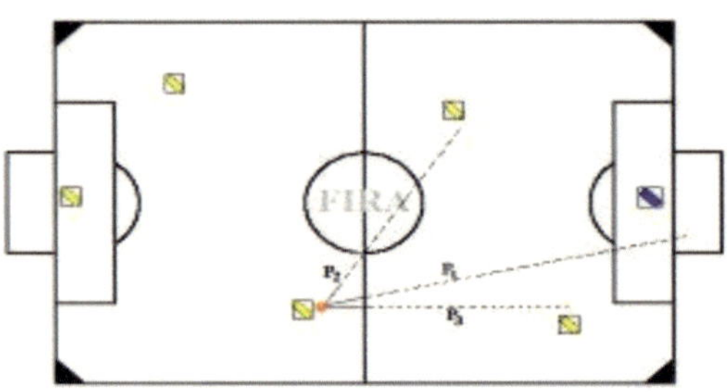

Fig. 40a. Basic offense formations of robots. (Yellow-home robots; Blue-opponent)

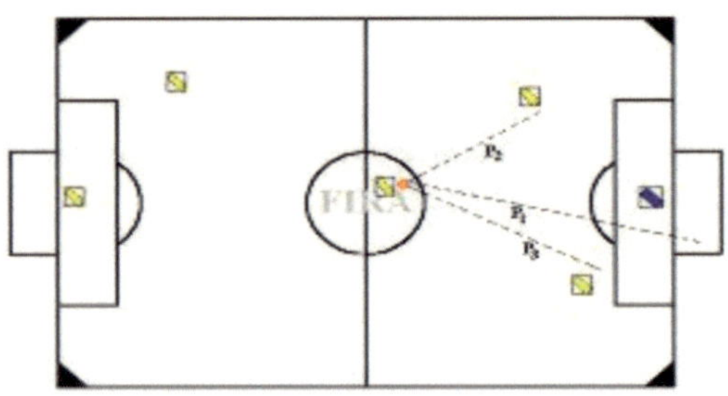

Fig. 40b. Basic offense formations of robots. (Yellow-home robots; Blue-opponent)

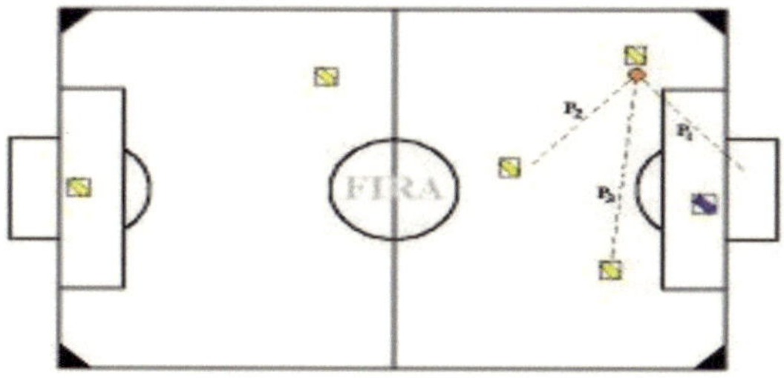

Fig. 40c. Basic offense formations of robots.(Yellow-home robots;Blue-opponent)

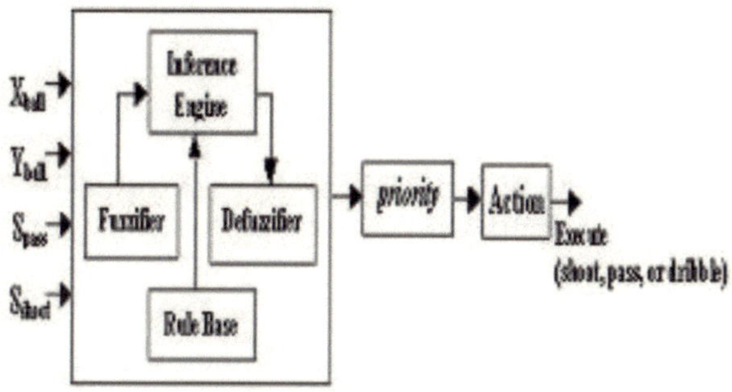

Fig. 41. Fuzzy logic arbiter of the ball handler.

5.5 Membership Functions Of The Output And Input Entities To The Fuzzy Logic Arbiter Of The Ball Handler

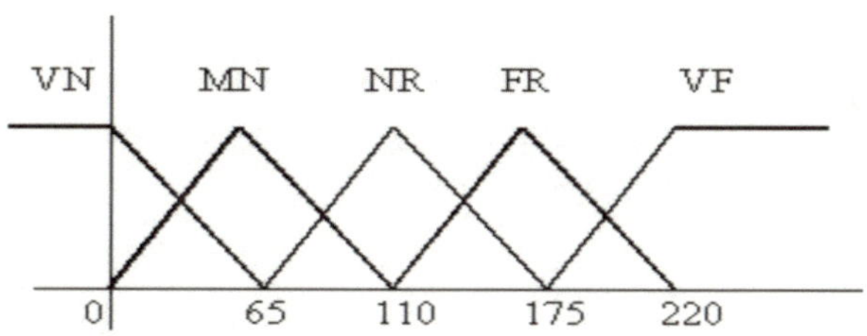

Fig. 42. Membership functions of ball's x-coordinate: VN (very near), MN (moderately near), NR(near), FR(far), VF(very far).

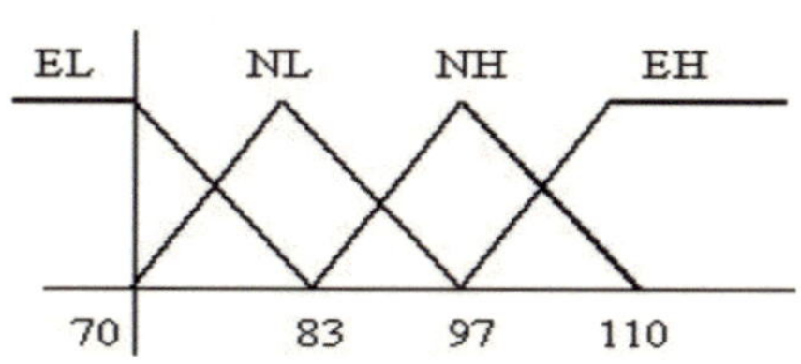

Fig. 43. Membership functions of ball's y-coordinate: EL(extreme low), NL(near low), NH(near high), EH(extreme high).

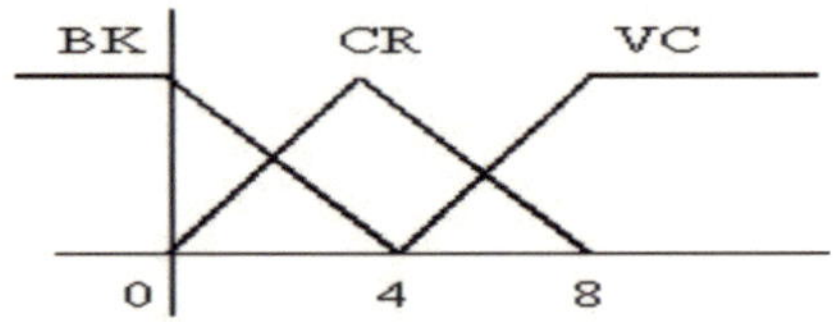

Fig. 44. Membership functions of shooting and passing lanes statuses: BK(blocked), CR(clear), VC(very clear).

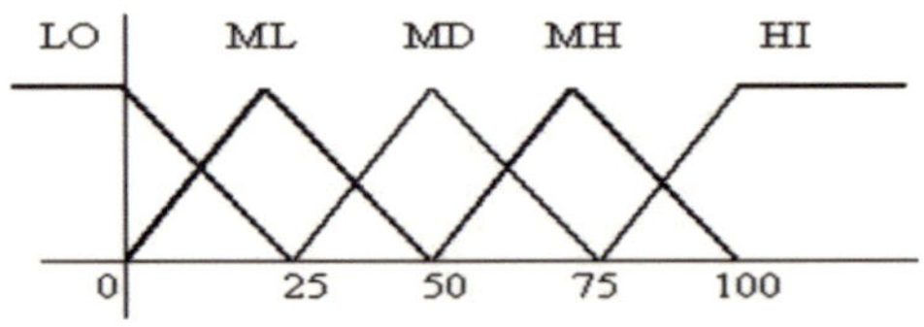

Fig. 45. Membership functions of priority variable: LO(low), ML(moderately low), MD (medium), MH(moderately high), HI(high).

5.6 Checking the Paths Lanes

Passing and shooting are two similar tasks that require lane checking, to determine whether the lane is clear or blocked, to ensure completion or success of said tasks. In avoiding obstacles alternate paths are first generated. These paths are then checked before they are being considered for inclusion in the pool of alternative paths. The same is done for the shooting and passing lanes. In shooting the target is any open part of the opponent goal, preferably the point that is farthest from the opponent goalie at the instant it is kicked. Passing is pushing the ball to a teammate robot so it can be controlled immediately.
To check the availability of the lane for passing, shooting, or for alternative paths is to determine the distances of all objects (teammates or opponents) from the line passing through the robot and the target using analytic geometry. The lane is free when the distance of the nearest object from the line exceeds a threshold value.

5.7 Defense Strategy

Figure 46 shows the three basic defense formations of robots with the ball located on different vertical coordinates of the playing field. At any instance two robots are assigned to block the ball and one of them is tasked to intercept it. The boundaries of the areas where the defender robots operate are flexible except on the home goal area where the goalie robot is in charge.

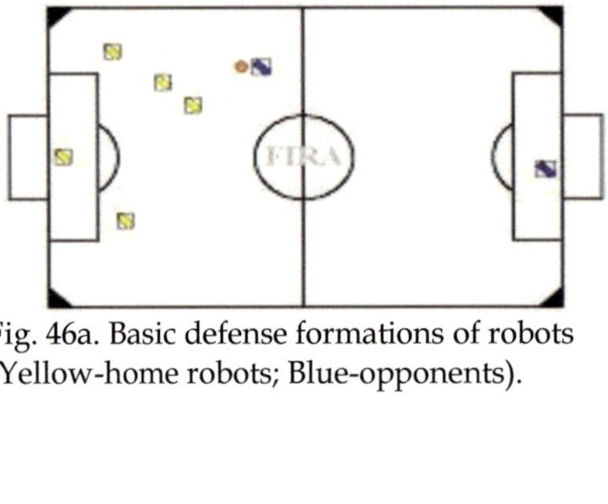

Fig. 46a. Basic defense formations of robots (Yellow-home robots; Blue-opponents).

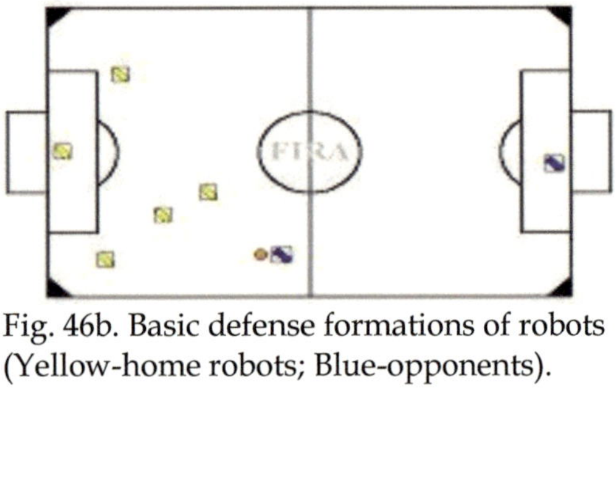

Fig. 46b. Basic defense formations of robots (Yellow-home robots; Blue-opponents).

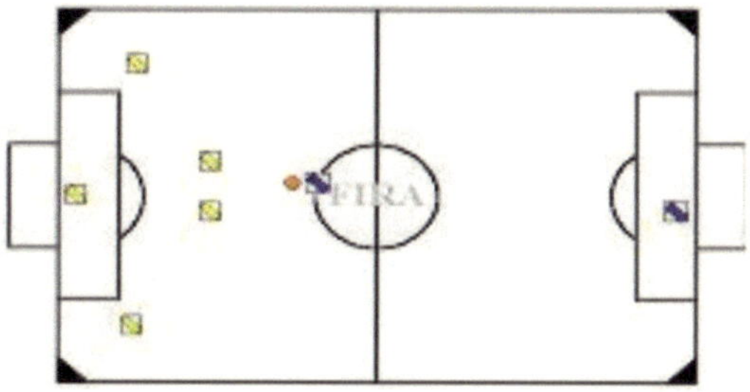

Fig. 46c. Basic defense formations of robots (Yellow-home robots; Blue-opponents).

5.8 Shooting

Shoot algorithm is implemented by attacking the ball with the intent of kicking it towards the opponent goal. In Figure 47, the robot moves toward point E, just behind the ball, while it constantly calculates φ (the angle between L_1 and L_2). When the value of φ falls below the set minimum value, the robot changes its direction by moving towards the ball kicking it directly to the opponent's goal. The algorithm requires a robot number and coordinates of the point where the ball will be kicked to.

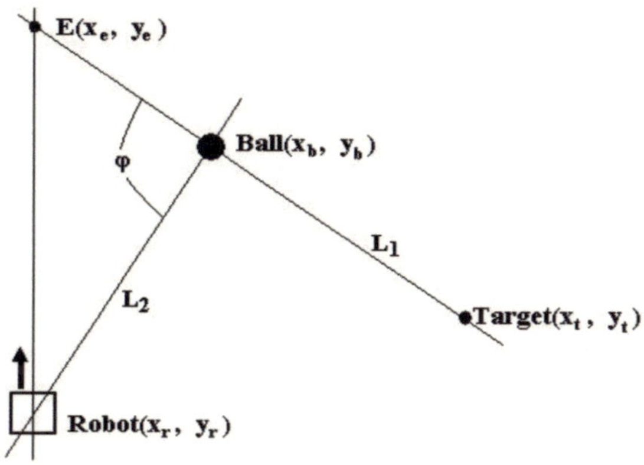

Fig. 47. Shoot algorithm.

Let m_1 = slope of L_1
m_2 = slope of L_2
φ = angle between L_1 and L_2
r_1 = distance of target from E
r_2 = distance of ball from E (arbitrary value)
r_3 = distance of ball from target

$\varphi = \tan -1 [(m_2-m_1)/(1+ m_1 m_2)]$; $r_1=r_2 + r_3$; $r_3 = sqrt((x_b-x_t)2 + (y_b-y_t)^2)$
$x_e = (r_2 x_t + r_1 x_b)/(r_1 + r_2)$; $y_e = (r_2 y_t + r_1 y_b)/(r_1+r_2)$ (1)

5.9 Experimental Results of Hierarchical Multi-agent Cooperative Control Structure

To illustrate the performance of the main task identifier fuzzy logic controller, let's say for example in Table 4, it has the following inputs: ND=-11, O=60, and D=15. When these values are fuzzified: ND will be LEW(leading widely) of degree=0.1 and LES(leading slightly) of degree=0.9 which means that the home robot is slightly nearer to the ball than its nearest opponent; O will be VG(very good) of degree=0.67 and GD(good) of degree=0.33 which means that the nearest home robot's front is adequately facing the ball; D will be MN(moderately near) of degree=0.91 and MD(medium) of degree=0.09 which means that the ball is quite near the opponent's goal. In other words, the situation can be translated into English as "a home robot can take possession of the ball near the opponent's goal". A rational agent given this situation will decide to assume the offensive task, since the chance of scoring is high, which is the actual output of the main task identifier fuzzy logic controller.

Likewise, to illustrate the performance of the ball handler fuzzy logic arbiter, let's say for example in Table 5, it has the following inputs: X_{ball}=170, Y_{ball}=73, S_{shoot}=6, and S_{pass}=3. When these values are fuzzified: X_{ball} will be NR(near) of degree=0.08 and FR(far) of degree=0.92 which means that the ball is slightly beyond the midcourt towards the opponent's goal; Y_{ball} will be EL(extremely low) of degree=0.77 and NL(near low) of degree=0.23 which means that the ball is quite near on the abscissa of the field; S_{shoot} will be CR(clear) of degree=0.50 and VC(very clear) of degree=0.50 which means that the shooting lane is half clear; S_{pass} will be BK(blocked) of degree=0.25 and CR(clear) of degree 0.75 which means that the passing lane is partially blocked. In other words, the situation can be translated into English as "the ball handler is on the lower part of the opponent's side, it's not okay to shoot, and the passing lane is partially blocked." A rational agent given this situation will decide to pass the ball, since the chance of scoring is low, which is the actual output of the ball handler fuzzy logic arbiter.

Figs. 48, 49, & 50 show how cooperative behaviors between autonomous mobile robots can produce positive results during experimentations on the actual robot-soccer system. In Figure 48 two- friendly robots display their passing abilities. Robot 2 goes to the ball and passes it to its teammate allowing the latter to shoot. Table 12 gives the performance of two cooperating robots in passing and scoring. Figure 49 shows a robot implementing the aforementioned shoot algorithm. In Figure 50 illustrates how the goalie robot blocks every attempt of the ball to enter the goal. This performance is tabulated in Table 13 which shows an overall success of 66.66%.

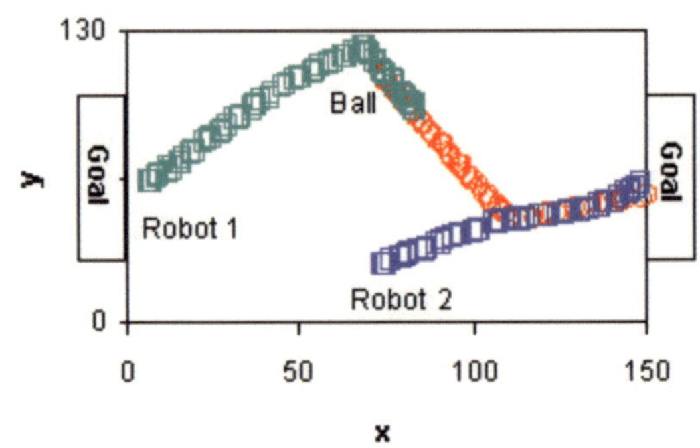

Fig. 48. A cooperative behavior through passing.

	Trials				
	1	2	3	4	Total
Intercepts	2	0	2	2	6
Goal scores	2	0	2	2	6
Total points					12
Success	0.25	0	0.25	0.25	75%
Failure	0	0.25	0	0	25%

Table 12. Performance evaluation of the robot for passing and scoring.

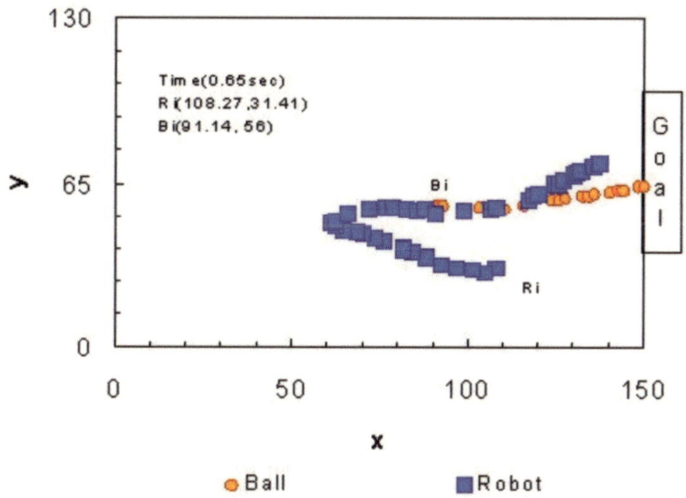

Fig. 49. Shoot algorithm performance.

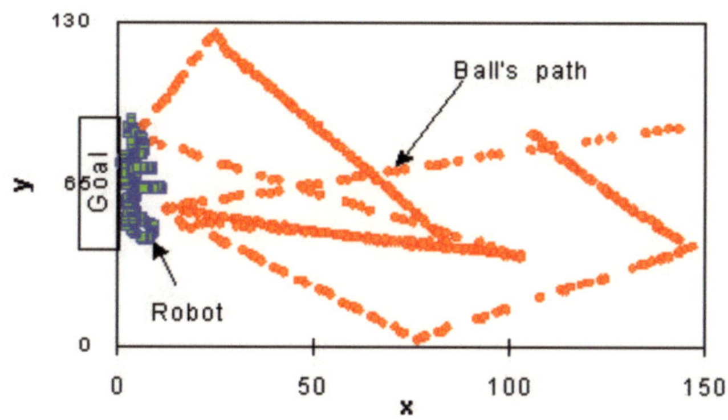

Fig. 50. Sample result of goalkeeper .

	Results	Rate
Success	4	66.66%
Failure	2	33.33%
Total	6	attempts

Table 13. The goalkeeper's performance with 6 trials conducted in real time. A "success" is a successful interception of the ball; "failure" is a failed attempt in blocking the ball.

Experimental results revealed that faster and swifter performance was achieved when more globally available information was given to individual robots. For example, a robot knows beforehand the presence or absence of obstacles leading to its destination and takes alternate path with the minimum cost. In addition, improved coordination between robots was realized resulting from the clear and specific instructions coming from the central controller. The performance of the fuzzy logic systems in general was satisfactory. While the improvement of their performance will depend more on fine-tuning of fuzzy set membership functions and accurate generation of the rule-base, the effects resulting from the unsatisfactory adjustments of these parameters was not seen as drastic that would in effect garble the overall performance of the robotic system. As Tables 4 and 5 show, a huge variation in the values of more than one input parameter did not have any effect on the outcome of the main task identifier's or ball handler's arbiter's action. This is a manifestation of the system's robustness.

INPUTS AND ANTECEDENTS															OUTPUT	
ND	FS1	Deg	FS2	Deg	O	FS1	Deg	FS2	Deg	D	FS1	Deg	FS2	Deg	SV	G.S.
-11	LEW	0.1	LES	0.9	60	VG	0.67	GD	0.33	15	MN	0.91	MD	0.09	8.18	OFF
15	LAS	0.5	LAW	0.5	10	GD	0.33	BD	0.67	75	VN	0.82	MN	0.18	4.02	DEF
4	SE	0.6	LAS	0.4	200	VG	0.78	GD	0.22	10	MF	0.36	VF	0.64	6.57	OFF
-18	LEW	0.8	LES	0.2	112	VG	1.00	-	0.00	0	MD	0.96	MF	0.04	9.04	OFF
-2	LES	0.2	SE	0.8	57	GD	0.98	BD	0.02	46	MN	0.96	MD	0.04	5.52	DEF
1	SE	0.9	LAS	0.1	178	VG	0.11	GD	0.89	40	MF	0.76	VF	0.24	5.79	DEF
-11	LEW	0.1	LES	0.9	30	VG	0.96	GD	0.04	2	VN	0.45	MN	0.55	8.58	OFF
9	SE	0.1	LAS	0.9	108	GD	0.22	BD	0.78	80	MN	0.04	MD	0.96	3.40	DEF
-20	LEW	1	-	0	2	VG	0.78	GD	0.22	10	VN	0.96	MN	0.04	8.34	OFF
-3	LES	0.3	SE	0.7	51	VG	0.29	GD	0.71	32	VN	0.07	MN	0.93	6.68	OFF

Table 14. Main Task Identifier Fuzzy Logic System Performance (**OFF**-Offense, **DEF**-Defense).

INPUTS AND ANTECEDENTS																				OUTPUT	
Ball's X-coord	Fuzzy Set 1	Degree	Fuzzy Set 2	Degree	Ball's Y-coord	Fuzzy Set 1	Degree	Fuzzy Set 2	Degree	SShoot	Fuzzy Set 1	Degree	Fuzzy Set 2	Degree	Spass	Fuzzy Set 1	Degree	Fuzzy Set 2	Degree	PRIORITY	ACTION
170	NR	0.08	FR	0.92	73	EL	0.77	NL	0.23	6	CR	0.50	VC	0.50	3	BK	0.25	CR	0.75	48.06	**PASS**
15	VN	0.77	MN	0.23	99	NH	0.85	EH	0.15	3	BK	0.25	CR	0.75	3	BK	0.25	CR	0.75	55.76	**PAS S**
67	MN	0.96	NR	0.04	108	NH	0.15	EH	0.85	5	CR	0.75	VC	0.25	7	CR	0.25	VC	0.75	79.95	**SHT**
33	VN	0.49	MN	0.51	85	NL	0.86	NH	0.1 4	5	CR	0.75	VC	0.25	6	CR	0.50	VC	0.50	79.78	**SHT**
217	FR	0.07	VF	0.93	85	NL	0.86	NH	0.14	2	BK	0.50	CR	0.50	1	BK	0.75	CR	0.25	22.30	**DRIB**
217	FR	0.07	VF	0.93	70	EL	1.00	-	0.00	7	CR	0.25	VC	0.75	7	CR	0.25	VC	0.75	43.26	**PASS**
43	VN	0.34	MN	0.66	84	NL	0.93	NH	0.07	2	BK	0.50	CR	0.50	7	CR	0.25	VC	0.75	80.96	**SHT**
43	VN	0.34	MN	0.66	84	NL	0.93	NH	0.07	2	BK	0.50	CR	0.50	4	CR	1.00	-	0.00	75.00	**SHT**
43	VN	0.34	MN	0.66	84	NL	0.93	NH	0.07	6	CR	0.50	VC	0.50	4	CR	1.00	-	0.00	75.00	**SHT**
43	VN	0.34	MN	0.66	84	NL	0.93	NH	0.07	7	CR	0.25	VC	0.75	3	BK	0.25	CR	0.75	63.22	**PASS**
136	NR	0.60	FR	0.40	96	NL	0.07	NH	0.93	7	CR	0.25	VC	0.75	3	BK	0.25	CR	0.75	50.49	**PASS**
115	NR	0.92	FR	0.08	108	NH	0.15	EH	0.85	7	CR	0.25	VC	0.75	5	CR	0.75	VC	0.25	64.17	**PASS**
115	NR	0.92	FR	0.08	108	NH	0.15	EH	0.85	7	CR	0.25	VC	0.75	8	VC	1.00	-	0.00	72.44	**SHT**

Table 15. Ball Handler Arbiter Fuzzy Logic System Performance (**SHT-** Shoot, **DRIB-** Dribble).

6. Conclusions

The advancement of industrial robots from mere Simple-Reflex-Agents to intelligent, learning, and autonomous agents has been accelerated by the integration of body, sensor, and AI-based software. This is also true to non-industrial robots [26]. This development will pave the way for the full automation and exclusion of human interventions in the robot soccer arena. This chapter presented some of the next great experiments in robot soccer, i.e. the development of more robust and intelligent robot soccer strategies (human-like behaviors) and an environment free of human entities, for scoring and refereeing, while the game is being played. All of these can be successfully done using AI-based paradigms, e.g. fuzzy logic algorithms.

7. References

[1] Dadios, E.P., et. al., "Neural Networks, Fuzzy Logic and Knowledge Base Hybrid Control System for Indoor AMR", ISARA 2000 Proceedings, pp. 9-15.

[2] Dadios, E.P., et. al., "Fuzzy Logic Controller for Micro-robot Soccer Game," Proceedings of the 27th IEEE Industrial Electronics Society Annual Conference, Denver, Colorado, USA.

[3] Jamshidi, M., et.al., "Applications of Fuzzy Logic: Towards High Machine Intelligence Quotient Systems", Prentice Hall PTR, Upper Saddle River, New Jersey, USA, 1997.

[4] Leinecker, R.C., "Visual C++ 6 Bible", IDG Books World Wide, Inc., Foster City, CA, USA, 1998.

[5] Dadios, E.P. and Maravillas, O.A.,"Fuzzy Logic Controller for Micro-robot Soccer Game", Proceedings: 27th IEEE Industrial Electronics Society Annual Conference, Denver, Colorado, USA.

[6] Thornton, J., et. Al.,"Shape Recognition and Enhanced Control Systems for Robot Soccer", Proceedings: 2002 FIRA Robot World Congress, Seoul, Korea, pp. 670-674.

[7] Kim, J.H., et.al.,"Nonlinear Field Based Path Planning and Petri-Nets Based Role Selection Mechanism w/ Q-learning for the Soccer Robot System", Proceedings: ISARA 2000 (International Symposium on Autonomous Robots and Agents), 26 May 2000, NUS,Singapore, pp. 86-100.

[8] Ong Chin Siong, et. al.,"Java-Based Implementation of Robot Soccer", Proceedings: ISARA 2000 (International Symposium on Autonomous Robots and Agents), 26 May 2000, NUS, Singapore, pp. 21-27.

[9] T. Simeon, S. Leroy, and J.-P. Laumond, "Path Coordination for Multiple Mobile Robots: A Resolution-Complete Algorithm", IEEE Transactions on Robotics and Automation, Vol. 18, No. 1, February 2002, pp. 42-48.

[10] K. Inoue, J. Ota, T. Hirano, D. Kurabayashi, and T. Arai, "Iterative Transportation by Cooperative Mobile Robots in Unknown Environment", Intelligent Autonomous Systems Y. Kkazu et. al. Eds., IOS Press, 1998, pp. 30-37.

[11] E. Pagello, A. D'Angelo, F. Montesello, C. Ferrari, "Emergent Cooperative Behavior for Multi-robot Systems", Intelligent Autonomous Systems Y. Kkazu et. al. Eds., IOS Press, 1998, pp. 45-52.

[12] L. Vlacic, A. Engwirda, M. Hitchings, and Z. O'Sullivan, " Intelligent Autonomous Systems", Y. Kkazu et. al. Eds., IOS Press, 1998, pp. 53-60.

[13] K.-H. Han, K.-H. Lee, C.-K. Moon, H.-B. Lee, and J.-H. Kim, "Robot Soccer System of SOTY 5 for Middle League Mirosot", 2002 FIRA Robot Congress, Seoul, Korea.

[14] H.-S. Shim, M.-J. Jung, H.-S. Kim, I.-H. Choi, and J.-H. Kim, "Development of Vision-Based Soccer Robots for Multi-Agent Cooperative Systems", 1997 Micro-Robot World Cup Soccer Tournament Proceedings, Taejon, Korea.

[15] K.M. Passino and S. Yurkovich, "FuzzyControl", Addison-Wesley Longman, Inc., 1998.

[16] Y.U. Cao, "Cooperative Mobile Robotics: Antecedents and Directions", Autonomous Robots, Special Issues on Robot Colonies, R.C. and G.A. Bekey Eds., Vol. No. 4, March 1997.

[17] E.P. Dadios, E.A. Maravillas, and N. Reyes, "Color-Based Fuzzy Vision System for the FIRA Robot Soccer Game", 2002 FIRA Robot Congress, Seoul, Korea.

[18] S. Marwaha and D. Srinivasan, Advances in Multi-Agent Learning Systems: A Survey, *Proceedings CIRAS 2001,* NUS, Singapore, pp. 188-189.

[19] K. Sugawara, I. Yoshihara, M. Sano, K. Abe, and T. Watanabe, Cooperative Behavior of Simple multi-Robot in a Clockface Arranged Foraging Field, *Proceedings CIRAS 2001*, NUS, Singapore, pp. 166-169.

[20] S. Premvuti, Aviation Knowledge Based Multiple Mobile Robot Systems: Consideration of Analogy between Air Traffic Control Systems and Multiple Mobile Robot Systems, *Intelligent Autonomous Systems, Y. Kakazu et. al. Eds.*, IOS Press, 1998, pp.38-44..

[21] J.H. Johnson and M.J. Booth, Robot football: emergent behavior in nonlinear discrete systems, *1997 Micro-robot World Cup Soccer Tournament Proceedings*, S3-5.

[22] Randy Sargent and Bill Bailey, Fast Vision Tracking and Coordinated Control for Soccer-Playing Robots, *1997 Micro-robot World Cup Soccer Tournament Proceedings*, S2-2.

[23] R.C. Gonzales and .E. Woods, *Digital Image Processing* (Addison-Wesley Publishing Company, Inc., 1993).

[24]Sun-Gi Hong, et. Al., "Designing Soccer-Playing Robot Team (F.B.I.) Based on the Centralized Approach", 1997 Micro Robot World Cup Soccer Tournament Proceedings, S4-4.

[25]Giridhar Rajaram, et. Al., "STRIKER - A vision based robot that learns reactive behaviors for scoring goals", 1997 Micro Robot World Cup Soccer Tournament Proceedings, S4-6.

[26] K. Kawamura, S. M. Gordon and P. Ratanaswasd, "Robotic Body-Mind Integration: Next Grand Challenge in Robotics", Industrial Robotics: Theory, Modelling and Control, pro literatur Verlag, © 2007 Advanced Robotic Systems International, www.ars-journal.com, pp.1-42.

9

Artificial Immune Systems, A New Computational Technique for Robot Soccer Strategies

Camilo Eduardo Prieto S., Luis Fernando Nino V. and Gerardo Quintana
Universidad Nacional de Colombia- National University of Colombia
Intelligent Systems Research Laboratory
Colombia

1. Introduction

Over last decades a computational intelligence technique has take more action field in engineering, this method based on biological systems provides new solutions to tackle a wide range of engineering problems. This technique is the Artificial Immune Systems (AISs) (DeCastro&VonZuben,2000), which utilizes metaphors from the immune systems in order to solve problems. From different studies and investigations, multiples techniques had surged such as negative selection, immune networks, clonal selection and others. Precisely a new technique thrives inside on AIS, its name: Danger Theory. In this chapter we used some methods for develop robot soccer game strategies.

Robot soccer presents a dynamic environment where it is possible to implement and test diverse and new computational designs. In robot soccer, specifically, SIMUROSOT from FIRA (website,2006), there are two leagues: middle and large leagues. In actual work, first one was used.

In this chapter, robot soccer strategies based on natural immune response are presented. Due to the adaptability of natural immune system with unknown pathogen, the algorithms presented carry on this feature in order to play a robot soccer game. As well as their biological inspiration, strategies based on natural immune response are exposed in several situations unknown that offer the Robot Soccer Game.

2. Background

Many fields of science have problems with systems and process identification become on inconvenient what need to be controlled. Most cases a complex process of control is used, but the best control in these cases is an adaptive control; for this reason new techniques are necessary to engage several dynamical environments, such as Robot Soccer. The researches trends towards biological inspiration because of its adaptability, some of these are: Neural networks, genetics algorithms, swarm and, recently, immune systems. Some characteristics of immune systems are learning, distributed process and memory; these features are ideals for Robot Soccer since they can provide a cognitive behaviour emergent where several agents form a team looking for same objective.

2.1 Some works related

There are several works on Robot Soccer; most of them are focused to generate highly competitive teams providing in this way more reactive teams than deliberative teams. Some works related with this chapter are described as follows. A work what develops its strategy with Artificial Immune Systems (AIS) is showed in (Guan-Chun et al 2006). Basically the strategy enables one robot in order to *select* one behaviour between pass, kick, shoot, follow and protect. For strategies evaluation, the middle league from FIRA is used; the player robot *takes a decision* by using an artificial neural network implementing on this way an action to execute. In (Guan-Chun et al 2006) an antigen is represented by environmental information, besides each robot had 6 antibodies (see figure 1) that they correspond to actions or behaviours mentioned before. In order to calculate Antigen-Antibody affinity a fuzzy logic system is used. This combination AIS-Fuzzy logic presents good results and it offers a interesting tactic for robot soccer games.

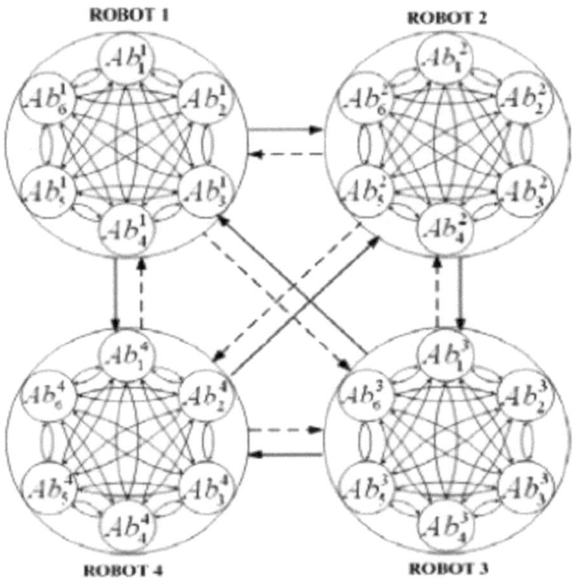

Fig. 1. Immune network used in (Guan-Chun et al 2006)

Neuronal networks, in special, a Multilayer Perceptron –MLP- is used in (Hwan at al,1997) to make learning process; however a Action Selection Mechanism (ASM) is in charged of *make* an action according to play role (see figure 2). The play roles used in that work are goalkeeper, back –defense- and forward. As well as on this work, research is focused into adaptation in dynamical systems.

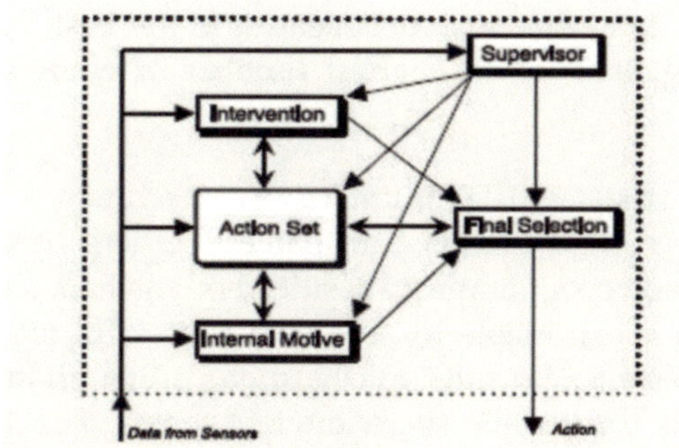

Fig. 2. ASM structure used in (Hwan at al, 1997)

In (Yu-Hwan,2005) fuzzy logic is utilized to control speed and trajectory of each robot. It also implements a "genetic regulatory network" which uses a concept of the bioinformatics field based on how genes are involved in controlling intracellular and intercellular processes. In this way play roles are assigned dynamically. Its biological inspiration for the robot soccer environment showed good results. In Figure 3 shows the system architecture implemented. Like the previous work, the comparison is restricted to the computational model used.

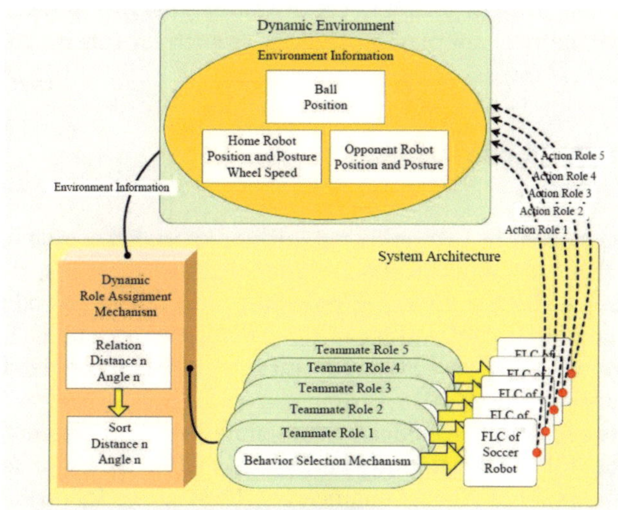

Fig. 3. System architecture used in (Yu-Hwan,2005)

In project FIBRA (Armagno at al, 2006) the ant colony technique is used in ordet to identify opponent's trajectories. Although good results are presented admits it has the disadvantage of requiring much processing time (about 4000 iterations). For decision making by using fuzzy logic like previous computational technique also has problems with the time needed for decision making, in addition, the large number of variables. In the evaluation process the following measures were used: percentage of time what the ball is played on his own or opponent's field, distributions of players on the field, number of goals, efficiency and effectiveness of attack to the opponent. At the beginning of that project was thought to an extent similar to the percentage of time the ball is played on each field, but based on human showed that football is not a very relevant, because even if a team has the ball in their dominance by one more time, this does not mean it is the best, because what counts in a championship is winning, that is, the highest number of goals scored and fewest goals received.

In work presented in (Farris et al, 1997), unlike other techniques what they develop low-level behavior such as pass and intercepted the ball and others, here high-level coordination is evolutioned using genetic programming; besides homogeneous and heterogeneous team are developed. Although it is possible to "evolve a successful team" this requires a lot of time in order to obtain reasonable results. Furthermore, although the evidence was obtained partial success (although the details were omitted here). For the evolution of team's strategies a mixed team is used because of the large number of iterations and mutation to take place, instead a pseudo-heterogeneous is implemented.

In (Thomas, 2005) the strategies are based on fuzzy logic with trajectory evolution. A model of 3 inputs - 2 outputs is developed, where the inputs are two angles and a distance, and the two outputs correspond to two wheel speeds in order to control the robot. The convergence of the algorithm of evolution is slow. Another controller 5 inputs (3 previous +2 current for each wheel) - 2 outputs is implemented. This last driver introduced improvements in behavior, but the computing time increases greatly due to the increased number of dimensions. Unlike the present chapter, control strategies are aimed at navigation and trajectory of the robot, more is not playing strategy as such for this reason is not comparable with this chapter work.

3. Immunology: Natural and Artificial

3.1 Fundamentals

The natural immune system is a complex network of specialized cells and organs that has evolved to protect the body against attacks from "outside invaders, defending infections caused by agents such as bacteria, viruses, parasites" and others (Woods, 1991). The ability of recognition is almost unlimited and it can remember past infections. In this way, when a pathogen attacks again the response is so efficient since it was recognized previously, which is known as secondary response.

Natural Immune System (NIS), specifically vertebrate immune systems has been taken as biological inspiration for Artificial Immune System (AIS). Different features of NIS are highly appealing from point of view of engineering, these are as follows:

- ✓ Uniqueness
- ✓ Pattern recognition
- ✓ Autonomy
- ✓ Diversity
- ✓ Multilayered
- ✓ Anomaly detection
- ✓ Distributivity
- ✓ Noise tolerance
- ✓ Robustness
- ✓ Learning
- ✓ Memory
- ✓ Self-organization

3.2 Natural Immune System

As mentioned above, NIS has the ability to distinguish foreign molecules or elements that can damage the body, this is known as the distinction between self and non-self. In normal situations the NIS may mistakenly identify itself as a non-cell itself and execute an attack, this is called auto immunity.

Although in NIS there are a variety of cells, some lymphocytes (type of white blood cell) are essential to mount an immune response, some lymphocytes featured are B and T; lymphocytes B complete their maturation in the bone marrow, while T cells migrate to the thymus. On the other hand, also exits dendritic cells and macrophages, first ones are found mainly in the skin, mucous membranes, lungs and spleen. Macrophages are specialized cells on to phagocyte (swallowing) large particles (e.g. bacteria) to decompose and then present them to lymphocytes. There is another type of cells with granules containing potent chemicals that kill other cells to be marked for elimination; these are known as natural killer cells (NKC). The figure below presents a classification of immune cells.

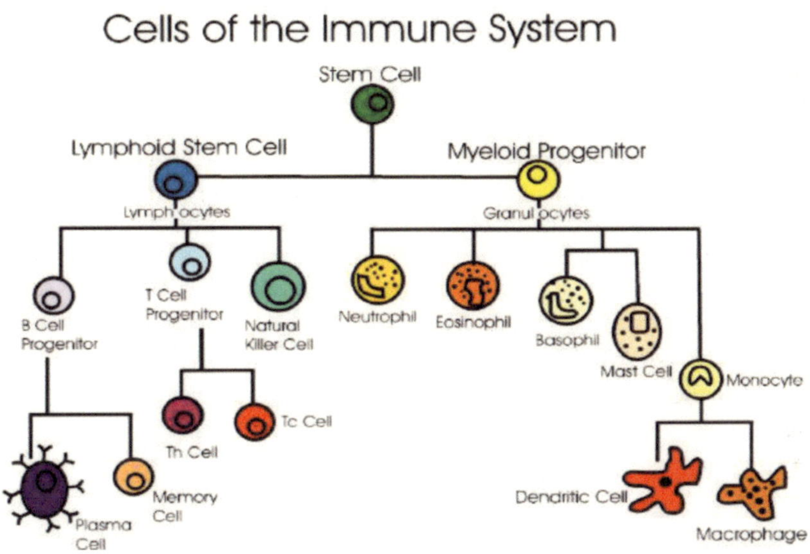

Fig. 4. Immune system cells.

Any substance or agent capableto produce an immune response are called antigens. An antigen may be a virus, bacterium or fungus, for example. When an antigen is presented, different cells can serve as antigen presenting cells (APC), these include B cells, T cells, macrophages and dendritic cells. APC role is to process a suspect foreign particles, broken down into peptides and after presente them on their surface to T cells in order to recognize antigen. Molecules that marks a cell as own are coded by a group of genes that is contained in a specific section of a chromosome, called Major Histocompatibility Complex or MHC. The MHC plays an important role in antigen identification, in particular for immune defense. Other specials molecules for immune defense are the antibodies, which belong to the family of molecules called inmunoglobinas and they are produced by B cells. Antibodies are composed of polypeptide chains which form a region V which is the area of coupling with the antibody (see Figure 5).

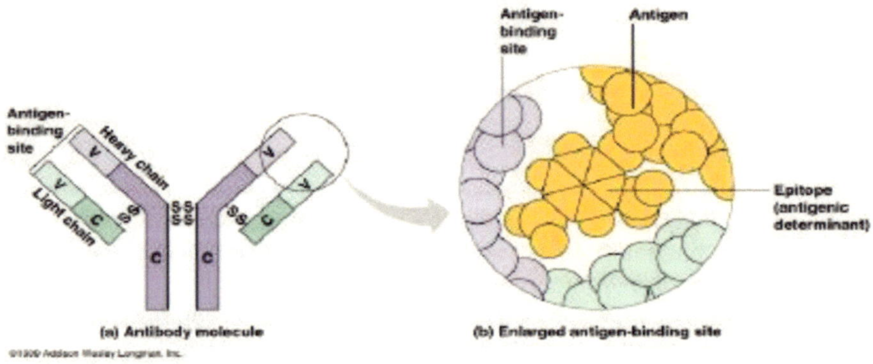

Fig. 5. Antibody structure.

The adaptive immune response is when the immune system can recognizes and selectively eliminates foreign molecules. This type of response may occur in two forms: Cellular and Humoral immunity. For development of actual chapter only the last one is taken into account. For details on the cellular immune response consult (Woods, 1991). Humoral immunity consists of the following phases: Macrophage swallows an antigen and becomes an APC, this APC divides the antigen in peptides, peptides and MHC join together to form an internally MHC-peptide molecule which one is presented on the APC surface. This stimulates to Helper T cells which recognizes the antigen through its TCR (T-Cell Receptor). Once recognition by the Helper T-cell is made, T-Cell emits two kinds of signals: CD40L and cytokines, for the purposes of this study only takes into account the last ones, since cytokines cause cell B proliferation and differentiation as clonal plasma cells and memory cells. Plasma cells secrete antibodies which are attached to the antigen in order to the natural killer cells (NKC) can be identify and eliminate them.

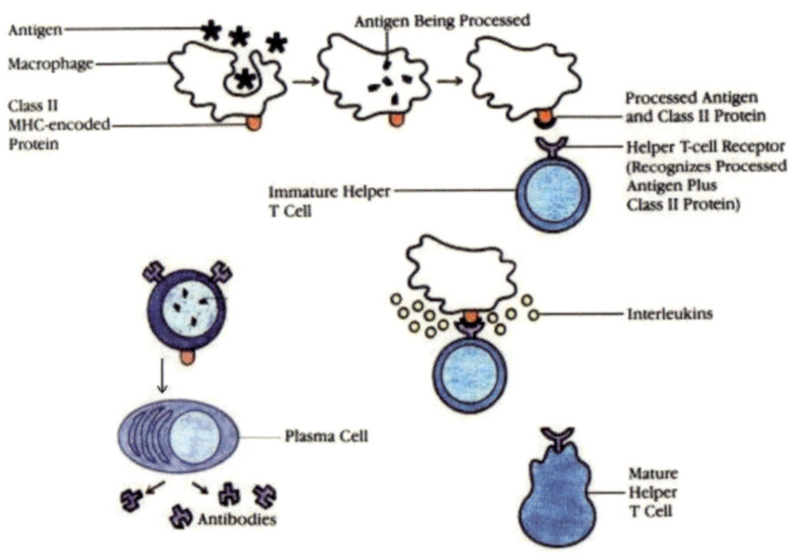

Fig. 6. Humoral Immune response process.

3.3 Artificial Immune Systems

Artificial immune systems (AIS) are adaptive systems inspired by immunological theories and observed immune functions, principles and models that are applied to solve problems (DeCastro&Timmis, 2002). Although this area of research is relatively recent, highlights several algorithms and models (website, 2009), some of them are:

✓ *Negative Selection Algorithm*: it is inspired mainly by the mechanism in the thymus that produces a set of T cells what are able to bind only to antigens (non-equity), items on the basis only of their own.

✓ *Clonal Selection Algorithm*: it is based on affinity maturation of B cells basically, this algorithm extracts two fundamental characteristics: proliferation of B cells proportional to the affinity with antigen (higher affinity, greater is number of clones produced) and the mutation undergoes the antibody (lower affinity, greater is mutation).

✓ *Immune Networks*: these models are based on the fact that any cell receptor can be recognized by a receptor repertoire, recognizing each other, ie, B cells are stimulated by not only antigens but also by other B cells, and this feedback mechanism leads to a memory.

3.3.1 Humoral Response Algorithm (HRA)

Here, an algorithm inspired by Humoral Immunity is developed in order to implement behavior of some robot soccer players. This algorithm uses some features of Humoral response described previously. The artificial immune response of HRA is considered in two stages: Stage or phase of activation and effector phase. In the next figure , these phases (natural immune response) are presented.

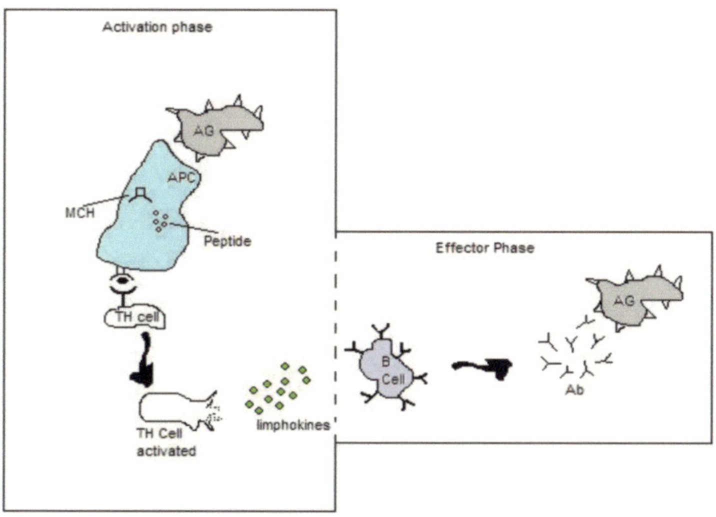

Fig. 6. Activation and effector phases.

(a) *Activation Phase:* At this stage the process of antigen identification is made, whose biological inspiration can be summarized as follows: Once reaching an antigen, an antigen presenting cell -APC- engulfs and processes it, once the molecular material is processed, it is presented on antigen surface joined to a molecule of the largest histocompatibility complex for subsequent recognition by a Helper T-lymphocyte which sends limphokynes to activate the B-cells. This stage is modeled as follow:

PRE-HRAALGORITHM

```
Begin
    For each antigen
        divide in peptides
        show MCH/peptide
        T-Helper sends limphokynes
    End For
End
```

Each time that an antigen arrives, breaks down elements that make up this antigen, for this chapter the opponent's strategy represents an antigen; peptides represent the coordinates' opponents players with respect to the ball. In order to process information of the opponent, information from the local team represents MCH and form MCH / peptide (biological point of view). According to this information the opponent's strategy is identify. Although in the biological process there is a mutation from T Helper cells, the model proposed in this work was not taken into account since it represents a high computational cost.
(b) *Effector phase:* At this stage the process of antigen elimination is carries out. The biological process modeled is showed as follows.

HRA ALGORITHM

```
Begin
 While receives limphokyne Do
  If Ab exists in memory Then
        Use Ab
  Else
        Generate Abs
        For each Ab Do
              Calculate affinity
              Choose Abs with best affinity
              If exist bests Abs Then
                    Choose the best
              Else
                    Mutation of  Antigen Image
              End If
        End For
  End If
 End While
End
```

Once the limphokines are received from earlier stage, the algorithm verifies if an antibody (Ab) exists into memory then it is used, otherwise generates Abs, calculates his affinity with antigen for each one and chooses the best Ab. In some cases the affinities are not the best, and then a mutation of antigen image is necessary.

3.4 Danger Theory

In 1994, Polly Matzinger proposed a theory that tries to explain why, in some cases there is no distinction between self and strange, for example, why does not the immune system reacts to foreign bacteria in food?. Matzinger proposes to change the classical theory of self/ non-self by dangerous/harmless on a new thoery called the Danger Theory. The central concept of this theory is what the immune system reacts to danger, the danger is measured by damage to cells as indicated by stress signals that are sent when the cells die so unnatural way (Aickelin&Cayzer,2002). In detail, when a cell dies in unnatural conditions it sends a alarm or danger signal that establishes a zone of danger around it (see Figure 7). In this context, B cells are responsible to produce antibodies that detect these antigens. If this detection occurs within danger zone, antibodies are stimulated and thus are activated. However, if detection is outside the danger zone, then the antibodies are not stimulated. Even if there is an affinity between antigen-antibody by outside the danger zone will not develop a cellular activation.

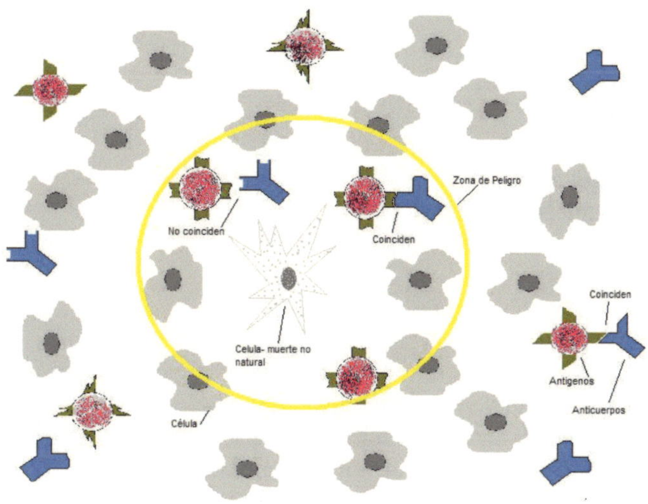

Fig. 7. Danger zone.

According to the two-signal model proposed by (Brester and Cohn,1970), antigen recognition is accomplished through two signals: signal recognition (signal one) and signal of co-stimulation (signal two), the last one really means dangerous (Aickelin&Cayzer,2002). Under the Bretscher-Cohn model, Danger Theory operates with the laws of lymphocytes, which are (Matzinger,2001):

1) A lymphocyte requires two signals to be activated, Signal One (signal recognition) comes from the junction of TCR (T cell receptor) and MHC-peptide. Signal Two comes from an APC.
2) A lymphocyte only accepts Signal Two from APC.
3) The activation phase delays a certain time and after activated, the lymphocytes not need Signal Two.

3.4.1 Danger Theory Algorithm

In order to make an abstraction of model proposed by Matzinger, it is necessary to identify certain characteristics to be implemented on this model, which one has been called DTAL (Danger Theory Algorithm). Here are the features modeled:

- ✓ Danger Zone
- ✓ Signal One
- ✓ Siganl Two
- ✓ Antigen
- ✓ Lymphocyte
- ✓ APC

Following the characteristics from model proposed by Matzinger an algorithm is generated. The proposed algorithm is presented below.

Algorithm DTAL

```
1 - Start
2 - for each antigen
3 -     detect alert (Signal One)
4 -     monitor antigen
5 -     if you receive signal then Two
6 -             danger zone set
7 -             activate lymphocyte NKC
8 -     end if
9 - end for
10 - end
```

The flexibility of this algorithm is to define a danger zone that could be placed where is needed, since there are systems which require what certain area or subsystem always is at same place in order to be monitored for abnormalities; in these cases line 6 of algorithm can be located at 1.5 to establish the danger zone. This flexibility makes the algorithm can be applied in different environments (Prieto et al, 2008).

4. Robot Soccer Strategies Inspired on Immunology

Taking the concepts of immunology treated previously strategies for robot soccer are developed, both from a classical viewpoint and from the viewpoint of danger theory.

In many cases, the strategy used by a robot soccer team is treated globally. However, this may not be appropriate since the goalkeeper carries out other functions and can be viewed as a special player. Therefore, goalkeeper strategy can be separated from the rest of team, but still must be coordinated or linked to the rest of team to achieve objective: win the game. Therefore, in this chapter we define the strategies of the players as well:

1) Goalkeeper Strategy: We propose the use danger theory to develop the goalkeeper strategy.
2) Strategy Team: For the average league soccer robots (SIMUROSOT), players, excluding the goalkeeper, is four. With these players can deploy multiple roles football game. In this case, we propose the use of the humoral response theory to develop team strategy.

4.1 Goalkeeper Strategy

In this study, APC cell will be a B cell, the lymphocyte to be activated will be a NKC (natural killer cells), antigen is the strategy of the opponent's attack and signals One and Two will be which indicate to the goalkeeper that its bow is on danger. This analogy between elements of soccer robots and immunology is detailed in Table 1.

Danger Theory Elements	Robot Soccer
APC	Goalkeeper ID-Strategy
Tissue	1/3 Home Side
Antigen	Opponent and Ball at home side
Signal One	Ball at home side
Signal Two	Opponent with ball close to penalty area
Danger Zone (Fixed)	Penalty area
Lymphocyte (NKC)	Clear Strategy-AIKIDO

Table 1. Analogy between danger theory an robot soccer.

Goalkeeper identifies if there are opponent plays that can become dangerous and end up in goal, all of these by taking into account the signals of the model. When in the course of game, the ball is on final third side of the stadium itself (see Figure 8), will trigger an alert, this is sign one, since ball position can become goal. That is why active surveillance opponent's moves in order to detect if there is danger. The goalkeeper is so attentive to receive signal Two or co-stimulation, which indicates whether the situation is dangerous or not, that signal will be present when opponent has the ball and is in proximity to penalty area.

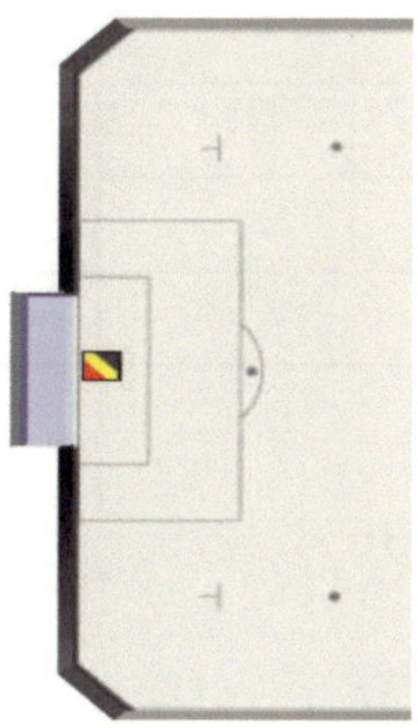

Fig. 8. Danger zone on game field.

As human soccer, goalkeeper gets alerted when an opponent player with the ball is coming. Usually, goalkeeper is located close to the vertical post nearest where is dangerous situation. This work takes such human reaction in order to do monitoring process. In robot soccer case a fixed danger zone is used, which corresponds to the penalty area. When goalkeeper receives the two signals lymphocyte NKC is activated, which is clearance strategy, disarming the opponent's goal play. As in human and robot soccer, clearance does not guarantee full disarmament of opposing team's attack, in many cases it needs the intervention of other elements to make the response effective. The clearance strategy used by goalkeeper is taken from aikido art, a strategy proposed by (Fredman&Mon,2004). This technique uses the force that is blowing to deflect it. The strategy of goalkeeper is uses angle and speed that comes with the ball and deflected its course, away from the arc as far as possible.

4.2 Team Strategy

Since the dynamics of soccer game is necessary that resident system; in this context, local team will be capable adapting to game schema of opponent in order to win soccer match. For team strategy development, information provided by simulator is used and putted into a vector of 22 elements, which result from to combine positions *(x, y)* of all players (home and visit) and ball.

LX^0_t	LY^0_t	LX^1_t	LY^1_t	...	LX^4_t	LY^4_t	OX^0_t	OX^0_t	OX^1_t	OX^1_t	...	OX^4_t	OX^4_t	BX_t	BY_t

Where LX^i_t and LY^i_t represent local player i coordinates at t time instant. For the team strategy, an antigen represents opponent's strategy, noting that the concept of strategy used for the team is defined as formation of the opponent with regard to both ball and field game into a window of time. To determine a strategy is necessary to use a sequence of movements of the opponent into a period of time. Because of need to sample the opponent's strategy, it is essential to have a history of opponent's movements (see Table 2). For this reason, when opponent player location is needed does not take the current position but for that player's position corresponds to predicting the next move according to that history (see equations 1 and 2).

OX^0_t	OY^0_t	OX^1_t	OY^1_t	...	OX^4_t	OY^4_t
OX^0_{t-1}	OY^0_{t-1}	OX^1_{t-1}	OY^1_{t-1}	...	OX^4_{t-1}	OY^4_{t-1}
⋮	⋮			⋮	⋮	⋮
OX^0_{t-4}	OY^0_{t-4}	OX^1_{t-4}	OY^1_{t-4}	...	OX^4_{t-k}	OY^4_{t-4}

Table 2. History opponent's moves.

$$OX_{t+1} = OX_t + \frac{1}{4}\sum_{k=0}^{3}\left(OX_{t-k} - OX_{(t-k)-1}\right) \tag{1}$$

$$OY_{t+1} = OY_t + \frac{1}{4}\sum_{k=0}^{3}\left(OY_{t-k} - OY_{(t-k)-1}\right) \tag{2}$$

On this way, an antigen can be represented as follows:

OX^0	OY^0	OX^1	OY^1	...	OX^4	OY^4	BX	BY

Positions XY of opponent players — Ball Position XY

Activation stage is responsible for make acknowlegde. To perform this process is necessary take this information and process it with the Major Histocompatibility Complex-MCH- (biologically speaking). From Robot soccer point of view, MCH can be represented by XY positions from local team (10 data). Using all the information (antigen and MCH) is necessary to find the distance between each local player and each opponent player with respect to ball, thus finding opponents who are an active participation on game and what players can participate in current move, in other words, we know the opponent's strategy by the players directly involved in action game, getting an image of the antigen. This process is analogous to the decomposition of such a peptide antigen in the biological.

5. Experimentation and Results

Due to Robot Soccer environment the experiments are based on matches of 5 minutes each, depending on the category SIMUROSOT. To test the proposed strategies, it is necessary to match the team in which strategies were implemented (local) with other teams, that is, other gaming strategies. One difficulty in the robot soccer environment -FIRA- lies in the unavailability of reference test strategies or benchmark for evaluation. However, different strategies developed in the work of (Sotomonte, 2005) and (Kogan and Parra,2006) were used in experimentation. That is, used 4 strategies:

- ✓ H01-heterogeneous system model 1, M04-homogeneous system with knowledge of rules and collision detection. Both were designed by (Sotomonte, 2005).
- ✓ Rakiduam developed by Kogan and Parra in 2006. Participated in the Argentine Championship Robot Soccer (website,2008), earning fifth place among 10 participants.
- ✓ Strategy which has by default the simulator league official SIMUROSOT.

In addition, a random attack strategy is used.

5.1 Results

In order to do all experiments only game time where the ball is in action game was taken. For this reason, the number of matches is not bigger in comparison with others investigations, but inside Robot Soccer context is enough in order to prove the computational intelligence.

5.1.1 DTAL Algorithm

To determinate effectiveness of this strategy, 15 matches were carried out. Time used on these tests was 15 minutes nets -no dead times were taken into account-. In order to evaluate the strategies developed two primordial features were used: Goal Options and Goal Annotations. The first ones are which Signal Two was present, it means, antigen was recognized as dangerous.

Match	Goal Option	Goal Annotation	Effectiveness Goalkeeper (%)
1	26	2	92.30
2	52	6	88.46
3	26	0	100
4	44	6	86.36
5	42	4	90.47
6	46	8	82.61
7	56	11	80.36
8	58	10	82.76
9	58	12	79.31
10	43	9	79.07
11	36	7	80.56
12	44	10	77.27
13	60	10	83.33
14	62	12	80.65
15	47	8	82.98

Table 3. Results for DTAL algorithm

It is important highline the fact that matches use 5 opposite player vs. goalkeeper, and so the effectiveness obtained was 84.43% with a standard deviation of 6.10%.

5.1.2 HRA Algorithm

Difference between this test and before test is the combination of 2 algorithms is into this algorithm. So, following its biological inspiration the HRA algorithm use memory in order to adaptation will be successfully. Into the next figure a goal tendency is showed; this represents its adaptation a different opponents.

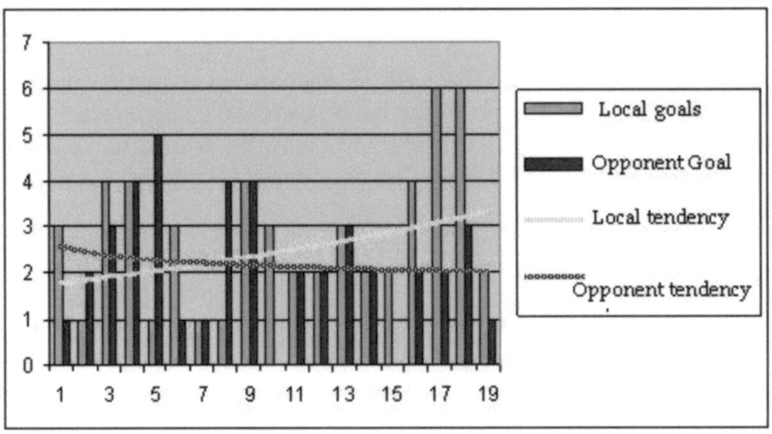

Fig. 9.

5.1.3 Analysis

In experiments for algorithm DTAL, despite the fact that goalkeeper played only against the rest of the opposing team (1 vs. 5), a high rate of effectiveness tackling opposing moves that may become a goal, was achieved. Although some aspects may improve the prediction of the position of the ball, the algorithm has many qualities to be implemented in other engineering fields. Those characteristics are the simplicity of implementation, speed of response and security system applications.

When algorithm DTAL is combined with algorithm HRA to develop the equipment, the team has characteristics of cooperation that was not formally designed, but its inspiration immune system makes this system suitable for multi-agent dynamic environments. The HRA algorithm can be implemented in other fields of action, preliminary interpretation of the environment so that its effectiveness is reflected in the particular application.

Even though the team was able to face different game situations, the method of navigation could be improved to make it much faster and generate movements that the opponent can not respond optimally, and thus will find a game much more competitive. However, in several previous works are presented various forms of navigation, but in the present study opted for a simple and effective way of navigation, since the focus of research was the application of artificial immunology concepts for a multi-agent system in highly dynamic environment.

6. Future Research

As future work, it is worthwhile to deepen some aspects of this work, besides continuing some work done. These aspects are:

- ✓ Because of work focused on a high level of abstraction, namely the implementation of strategies in play, a task ahead is to strengthen the players' actions to be implemented in a competition either domestically or internationally.
- ✓ There should be testing and improving, if necessary, the navigation system in order to be faster, since in gaming systems this is a very important feature in a competition.
- ✓ Perform other hybrid models involving computer techniques bio-inspired such as neural networks and genetic algorithms, in order to find a very competitive internationally.
- ✓ Using other platforms to interact with official simulator from FIRA to run faster actions, besides being able to implement different programming structures for the development of strategies.

7. Conclusions

Algorithms based on immunology concepts are presented; these features are used into a computational system in this case a robot soccer team in order to learn the game situations. Through interaction with the rating system, a memory is built so that its response is fast growing and adapts its behavior to different game situations regardless of the opponent.

It is important highlight that although the algorithms developed in this work, initially did not schedule for explicit communication between players (ie, between the goalkeeper and other players), thanks to the biological inspiration in immunology surges a collaborative relationship between the players in order to give a response to actions of the opponent who has not been previously scheduled. This implies that intelligent behavior emerges making results expected from these strategies developed meet the expectations raised initially.

8. References

De Castro Leandro, Von Zuben Fernando. *Artificial Immune Systems: A Survey Of Applications*. Thechnical Report, February 2000.

Lee Dong-Wook, Sim Kwee-Bo. *Artificial Immune Network-Based Cooperative Control In Collective Autonomous Mobile Robots*. IEEE Intenational Workshop On Robot. 1997.

De Castro Leandro. *Immune Cognition, Micro-Evolution, And A Personal Account On Immune Engineering*. Graduation And Research Institute. Catholic University Of Santos, Brazil. 2004

Kim Jong-Hwan, Shim Hyun-Sik, Jung Myung-Jin, Kim Heung-Soo And Vadakkepat Prahlad. *Cooperative Multiagent Robotic Systems: From De Robot Soccer Perspective*. 1998.

Sotomonte Wilson. *Estrategias De Sistemas Inteligentes (Simple Y Multiple). Caso De Estudio: Fútbol De Robots*. Universidad Nacional De Colombia. 2005.

Alonso Oscar, Niño Fernando, Velez Marcos. *A Robust Immune Based Approach To The Iterated Prisoner's Dilemma*. ICARIS 2004.

Romero Diego Andres, *Simulación De Un Agente Móvil Autónomo Basado En Sistemas Inmunes Artificiales*. Universidad Nacional De Colombia.2005.

Cortes Rivera Daniel. *Un Sistema Inmune Artificial Para Resolver El Problema Del Job Shop Scheduling*. Cinvestav-IPN. 2004.

Tomoharu Nakashima, Masahiro Takatani, Naoki Namikawa, Hisao Ishibuchi, Manabu Nii. *Robust Evaluation Of Robocup Soccer Strategies By Using Match History*. CEC 2006.

Gonzalez Fabio. *A Study Of Artificial Immune Systems Applied To Anomaly Detection*. University Of Memphis. 2003.

Página oficial de Robocup Soccer. Www.Robocup.Org – visitada en Octubre de 2006

Página oficial de Federation Of International Robot-Soccer Association www.Fira.Net. 2006.

Farías Terrens, Damián Gustavo, Pérez Orozco, Adith Bismarck, González Guerrero, Enrique. *Cooperación En Sistemas Multiagente: Un Caso De Estudio ROBOCUP*. Pontificia Universidad Javeriana. 2002.

Cecchi Laura, Parra Gerardo, Vaucheret Claudio. *Aspectos Cognitivos En El Desarrollo De Un Equipo De Futbol De Robots*. Universidad Nacional De Comahue. 2004.

Castro Leandro and Von Zuben. An evolutionary network for data clustering. IEEE Brazilian Symppsium on Artificial Neural Networks. 2002.

Brooks R. How to build complete creatures rather than isolated cognitive simulators, in K. VanLehn (ed.), Architectures for Intelligence, pp. 225-239, Lawrence Erlbaum Assosiates, Hillsdale, NJ, 1991.

Brooks R. A Robust Layered Control System for a Mobile Robot. AI Memo 864, MIT AI Lab. (1985).
http://www.vaneduc.edu.ar/cafr/equipos_fixture.htm

Bretscher Peter and Cohn Melvin. A Theory of Self-Nonself Discrimination. Science 11 September 1970: Vol. 169. no. 3950, pp. 1042 - 1049.

Matzinger Polly. "The Danger model in its historical context," Scandinavian Journal of Immunology, 54, 2001.

Prieto Camilo, Niño Fernando, Quintana Gerardo. A goalkeeper strategy in Robot Soccer based on Danger Theory. Proceedings of 2008 IEEE Congress on Evolutionary Computation. 2008.

De Castro Leandro, Timmis Jo. Artificial immune systems: a new computational intelligence approach. Springer, 2002.

Galeano Juan, Veoza-Suan Angélica and Gonzalez Fabio. A comparative análisis of Artificial Immune Network Models. GECCO 2005. Washington DC, USA.

Jong-Hwan Kim, Hyun-Sik Shim, Heung-Soo Kim, Myung-Jin Jung and Prahlad Vadakkepat. Action Selection and strategies in robot soccer systems. Circuits and Systems, 1997. Sacramento, CA, USA.

Vargas Patricia, De Castro Leandro and Von Zuben Fernando. Artificial immune systems as complex adaptive systems. ICARIS, 2003.

Sathyanath Srividhya and Sahin Ferat. AISIMAN - An artificial immune system based intelligent multi agent model and its application to a mine detection problem. www.citeseer.ist.psu.edu/640818.html

Luh Guan-Chun, Wu Chun-Yin and Liu Wie-Wen. Artificial immune system based cooperative strategies for robot soccer competition. International Forum on Strategic technologic. Octubre 2006

Baxter, J.L., Garibaldi, J.M., Burke, E.K. and Norman, M. Statistical Analysis in MiroSot. *Proceedings of the FIRA Roboworld Congress,* ISBN 981-05-4674-2, Singapore. December 2005

Laurenzo Tomás and Facciolo Gabriele. Una herramienta de análisis de estrategias de fútbol de robots Middle league Simurosot. Instituto de computación, Facultad de Ingeniería, Universidad de la República. Montevideo, Uruguay. 2004.

Secker Andrew, Freitas Alex and Timmis Jon. A Danger Theory Inspired Approach to Web Mining. Springer Berlin / Heidelberg. ISBN 978-3-540-40766-9. 2003.

G. Sen Gupta and C.H. Messom. Strategy for Collaboration in Robot Soccer. IEEE International workshop on electronic design. 2002

Aickelin Uwe and Cayzer Steve. The danger theory and its application to artificial immune systems. Proceedings of the 1st International Conference on Artificial Immune Systems (ICARIS), pages 141--148, University of Kent at Canterbury, September 2002.

Lin Hong. A real-time dynamic danger theory model for anomaly detection in file systems. MSc Thesis, Department of computer science, University of york. 2005

Armagno Gustavo, Benavides Facundo and Rostagnol Claudia. Proyecto Fibra. Instituto de computación, Facultad de Ingeniería, Universidad de la República. Montevideo, Uruguay. 2006.

Aickelin Uwe, Bentley P, Kim Jungwon, Cayzer Steve and McLeod Julie. Danger Theory: the link between AIS and IDS. Proceedings ICARIS-2003, 2nd International Conference on Artificial Immune Systems, pp 147-155.

Anjum Iqbal. Danger theory metaphor in artificial immune system for system call data. PhD Thesis, Faculty of Computer Science and Information Systems, Universiti Teknologi Malaysia. 2006.

Hart Emma. Immunology as a metaphor for computational information processing: fact or fiction?. PhD Thesis, Artificial Intelligence Applications Institute, Division of informatics, University of Edinburgh. 2002.

Huang Yu-Huan. Study and Design of a two-stage control strategy for robot soccer competiton. MSc thesis, National Cheng Kung University, Taiwan. 2005

Cortés Daniel. Un sistema immune artificial para resolver el problema del Jb Shop Scheduling. Tesis de Maestría, Departamento de ingeniería eléctrica, Cinvestav, México. 2004.

Lai Chien-Hsin. Study of fuzzy control strategy for five-on-five robot soccer competition. MSc Thesis , National Cheng Kung University, Taiwan. 2005.

Stone Peter and Veloso Manuela. Multiagents systems: A survey from a machine learning perspective. Carnegie Mellon University. 2000

Yang TW, Chan-Tan YW, Lee HA, C Teoh EL, Jiang H and Sng HL. Dynamic Model and shooting algorithm on Simurosot. Second International conference on autonomous robots and agents, New Zeland. 2004.

Kogan Pablo, Yañez Jael, Campagnon Costanza, Cecchi Laura, Parra Gerardo, Vaucheret Claudio and Del Castillo Rodolfo. Aspectos de diseño e implementación del equipo de fútbol con robots RAKIDUAM. Grupo de investigación en Robótica inteligente, Universidad del Comahue, Argentina. 2006.

Kogan Pablo, Parra Gerardo. Diseño e implementación de un sistema Multiagente: un equipo de fútbol con robots. Tesis de licenciatura en ciencias de la computación. Universidad Nacional de Comahue, Argentina. 2006

Freedman Hernán and Mon Gonzalo. How Spiritual machine works. Buenos Aires, Argentina. 2004

Thomas Peter. Evolutionary learning of control and strategies in robot soccer. PhD thesis, Central Queensland University. 2003.

Lydia Woods Schindler. Understanding the immune system. US Department of health and human service. October 1991.

The Online Home of Artificial Immune Systems. http://www.artificial-immune-systems.org/. 2009

Farris Jonathan, Jackson Gary and Hendler James. Co-evolving soccer softbot team coordination with genetic programming. Proceedings on the first international workshop on robocup. Japan. 1997.

Cómo se juega al fútbol. www.supercampeonato.com/futbol/como_se_juega_al_fútbol.php. 2008

Campeonato Argentino Fútbol Robots. http://www.vaneduc.edu.ar/cafr/equipos_fixture.htm 2008

Adaptive immunity. http://textbookofbacteriology.net/adaptive.html. 2008

10

The Role Assignment in Robot Soccer

Ji Yuandong, Zuo Hongtao, Wang Lei and Yao Jin
Sichuan University
China

1. Introduction

Multi-agent system is an emerging cross-disciplinary, involving robotics, artificial intelligence, mechatronics, intelligent control and etc. To improve the collaboration capabilities in unpredictable environment is an important part of the development of robotics.

Robot soccer system is a typical and challenging multi-robot system. It is an emerging field of artificial intelligence research, combining the real-time vision system, robot control, wireless communications, multi-robot control, and many other areas of technology. Robot soccer provides an ideal platform for robot collaboration research in dynamic and unpredictable environment. Assigning the appropriate role for each robot in robot soccer under the fast-changing environment is a basis issue of the real time decision-making system and the key to victory.

Role assignment in robot soccer has a direct impact on the efficiency of the entire system, and influences the ability of each robot to finish their task. At present, role assignment in robot soccer is mainly based on behavior, fuzzy consistent relation, robot learning, and evolutionary algorithm, etc. Behavior-based approach is simple, but only has a local optimal solution and is poor of robot collaboration. Fuzzy consistent relation increases the flexibility of the team, but it is difficult to build the model. Machine learning and genetic algorithm adapt to the complex dynamic environment but need training process.

This chapter presents two new algorithms based on analytic hierarchy process and market mechanism.

Analytic Hierarchy Process (AHP), a method of decision theory in operational research, is used for the role assignment. Based on mathematics and psychology, Analytic Hierarchy Process was developed by Thomas L. Saaty in the 1970s and has been extensively studied and refined since then. The AHP provides a comprehensive and rational framework for structuring a decision problem, for representing and quantifying its elements, for relating those elements to overall goals, and for evaluating alternative solutions. It is used around the world in a wide variety of decision situations, in fields such as government, business, industry, healthcare, and education.

AHP can rank choices in the order of their effectiveness in meeting conflicting objectives. Logic errors can be avoided when policy makers facing complex structure and many programs. In this chapter, in order to select a suitable robot for a certain role, the hierarchy is built, the pairwise comparison matrices are constructed with the help of experts, the weight

vector and the combination weight vector are calculated, and their consistencies are measured according AHP.
Market mechanism which is introduced into traditional behavior-based allocation is also used for role assignment. Firstly the task in the method is divided into two parts, combinatorial task and single task. The former is allocated to the specified robot group by using auction method while the latter is allocated to the single robot. Then the Bipartite Graph's weight and maximum match are adopted to solve the role collisions brought by dynamic assigning.

2. Behavior-based Role Assignment

The most fundamental and simplest role assignment method is fixing the role of each robot. Fixed role method is designating a role for each robot, when designing the decision-making system. And the role of each robot will not change in a strategy. Since the role of each robot is fixed before the game beginning, the movement of each robot is coherent. But this fixed role method has an obvious shortcoming: the robot could not always do the most suitable task, and it can not change its task according the fast-changing environment. This may cause the wastage of resources and the weak control ability of the decision-making system.
Every robot has a weight of each task, and the role of the robot is assigned by the weights in behaviour-based method. According to the different weights which are different for different task, different robot, and different status of the game, the decision-making system could choose the best finisher for each task.
Behavior-based role assignment algorithm is generally divided into three steps:

Step1: For a task j, calculate the weight of each robot according some algorithm which is designed in the decision-making system.

Step2: Find out the robot which has the greatest weight.

Step3: Assigned the task j to the robot which has the greatest weight, and do not assign another task to this robot in the follow-up distribution.

Step4: Go to step1 until every task finds a robot to achieve.

The method is characterized by the simpleness of the calculation process, and it is real-time and fault-tolerance. But this method sometimes leads to inconsistent of one robot's role. That is the role of a robot may change fast. Behavior-based approach is simple, but it only has a local optimal solution and is poor of robot collaboration.

3. Role Assignment Based on Analytic Hierarchy Process

3.1 The Algorithm of AHP

Analytic Hierarchy Process (AHP) is one of Multi Criteria decision making method, and it allows some small inconsistency in judgment because human is not always consistent. The ratio scales are derived from the principal Eigen vectors and the consistency index is derived from the principal Eigen value. AHP has been widely used in economic planning and management, energy policy, military command, transportation, education, etc. The algorithm of AHP is as follows (more details are presented in Thomas L. Saaty's book: *The Analytic Hierarchy Process*):
1. Build the hierarchy.

Develop the hierarchy by breaking the problem down into its components. The three major levels of the hierarchy are the goal, objectives, and alternatives.

2. Construct the pairwise comparison matrices.
Begin with the second layer, use pairwise comparison and 1-9 scale to construct the comparison matrices. A comparison is the numerical representation of a relationship between two elements that share a common parent. Do until the last layer.

3. Calculate the weight vector and measure the consistency.
For each pairwise comparison matrix calculate the largest eigenvalue and the corresponding eigenvector. Calculate the CI (consistency index) and CR (consistency ratio) to text the consistency of the matrices. If CR < 0.1, the test passed, and the normalized eigenvectors can be seen as the weight vectors. If adopted, the pairwise comparison matrices would be restructured.

4. Calculate the combination weight vector and measure the consistency.
Calculate the combination weight vector and test the consistency. If the test passed, the decision can be made according the outcome expressed by the combination weight vectors. Otherwise, restructure the pairwise comparison matrices whose CR are larger. AHP tells us that the plan with the largest combination weight is the most suitable for the goal.

3.2 Role Assignment
We define four roles in this chapter: attacker, winger, assistant and defender, except the goalkeeper (fixed to be robot 0).
Different stadium situations mean different role assignment strategies. The position of the ball is used to describe the stadium situation, such as attack, defence and etc. And it will influence the values of pairwise comparison matrices. The stadium is divided into 7 regions (shown in Figure 1). Roles of robots need to be assigned in every region which is shown in Figure 1.
Now we take selecting the attacker when the ball is in region 4 for example to show how to assign the roles using AHP.

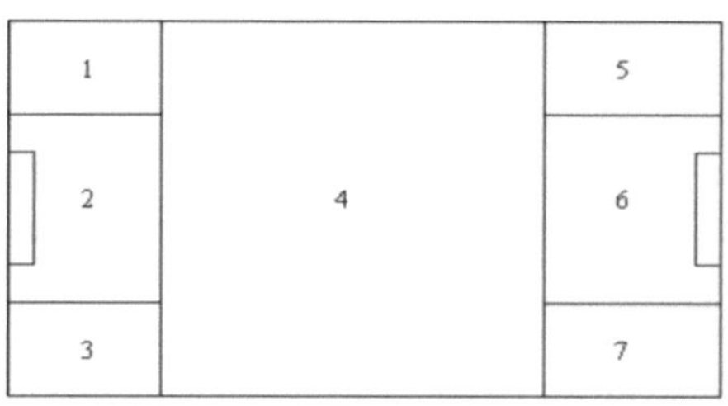

Fig. 1. Region division

3.2.1 Build the Hierarchy
There are three levels of the hierarchy: goal, objectives, and alternatives.

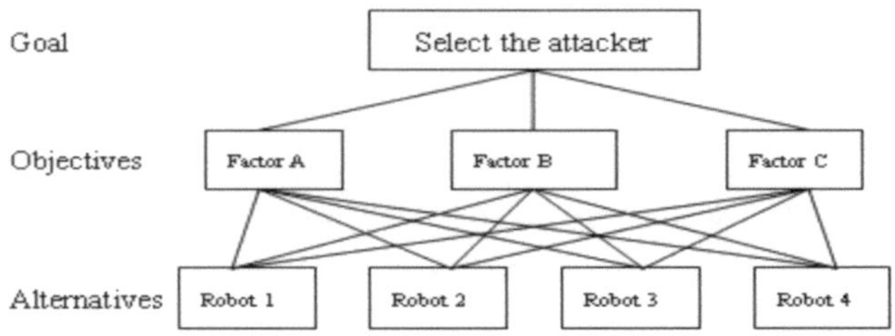

Fig. 2. Hierarchy diagram

1. Goal

The goal of role assignment is finding a suitable robot for a certain role. Selecting the attacker when the ball is in region 4 is for example in this section. So the goal is defined to be finding the attacker, shown in Figure 2.

2. Objectives

The factors which influence the assignment are as follows:

A: the distance between the ball and the robot.

B: the pose of the robot (here, we use the angle between the ball motion direction and the robot motion direction to describe it).

C: obstacles (we define the obstacle as this: robots in a particular fan region in the robot motion direction, and the bound is also an obstacle).

D: the position of the ball.

We do not take the position of the ball as an element of the objectives (shown in Figure 2). Because the role assignment and the importance of the factors are different when the ball is in different regions, so the pairwise comparison matrices may not be the same in each region. Now we only consider the situation when the ball is in region 4.

3. Complete the hierarchy

The goal and the objectives are described upper. And the alternatives are the four robots. The hierarchy with the goal, objectives, and the alternatives is shown in Figure 2.

3.2.2 Construct the Pairwise Comparison Matrices

When comparing the impact of two different factors for an upper layer factor, which relative measure scale is good? We use the 1-9 scale for the following reasons:

1. Conducting qualitative comparison, people usually have 5 clear hierarchies which can be easily expressed by 1-9 scale.

2. Psychologists believe, too many paired comparison factors will exceed peoples' judge ability. Using 1-9 scale to describe the difference is appropriate.

3. Saaty have experiment total 27 kinds of comparison scale such as 1-3, 1-5, 1-9, 1-27, etc[9]. The result showed that not only in the simple measures 1-9 scale is the best, but also as good as the complicated scales.

Currently, most people use 1-9 scale (shown in Table 1.) in their applications of AHP.

Intensity of Importance	*Definition*	*Explanation*
1	Equal Importance	Two activities contribute equally to the objective
2	Weak or slight	
3	Moderate importance	Experience and judgement slightly favour one activity over another
4	Moderate plus	
5	Strong importance	Experience and judgement strongly favour one activity over another
6	Strong plus	
7	Very strong or demonstrated importance	An activity is favoured very strongly over another; its dominance demonstrated in practice
8	Very, very strong	
9	Extreme importance	The evidence favouring one activity over another is of the highest possible order of affirmation
Reciprocals of above	If activity i has one of the above non-zero numbers assigned to it when compared with activity j, then j has the reciprocal value when compared with i	A reasonable assumption
1.1–1.9	If the activities are very close	May be difficult to assign the best value but when compared with other contrasting activities the size of the small numbers would not be too noticeable, yet they can still indicate the relative importance of the activities.

Table 1. The fundamental scale of absolute numbers(1-9 scales)

We first construct the pairwise comparison matrix of the second layer use the 1-9 scale. This work must be done by the experts, the opinion of each expert should be considered. We invited several experts in robot soccer to attend our work and the result is shown in Table 2. For example, B is moderate importance compared with A judged by the experts, so the value of matrix element a_{21} is 3 and the value of a_{12} is 1/3.
So the pairwise comparison matrix of the second layer is as follows:

$$A = (a_{ij})_{3\times 3} = \begin{pmatrix} 1 & \frac{1}{3} & 5 \\ 3 & 1 & 7 \\ \frac{1}{5} & \frac{1}{7} & 1 \end{pmatrix} \tag{1}$$

Factors	A	B	C
A	1	$\frac{1}{3}$	5
B	3	1	7
C	$\frac{1}{5}$	$\frac{1}{7}$	1

Table 2. Pairwise comparison of the second layer

When constructing the pairwise comparison matrices of the third layer, we need to compare the importance of the same factor among the four robots. But the values of the factors are not 1-9, this chapter establish a transformation using the idea of fuzzy mathematics to transform the factor value into 1-9 scale. The transformation method is shown in Table 3.

Grade	*A (distance)*	*C (difference of angles)*	*D (the number of the obstacles)*
1	>=90	>=100	>=6
2	80~90	90~100	
3	70~80	75~90	4~5
4	60~70	50~75	
5	50~60	40~50	2~3
6	40~50	30~40	
7	30~40	20~30	1
8	20~30	10~20	
9	<=20	0~10	0

Table 3. Measure transformation.

We use **A**(k), **B**(k), **C**(k) to express the grades of robot k according the factor A, B and C. Then we get the pairwise comparison matrices: $\mathbf{B}_1$, $\mathbf{B}_2$ and $\mathbf{B}_3$ as follows:

$$B_1 = (b_{ij}^1) = \begin{pmatrix} 1 & \frac{A(1)}{A(2)} & \frac{A(1)}{A(3)} & \frac{A(1)}{A(4)} \\ \frac{A(2)}{A(1)} & 1 & \frac{A(2)}{A(3)} & \frac{A(2)}{A(4)} \\ \frac{A(3)}{A(1)} & \frac{A(3)}{A(2)} & 1 & \frac{A(3)}{A(4)} \\ \frac{A(4)}{A(1)} & \frac{A(4)}{A(2)} & \frac{A(4)}{A(3)} & 1 \end{pmatrix} \quad (2)$$

$$B_2 = (b_{ij}^2) = \begin{pmatrix} 1 & \frac{B(1)}{B(2)} & \frac{B(1)}{B(3)} & \frac{B(1)}{B(4)} \\ \frac{B(2)}{B(1)} & 1 & \frac{B(2)}{B(3)} & \frac{B(2)}{B(4)} \\ \frac{B(3)}{B(1)} & \frac{B(3)}{B(2)} & 1 & \frac{B(3)}{B(4)} \\ \frac{B(4)}{B(1)} & \frac{B(4)}{B(2)} & \frac{B(4)}{B(3)} & 1 \end{pmatrix} \quad (3)$$

$$B_3 = (b_{ij}^3) = \begin{pmatrix} 1 & \frac{C(1)}{C(2)} & \frac{C(1)}{C(3)} & \frac{C(1)}{C(4)} \\ \frac{C(2)}{C(1)} & 1 & \frac{C(2)}{C(3)} & \frac{C(2)}{C(4)} \\ \frac{C(3)}{C(1)} & \frac{C(3)}{C(2)} & 1 & \frac{C(3)}{C(4)} \\ \frac{C(4)}{C(1)} & \frac{C(4)}{C(2)} & \frac{C(4)}{C(3)} & 1 \end{pmatrix} \tag{4}$$

3.2.3 Calculate the Weight Vectors and Measure the Consistency

First we calculate the largest eiginvalue $\lambda_{\max}^{A}$ of matrix **A** and the normalized corresponding eigenvector ω^{A}.

$$\lambda_{\max}^{A} = 3.0649 \tag{5}$$

$$\omega^{A} = \begin{pmatrix} 0.2790 \\ 0.6491 \\ 0.0719 \end{pmatrix} \tag{6}$$

Then CI (Consistency index) and CR (Consistency ratio) are calculated to test the consistency. And the RI (Random consistency index) is shown in Table 4.

n	1	2	3	4	5	6	7	8	9	10	11
RI	0	0	0.58	0.90	1.12	1.24	1.32	1.41	1.45	1.49	1.51

Table 4. Random consistency

Consistency index of matrix **A**:

$$CI^{A} = \frac{\lambda - n}{n - 1} = 0.0325 \tag{7}$$

Consistency ratio of matrix **A**:

$$CR^{A} = \frac{CI^{A}}{RI} = \frac{0.0325}{0.58} = 0.0560 < 0.1 \tag{8}$$

Because $CR^A < 0.1$, according the theory of AHP the consistency test is passed, and ω^{A} can be seen as the weight vector of the second layer.

Then we continue to calculate the third layer. λ_k^B (k = 1, 2, 3) which is the largest eigenvalue of matrix and the normalized corresponding eigenvector ω_k^B (k = 1, 2, 3) are calculated.

Consistency index of matrix $\mathbf{B}_k$:

$$CI_k^B = \frac{\lambda_k^B - 4}{4 - 1} \tag{9}$$

Consistency ratio of matrix $\mathbf{B}_k$:

$$CR_k^B = \frac{CI_k^B}{RI} = \frac{CI_k^B}{0.90} \tag{10}$$

The results of the experiments show that almost every CR can pass the test, and if there is some CR > 0.1, it will not bring very bad influence to the final role assignment. We did some experiments which we intentionally make some CR > 0.1. And the final results of the role assignment are also good judging by the experts, if we do not restructure the pairwise comparison matrices.

3.2.4 Calculate the Combination Weight Vector and Complete the Role Assignment

After the eigenvectors ω_k^B (k=1, 2, 3) are calculated, we use them to calculate the combination weight vector **W**. The formulas are as follows:

$$\omega = (\omega_1^B, \omega_2^B, \omega_3^B) \tag{11}$$

$$W = \omega \cdot (\omega^A)^T \tag{12}$$

The theory of AHP tells us that the k-th plan which the corresponding **W**(k) is the largest is the most suitable for the goal. So robot k is the most suitable to be the attacker.
Using the AHP we find out the most suitable robot for being the attacker. The same method can be used to find out the most suitable robots for being the winger, assistant, and defender. If a robot is the most suitable for two more roles, for example attacker and assistant, we will choose it to be the attacker, in accordance with the principle of good offensive.
Above, we use AHP to solve the role assignment. The hierarchy is built, the pairwise comparison matrices are constructed with the help of experts, the weight vector and the combination weight vector are calculated, and their consistencies are measured according AHP. We followed the algorithm of AHP to assign roles when the ball is in region 4. The role assignment when the ball is in the other regions (shown in Figure 1) can be completed using the same algorithm.

3.3 Experiment

The simulation is done on Robot Soccer V1.5a (it can be downloaded on http://www.fira.net/soccer/simurosot/R_Soccer_v15a_030204.exe). In order to examine the role assignment method based on AHP, this chapter designs two strategies to compete with the own strategy of the platform for 50 matches. One use the role assignment method based on AHP, and the other use the role assignment method based on behavior. Figure 3 shows the goal difference when the two strategies compete with the platform's own strategy separately. The average goal difference of the strategy using behaviour-based role assignment algorithm is 1.28, and the average goal difference of the strategy using AHP-based role assignment algorithm is 3.06.
The results of experiments on SimuroSot 5vs5 show that the method is feasible and effective. Though there is probable that the consistency test may not be passed. Our experiments

show that if we do not restructure the pairwise comparison matrices, the final results of the role assignment are also good judging by the experts.

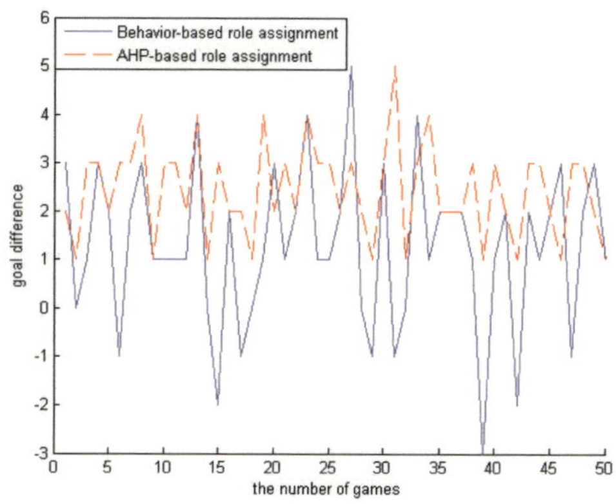

Fig. 3. Goal difference of Behavior-based role assignment and AHP-based role assignment

AHP can rank choices in the order of their effectiveness in meeting conflicting objectives. Logic errors can be avoided when policy makers facing complex structure and many programs. The success of the algorithm depends on the comparison matrices which are decided by the experts. Senior experts are the keys to this algorithm.

4. The Role Assignment Based on Market Mechanism

This section focuses on the role assignment based on market mechanism. Firstly define the tasks and dynamic task allocation in robot soccer using the concept of set theory. Then the basic idea and elements of the market mechanism, the solution process of the algorithm are presented. The problem of role conflict is solved using the maximum matching algorithm of bipartite graph. Experiments on the SimuroSot (5vs5) game platform show that the method is effective and convenient.

4.1 Robot Soccer Task and Task Allocation Dynamically

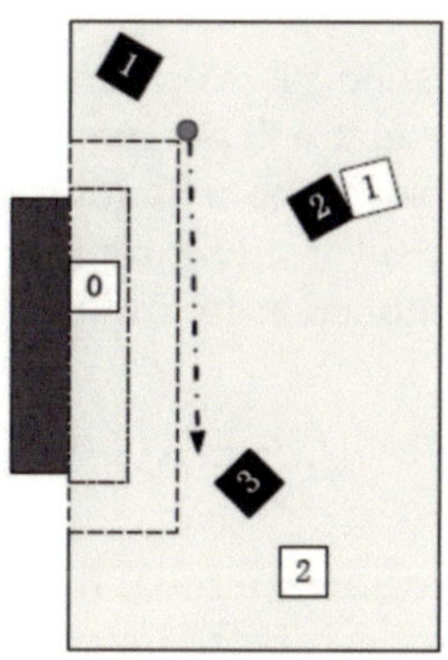

Fig. 4. Attack cooperate

Now assume a set of R consists of n robots $R = \{r_i \mid 1 \le i \le n\}$ and a set of T consists of n tasks $T = \{t_k \mid 1 \le k \le n\}$.

According to the situation of the game, this role assignment method will abstract several tasks from the formation of the team such as shooting, guarding and so on. Then using some algorithm to get a reflection f from the robots set to the tasks set ($f: R \rightarrow T$). In some statuses, it requires several robots to finish the task cooperatively.

In Figure 4 (The white is opponent, attack from left to right), it is obviously that the effect of only one robot shooting the gate is not good, and a better way is as follow: robot 1 move along the dotted line, pass the ball to robot 3, and robot 3 shoot the gate.

This chapter use the following definition: cooperative task M ($M \subseteq T$) is the task which requires several robots cooperatively to finish it, atom task S ($S \subseteq T$) is the task which requires only one robot to finish it. And all the cooperative tasks ($M_1, M_2, \ldots M_i$) and atom tasks ($S_1, S_2, \ldots S_j$) should satisfy the following condition: $\forall i, j \quad M_i \cap S_j = \varnothing$ and $M_1 \cup M_2 \cup \ldots M_i \cup S_1 \cup S_2 \cup \ldots S_j = T$. In this chapter, atom task is treated as cooperative task since it could be treated as special cooperative task which require only one robot.

4.2 The Basics of Role Assignment Based on Market Mechanism

4.2.1. The Cost of Task, Reward and Income

1. *Cost of the task*

The cost of a robot achieving a task in robot soccer is the time. In this chapter, use $cost(r_i, t_k)$ to mark the cost of robot r_i achieving task t_k. In robot soccer games, the cost will increase with the distance from the current position of the robot to the expected position, as well as the rotation angle and the difficulty of current status transforming to expected status. So the cost can be defined as follow:

$$\cos t(r_i, t_k) = f(d, \theta, \bar{v}) \tag{13}$$

d is the distance from the current position of the robot to the expected position, θ is the rotation angle, $\bar{v}$ is the difference between current velocity and the expected velocity.

2. *Reward*

Each robot will get reward if it finished its task. Robot r_i will get reward after it finishing task t_k: $reward(r_i, t_k)$. For offensive one, it will get more rewards if it runs toward the middle of the enemy's gate while the defensive one will get more rewards if it intercepts the ball run toward its gate. When one robot is at mid-ground, it will get rewards from attacking and defending. The $reward(r_i, t_k)$ is defined as follows:

$$reward\left(r_i, t_k\right) = g\left(Attack, Defend\right) \tag{14}$$

Attack indicates the ability of attacking after a robot finished its task and *defend* indicates the defending one. In a real match, they can be represented by the position of robot and the velocity of the ball and so on.

3. Income

The income of one robot after it finishing its task can be defined as followed:

$$income(r_i, t_k) = reward(r_i, t_k) - \cos t(r_i, t_k) \tag{15}$$

4.2.2 Auction and Combinational Task Auction

Market mechanism is based on auction theory. Generally, there are four types of auction: English auction, Dutch auction, sealed first-price auction, sealed-bid second-price auction. All the above auctions have the same important steps: auction announcement, bidding, contract. First the auction announcement will inform the entire bidder that what is going to be auctioned and the bidder will bid it and only one bidder will get the contact. Then the auction comes to the end.

Now the combination task auction in this chapter has the following assumption:

1. Auction is done by cooperative task.
2. Robots could be united as one union to bid one combinational task, and one robot could be in different unions.
3. Auction is done as sealed first-price auction. The bidder which gives the highest price will be the winner in the auction.

One robot union should bid according to the above rules and distribute the tasks to the robot in the union. It is allowed that the robots make a bargain with the others.

4.3 The Algorithm of Role Assignment

4.3.1 Initial Task Allocation

At the start of one match, the system should allocate the initial task to each robot. The task is divided into two layers: the first layer consists of cooperative tasks and the second layer consists of single tasks. Firstly cooperative tasks are allocated to robot unions by auction and single tasks are allocated to individual robots by behavior-based. The specific description of steps is as follow:

Step1: Generate the task set *T*, and divide it into several cooperative tasks $M_1, M_2, \ldots, M_k$ and assign constraint to each cooperative task.

Step2: According to the number of robots (mark with x) in the cooperative task, choose x robots in the entire robots R, which has C_n^x unions, calculate the income for each union and delete the unions which are not suitable to the constraints.

Step3: For cooperative task M_i, back to step 1 if there is not suitable bidder, else select the union which gives the highest price.

Step4: The union allocate the task to each robot by behavior-base method, and if there is not any task-conflict and go to *step6*.

Step5: Solve task-conflict.

Step6: Output the result.

4.3.2 Task Re-allocation

To avoid that the roles of robots change too frequently, the system needs to adjust each robot's task to get a consistent result. The following are the differences from initial task allocation:

1. The addressing of allocation

The task-allocation is executed forcedly by the decision-system at the start of the match, but each robot has to calculate the income from current task periodically. Robot requests task-allocation if the income is lower than its initial value. And the decision-making system will start task-allocation when receiving enough requests.

2. Whether receive new task or not

After one robot receiving a new task, firstly it will compare the status of the old one with the income of new one. It will not receive the new task if the old one is not done and the income of the old one is larger.

4.3.3 Removal of Role-conflict

Since one robot maybe belongs to different robot unions when bidding the cooperative tasks, it may get several tasks after allocation while some robots have no tasks, which is defined as task-conflict. In Figure 5, circles represent task, squares represent robots which having been allocated task, triangles represent robots which did not allocated yet and the line is the reflection of robot to task. If all the robots in the set X and tasks in the set T are complementary subsets of bipartite graph $G=(V,E)$, and $X=\{r_1,r_2,\ldots,r_n\}$, $Y=\{t_1,t_2,\ldots,t_n\}$ (n is the number of robots), $V=X\cup Y$. In Figure 5, *income*(r_i, t_j) ($1\le i,i\le n$) represents the weight of edges $w(r_i,t_j)$. So the method to solve the task-conflict is the way to find the max-match of the sum of weight in this non-complete bipartite graph. The steps are as follows:

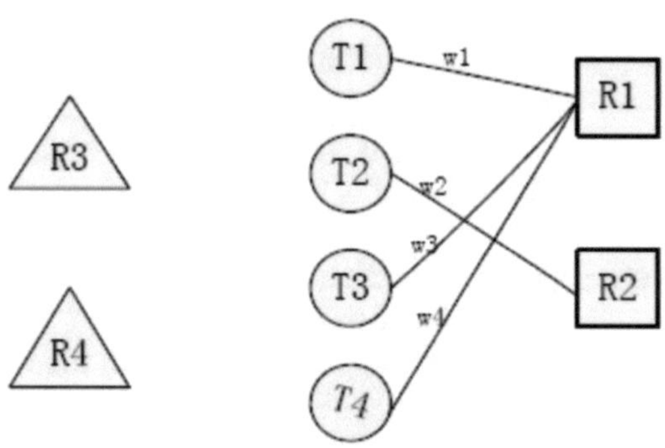

Fig. 5. Role Conflict

Step 1:

In graph G, give a real number $l(r)$ to each saturated point $r\in X$, called as flag of *r*, and *l* is as follows:

$$l(r)=\max_{t\in Y}\{w(rt)\} \tag{16}$$

For any $t\in Y$, there is $l(t)=0$.

Get the flag $l(r)$'s subset E_l :

$$E_l=\{rt\mid rt\in E\wedge l(r)+l(t)=w(rt)\} \tag{17}$$

And get M which is a match of $G(E_l)$.

Step2:
M is the solution if *M* has satisfied the point in subset *X*, and this algorithm comes to the end. Otherwise, select one non- saturated point $x \in X$ and generate set $S = \{x\}, T = \varnothing$.

Step3:
Calculate the adjacent point set $N(S)$:

$$N(S) = \{r \mid (\exists t \in S) \wedge (rt \in E)\} \tag{18}$$

In $G(E_l)$, if $N(S) \neq T$, go to step 4. Otherwise calculate follows:

$$\Delta = \min_{\substack{ri \in S, \\ tj \in Y-T}} \{l(r_i) + l(t_j) - w(r_i t_j)\} \tag{19}$$

$$l'(r) = \begin{cases} l(r) - \Delta, & r \in S \\ l(r) + \Delta, & r \in T \\ l(r), else \end{cases} \tag{20}$$

Then calculate E_l', use E_l' to substitute E_l, and use l' to substitute *l*.

Step4:
In $G(E_l)$, select one point *t* in $N(S) - T$. If *t* is *M*- saturated and it satisfies $tz \in M$, let $S \cup \{z\} \to S$, $T \cup \{t\} \to T$, and go to *step3*. If not, let $M \oplus E(P)$ substitute *M*, then go to *step2*.

4.4 Experiment and Analysis

50 matches using the strategy which contains the role assignment method based on market mechanism against the own strategy of the platform is shown in Figure 6.The results indicates that this algorithm improve the goal rate and increases the efficiency. The average goal difference of the strategy using Market-based role assignment algorithm is 2.84.
Dividing the tasks into cooperative tasks and atom tasks, using the auction theory to allocation in cooperative task layer, improve the cooperation between robots. This model can obviously increase the rate of successful and efficient of shooting.

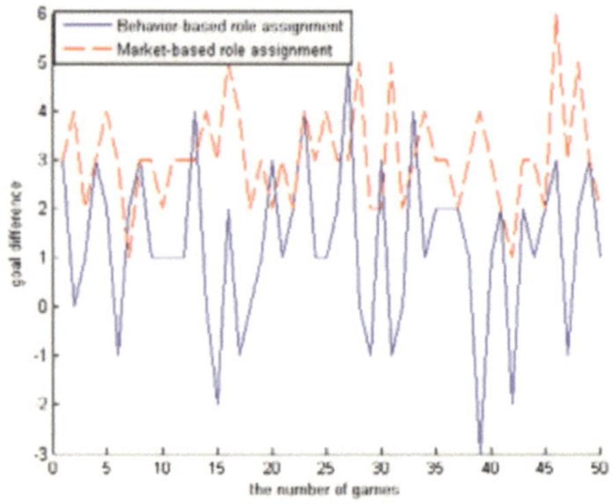

Fig. 6. Goal difference of Behavior-based role assignment and Market-based role assignment

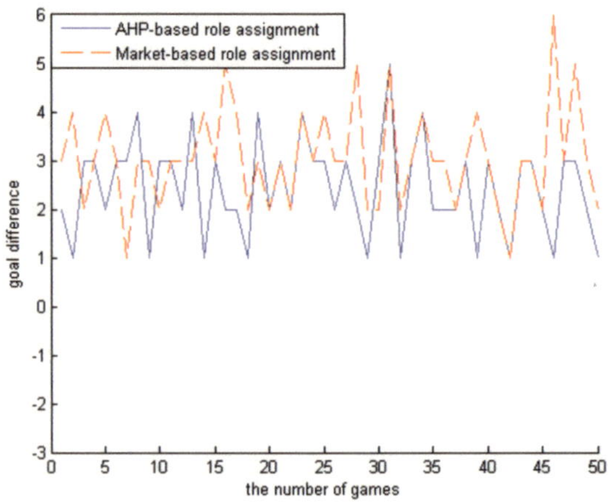

Fig. 7. Goal difference of AHP-based role assignment and Market-based role assignment

5. Conclusions and Future Research

Figure 7 shows the goal differences using the strategies which contain AHP-based role assignment and Market-based role assignment. The success of the AHP-based method depends on the comparison matrices which are decided by the experts. And the modelling of the Market-based method also needs the experience of the experts. Senior experts are the keys to these two algorithms. One of the future researches is using self-learning to improve these algorithms.

Robot soccer is a dynamic, uncertain, and difficult predicting system. Multi-robot role assignment has become a hot spot in the current research. Much work has been done in role assignment, many new algorithms and theories are introduced into this area. But the study of robot role assignment is still in its early stage, there is still a long way to go.

6. References

Cheng Xianyi, Yang Changyu(2007). Research Status and Development of Robotsoccer's Learning Mechanism. *Journal of Jiangnan University(Natural Science Edition),* Vol.6 No.6 Dec.2007:643-647.1671-7147

Daniel Playne(2008). Knowledge-Based Role Allocation in Robot Soccer. 2008 10th Intl. *Conf.on Control,Automation,Robotics and Vision*. Hanoi, Vietnam, Dec.2008: 1616-1619

Du Xinquan, Cheng Jiaxing(2008). Study on Soccer Robot Strategy Based on Annealing Evolution Algorithm. *Computer Technology and Development*. Vol.18 No.2. Feb. 2008:101-103.1005-3751

Emest Forman, Dsc. Mary Ann Selly(2002). *Decision by Objectives,* 2002:43-126, World Scientific Publishing Company, 9810241437, Hackensack., NJ, USA

Fu Haidong, Lei Dajiang(2006). Robot soccer role assignment system based on fuzzy consistent relation, *Computer Applications*. Vol.26 No.2.Feb.2006:502-504, 1001 -9081

Gerkey B P, Matarjc M j(2002). Sold!: auction methods for multirobot coordinnation. *Processdings of the IEEE Transactions on Robotics and Automation*. vol.18 No.5. May 2002:758-768.1042-296x

Gerkey B P, Maja J.M(2003). Multi-Robot task Allocation:Analyzing the Complexity and Optimality of Key Architectures. *Processdings of the IEEE International Conference on Robotices and Automation(ICRA2003)*2003:3863-3868,1050-4729 Taiwan.China, Org.2003, IEEE

Hong Bingrong. The Final Objective of the Robot Soccer and it's Realization Course. *J.Huazhong Univ. Of Sci.&Tech.(Nature Science Edition).* Vol.32 Sup.Oct.2004: 20-23.1671-4512

Ji Yuandong, Yao Jin(2008). Role Assignment in Robot Soccer Based on Analytic Hierarchy Process. *Journal of Harbin Institute of Technology(New Series).* Vol.15 Sup.2,Jul.2008.137-140

Kao-Shing Hwang, Shun-Wen Tan, Chien-Cheng Chen(2004). Cooperative Strategy Based on Adaptive Q-Learning for Robot Soccer Systems. *IEEE Transactions on Fuzzy System,* Vol.12 No.4, Aug 2004:569-576.1063-6706

Kardi Teknomo, PhD. Analytic Hierarchy Process(AHP) Tutorial, http://people. revoledu. com /kardi/tutorial/ahp/, visited on 10/3/2008.

Li Ping, Yang Yiming(2008). Progress of Task Allocation in Multi-robot System. *Computer Engineering and Applications,* Vol.44 No.17.2008:201-205.1002-8331

Liu Lin, Ji Xiucai, Zheng Zhiqiang(2006), Multi-robot Task Allocation Based on Market and Capability Classification. *Robot*. Vol.28 No.3 2006:337-343.1002-0446

Liu Shulin, Wang Shouyang, Li Jianqiang(1998). Decision Theoretic Approaches toBidding and Auctions. *Studies in Interntonal Technology and Economy.* Vol.1 No.2.May. 1998:20-33.1671-3435

Liu L, Wang L, Zheng Z Q, A learing market based layered multi-robot architecture. *Proceeding of the IEEE International Conference on Robotics and Automation*. Piscataway, USA, IEEE, 2004:3417-3422

Liu Wei, Zhang Cheng, MaChenwei, Han Guangsheng(2004). Decision-making and Role Allocatiing System for Robot Soccer. *Journal of Harbin Institution of Technology* . Vol.36 No.7 July 2004:966-968. ,0367-6234.

Robert Zlot, Anthony Stentz(2006). Market-based Multi robot Coordination for Complex Tasks. *International Journal of Robotics Research*, Vol.25 No.1 Jan.2006:73-101. 0278-3649

Thomas L.Saaty(2008). Decision making with the analytic hierarchy process, *International Journal of Services Sciences*. Vol.1 No.1,Jan 2008:83-98,1753-1454

T.L.Saaty(1980). The Analytic Hierarchy Process ,*McGraw Hill Intermational*, 1980

Wang Jinge, Wang Qiang, YaoJin(2005). Cooperation strategy for robot soccer based on defuzzification, *Journal of Harbin Institution of Technology*, Vol.137 No.7, July 2005: 943-946,0367-6234.

Wang Xiangzhong, Yu Shouyi, Long Yonghong(2004). Dynamic role assignment stategy in robo soccer competition, *Jourmal of Harbin Institute of Technology*, Vol.136 No.17, July 2004:943-945, 0367-6234

Wu Lijuan, Zhai Yuren, XuXinhe(2000). A Role Assignment Method of Multi- Intrlligence Agent Cooperation Based on Robot Soccer Match, *Basic Automation*, Vol.7 No.1 Feb 2000:4-6,1005-3662

Wu Zihua, Zhang Yili, Tang Changjie(1999). *Discrete Mathemaatics*, Chen Zhaolin, 217-221, SiChuan University Press, 7-5614-0175-2. Chengdu. China

Xue Fangzheng, Cao Yang, Xu Xinhe(2004), NEU Robot-Soccer Decision-making System Design. *Programmable controller& Factory Automation(PLC&FA)*, Vol.39 No.11 Nov. 2004:107-110,1606-5123

Yang Linquan, Lv Weixian(2005). A Role Assignment Method for Robot Soccer Based on Combinatorial Optimization, *China Robot Competition*, 31-33, Chang zhou .China, July 2005

Yang Lunbiao, Gao Yngyi(2006). *Fuzzy Mathematics:Principles and Applicatons(fourth edition)*, South China University of Technology Press,9787562304401,Guangzhou. China

Zhang Jie, Li Xiujuan, Zhang Xiaodong(2008). Application of Robot Soccer Role Assignment Based on Q Learning in Multi Agent Cooperation. *CD Tecnnology*, No.8 2008:45-47.1004-0447

Zhang Yu, Liu Shuhua(2008). Survey of multi-robot task allocation.*CAAI Transactions on Intelligent Systems*, Vol.3 No.2 Apr.2008:115-120.1673-4785

Zhong Biliang, Chen Chengzhi, Yang Yimin(2001). Study of Role Switching and assignment for Robot Soccer Based on Petri-net, *Computer Engineering and Applications*, Vol.37 No.20 Oct 2001:14-15,1002-8331

Zuo Hongtao, Lu Jinbo, Yao Jin(2009). Task Allocation Based on Market in Robot Soccer. *Journal of Harbin Institute of Technology*, Vol.41 Sup.2,July 2009:249-453. 0367-6234

Multi-Robot Systems: Modeling, Specification, and Model Checking

Ammar Mohammed and Ulrich Furbach
University of Koblenz-Landau, Department of Computer Science
Germany

Frieder Stolzenburg
Harz University of Applied Sciences, Department of Automation and Computer Sciences
Germany

1. Introduction

Specifying behaviors of physical multi-agent systems (MAS) – also called multi-robot systems – is a demanding task, especially when they are applied in safety critical applications. For this, formal methods based on mathematical models of the system under design are helpful. They allow us not only to formally specify the system at different levels of abstraction, but also to analyze the consistency of the specified systems before implementing them. The formal specification aims at both, a precise and unambiguous description of the behavior of MAS, and a formal verification whether a given specification is satisfied. For example, it should be possible to show that unsafe regions of the state space cannot be reached, or that a particular property is satisfied.
Generally, the behavior of an agent in MAS can be driven by external events and internal states. Therefore, an efficient way to model such systems is to use state transition diagrams, which are well established in software engineering. A state transition diagram describes the dynamic behaviour of an agent in terms of how the agent acts in certain scenarios of the

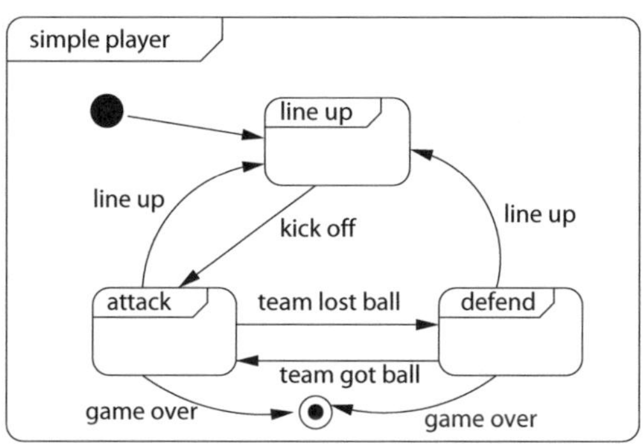

Fig. 1. A description of a simple agent in robotic soccer as a transition system.

environment. It aims at defining the behavior rules for the agents of the system. For example, Fig. 1 shows the behavior of an abstract simple agent/player in robotic soccer modeled as a state transition diagram. The agent may either *defend* or *attack*, depending on which team is controlling the ball. All changes of such global behaviors happen in response to one or more external events. Generally, state transition diagrams have been applied successfully for MAS, particularly in the RoboCup, a simulation of (human) soccer with real or virtual robots (cf. Arai & Stolzenburg, 2002; da Silva et al., 2004), in particular for the teams *RoboLog Koblenz* (two-dimensional simulation league) and *Harzer Rollers* (standard four-legged league) (Murray et al., 2002; Ruh & Stolzenburg, 2008).

In realistic physical environments, it is necessary to consider continuous actions in addition to discrete changes of the behaviors. Take for example, the movement of a soccer agent to kick off or to go to the ball, the process of putting out the fire by a fire brigade agent in a rescue scenario, or any other behaviors that depend on any timed physical law. Hybrid automata (Henzinger, 1996) offer an elegant method to model such types of behaviors. Basically, hybrid automata extend regular state transition diagrams with methods that deal with those continuous actions. The state transition diagrams are used to model the discrete changes of the agents' behavior, while differential equations are used to model the continuous changes. The semantics of hybrid automata make them accessible to a formal validation of systems, especially for those systems which are situated in safety critical environments. Model checking can be used to prove desirable features or the absence of unwanted properties in the specified systems (Clarke et al., 1999). Specifying and verifying behaviors of MAS by means of hybrid automata is challenging for many reasons. One of those is a state space problem: Essentially, MAS are specified as concurrent automata that have to be composed in parallel. The result of this composition captures all possible behaviors that may occur among the agents, which can be checked by hybrid automata verification tools (Behrmann et al., 2004; Frehse, 2005; Henzinger et al., 1995b). Obviously, this composition process may lead to a state explosion. Another problem is that hybrid automata describe not only the internal behaviors of agents, but also the external interaction among agents. This definitely adds complexity, which demands for structured and systematic methods for the specification of MAS. We propose to combine hybrid automata with software engineering methods to overcome these problems.

In this chapter, we provide a framework based on hybrid automata, which allows us to conveniently specify and verify physical MAS situated in a continuous dynamic environment. We will address the state space complexity raised from composition of agents, by composing the automata dynamically during the verification phase. This can relieve the problem in such a way that only the exact reached parts of the state space are activated, instead of activating all the entire state space at once.

Furthermore, in order to cope with complex multi-agent structures, we combine hybrid automata with hierarchical UML statecharts, which allows MAS specification with different levels of abstraction. We also give a formal semantics for this combination, and show how to analyze the dynamic behaviors of MAS. In principle, a straightforward way to analyze a hierarchical machines is to flatten them and to apply verification techniques to the resulting ordinary finite state machines. We show how this flattening can be avoided.

2. Hybrid Finite State Machines

Originally, hybrid automata (Henzinger, 1996) have been proposed as formal models for describing hybrid systems. They have been built as a generalization of timed automata (Alur & Dill, 1994), which have been used successfully as a standard framework to specify real-time systems. In addition to their mathematical models to formally specify and verify systems, the underlying mathematical models of hybrid automata can be represented graphically as a finite state machine (FSM). There are several approaches to apply this framework to MAS (see e.g. Egerstedt, 2000; Furbach et al., 2008; Mohammed & Furbach, 2008a).

In order to specify MAS by means of hybrid automata, the team of agents is described as concurrent automata, which in turn are combined via parallel composition into a global automaton, in order to coordinate their behaviors for reaching a common goal. It is well known that the major problem in applying model checking to analyze concurrent systems is the potential combinatorial explosion of the state space arising from parallel composition. Typically the state space of the parallel composition of an agent with K_1 states and another agent with K_2 states leads to a state space of $K_1 \times K_2$ states. Accordingly, the parallel composition of N agents, each with a state space of K states, leads to a state pace of K^N states. Even for small systems this may easily run out of control. Additionally, the state explosion problem is even more serious in verifying continuous dynamic systems. As such systems must satisfy certain timing and continuous constraints on their behaviors, a model checker must keep track not only of the part of the state space explored, but also of timing and continuous evolution information associated with each state, which is both time and space consuming. Traditionally, global state-space representations are constructed without regard to whether the states are reachable or not. In this section we will give a framework where the state space is built on the fly during the execution of the concurrent MAS. This can relieve the complexity in a sense that only the active parts of the state space will be taken into consideration during the run, instead of producing the composition prior to the verification phase.

In this section we will define the syntax and the semantics of our framework. Additionally, we will show how the composition of automata can be formally constructed. Finally, we will use constraint logic programming to implement the proposed framework. All this will be exemplified by a simple rescue scenario.

2.1 Rescue Scenario: Example

In the RoboCup rescue simulation league (Tadokoro et al., 2000), a large scale disaster is simulated. The simulator models part of a city after an earthquake. Buildings may be collapsed, or are on fire, and roads are partially or completely blocked. A team of heterogeneous agents consisting of police forces, ambulance teams, a fire brigade, and their respective headquarters is deployed. The agents have two main tasks, namely finding and rescuing the civilians and extinguishing fires. An auxiliary task is the clearing of blocked roads, such that agents can move smoothly. As their abilities enable each type of agent to solve only *one* kind of task (e.g. fire brigades cannot clear roads or rescue civilians), the need for coordination and synchronization among agents is obvious in order to accomplish the rescue tasks.

Now, consider the following simple scenario. If a fire breaks out somewhere, a fire brigade agent is ordered by its headquarters to extinguish the fire. The fire brigade agent moves to the fire and begins to put it out. If the agent runs out of water, it has to refill its tank at a supply station and return to the fire to fulfill its task. Once the fire is extinguished, the fire brigade agent is idle again.

An additional task of the agent is to report any injured civilians it discovers. In addition to the fire brigade agent, the model should include a fire station, fire, and civilians in the environment. A part of this scenario, specified as hybrid automata, is depicted in Fig. 2. The complete description and specification of the scenario will be shown in Sec. 3 (cf. Fig 4).

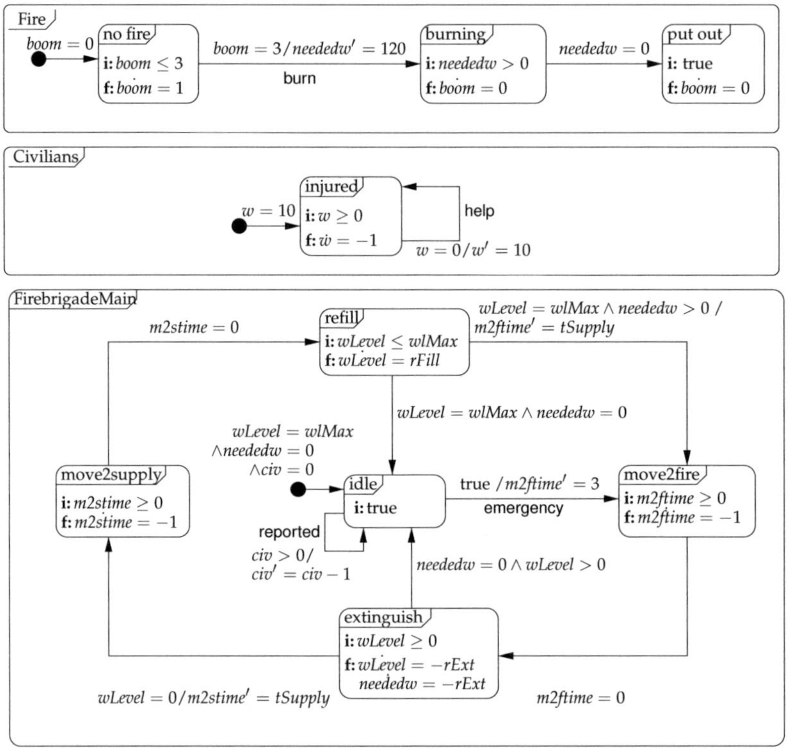

Fig. 2. A part of the RoboCup rescue scenario specified as hybrid automata.

As depicted in Fig. 2, the behavior of the agent *FirebrigadeMain* consists of five states corresponding to movements (*move2fire*, *move2supply*), extinguishing (*extinguish*), refilling the tank (*refill*), and an idle state (*idle*). It can report the discovered civilians when it is in its idle state. Details of this figure will be explained in details during this chapter.
It should be obvious that even in this simple case with very few components, it is difficult to see if the agent behaves correctly. Important questions like:

- Does the fire brigade agent try to extinguish without water?
- Will every discovered civilian (and only those) be reported eventually?

depend on the interaction of all components and cannot be answered without an analysis of the whole system.

2.2 Syntax

Since we intend to specify a multi-agent system with hybrid automata, the intuitive meaning of an agent is a hybrid automaton, which is represented graphically as a finite state machine, augmented with mathematical formalisms on both transitions and control states. Formally speaking, a hybrid automaton (agent with continuous actions) is defined as follows:

Definition 2.1 (basic components). *A hybrid automaton is a tuple* $H = (Q, X, Inv, Flow, E, Jump, Reset, Event, \sigma_0)$ *where:*

- *Q is a finite set of locations which defines the possible behaviors of the agent.*
  For example, in Fig. 2, the *FirebrigadeMain* agent has the locations *move2fire*, *move2supply*, *extinguish*, *refill*, and *idle* as possible behaviors. On the other hand, the *Fire* has *no fire*, *burning* and *put out* as its locations. It should be mentioned that we use the concept location instead of state, because an agent possesses different states inside each location, which are raised as a reason of continuous evolution. This will be described later in more details.

- *$X = \{x_1, x_2, ..., x_n\}$ is a finite set of n real-valued variables, including the variable t that denotes the time.*
  These variables will be used to model the continuous dynamics of the automaton with respect to t. For example, the variable *wLevel* represents the amount of water of the fire brigade, and it can be used to model the rate of change to refill or flow the water with respect to the time inside the tank. On the other hand, the variable *m2ftime* represents the distance to the fire, and its rate of change with respect to the time models the speed of the fire brigade agent.

- *Inv is a mapping which assigns an invariant condition to each location $q \in Q$. The invariant condition $Inv(q)$ is a predicate over the variables in X.*
  The control of a hybrid automaton will remain at a location $q \in Q$, as long as $Inv(q)$ holds. In the graphical representation, the invariant is tagged with the symbol *i:*. For instance the invariant $wlevel \leq wlMax$ inside the location *refill* of *FirebrigadeMain* shows that the fire brigade fills the water as long as the water level does not reach the maximum level represented by the *wlMax*. Conventionally, writing $Inv(q)[v]$ means that the invariant condition inside the location q holds, whenever the valuations of variables inside q are v.

- *$Flow$ is a mapping, which assigns a flow condition to each control location $q \in Q$. The flow condition $Flow(q)$ is a predicate over X that defines how the variables in X evolve over the time t at location q.*
  In the graphical representation, it is tagged with the symbol *f:*. A flow of a variable x is denoted as $\dot{x}$. In our example, the dotted variable $w\dot{L}evel$ describes the change of the water level in the location *refill*. The flow inside locations may be empty and hence omitted, if nothing changes continuously in the respective location.

- *$E \subseteq Q \times Q$ is the discrete transition relation over the control locations.*

- *$Jump$ is a mapping which assigns a jump condition (guard) to each transition $e \in E$. The jump condition $jump(e)$ is a predicate over X that must hold to fire e.*
  Omitting a jump condition on a transition means that the jump condition is always true and it can be taken at any point of time. In the rescue example Fig. 2, the jump condition between the locations *extinguish* and *move2supply* is given as *wLevel=0*, which means that the transition between these locations can be taken whenever *wLevel* reaches to 0. Conventionally, writing $Jump(e)[v]$ means that the jump condition on a transition e holds, when the valuations of variables on the transition are v.

- *$Reset$ is a mapping, which assigns values to variable to each transition $e \in E$. $Reset(e)$ is a predicate over X that defines how the variables are reset.*
  In the graphical representation, resetting a variable $x \in X$ is denoted as x'. For example, when the transition between location *refill* and *move2fire* holds, the action $m2ftime' = tSupply$ is executed, which means that the variable *m2ftime* is reset to the value *tSupply*.

Resetting variables are omitted on transition, if the values of the variables do not change before the control goes from a location to another.

- *Event is a mapping which assigns an event to each transition $e \in E$ from a set of events $Event_H$.* For instance, the transition between the locations *idle* and *move2fire* in *FirebrigadeMain* has *emergency* as its event. As we will see later, $Event_H$ is used to synchronize the automaton H with any other automata that share the same common events. It should be noted that in the graphical diagrams, an event $Event(e) \in Event_H$ is implicitly omitted, if it is not shared among any automata.
- *σ_0 is the initial state of the automaton. It defines the initial location together with the initial values of the variables X.*
  For example, the initial state of the agent *FirebrigadeMain* is the location *idle* with initial valuations $wLevel = wlMax$, $neededw = 0$, and $civ = 0$ to its variables $wLevel$, $neededw$, and civ respectively.

Before describing the semantics of a hybrid automaton, it should be mentioned that the hybrid automata are classified according to the type of continuous flow:

- If $\dot{x} = c$ (constant), then the hybrid automaton is called *linear* hybrid automaton. A special case of linear hybrid automata are *timed* automata (Alur & Dill, 1994), where $c = 1$).
- If $c_1 \leq \dot{x} \leq c_2$, then the hybrid automaton is called *rectangular* hybrid automaton.
- If $\dot{x} = c_1 x + c_2$, then the hybrid automaton is called *non-linear* hybrid automaton.

2.3 Semantics

Intuitively, a hybrid automaton can be in exactly one of its control locations at each stage of it computation. However, knowing the present control location is not enough to determine which of the outgoing transitions can be taken next, if any. A snapshot of the current state of the computation should also remember the present valuation of the continuous variables. Therefore, to formalize the semantics of a hybrid automaton, we first have to define the concept of a *state*.

Definition 2.2 (State). *At any instant of time, a state of a hybrid automaton is given by $\sigma_i = \langle q_i, v_i, t \rangle$, where $q_i \in Q$ is a control location, v_i is the valuation of its real variables, and t is the current time. A state $\sigma_i = \langle q_i, v_i, t \rangle$ is admissible if $Inv(q_i)[v_i]$ holds (i.e., the valuations of the variables satisfies the invariant at location q_i.*

A state transition system of a hybrid automaton H starts with the *initial state* $\sigma_0 = \langle q_0, v_0, 0 \rangle$, where the q_0 and v_0 are the initial location and valuations of the variables respectively. For example, the initial state of the *Civilians* (see Fig. 2) can be specified as $\langle injured, 10, 0 \rangle$.
Informally speaking, the semantics of a hybrid automaton is defined in terms of a labeled transition system between states. Generally, transitions between states are categorized into two kinds of transitions: continuous transitions, capturing the continuous evolution of states, and discrete transitions, capturing the changes of location. More formally, we can define hybrid automaton semantics as follows.

Definition 2.3 (Operational Semantic). *A transition rule between two admissible states $\sigma_1 = \langle q_1, v_1, t_1 \rangle$ and $\sigma_2 = \langle q_2, v_2, t_2 \rangle$ is*

discrete *iff* $e = (q_1, q_2) \in E$, $t_1 = t_2$ *and* $Jump(e)[v_1]$ *and* $Inv(q_2)[v_2]$ *hold. In this case an event* $a \in Event$ *occurs. Conventionally, we write this as* $\sigma_1 \xrightarrow{a} \sigma_2$.

continuous(time delay) *iff* $q_1 = q_2$, *and* $t_2 > t_1$ *is the duration of time passed at location* q_1, *during which the invariant predicate* $Inv(q1)[v_1]$ *and* $Inv(q1)[v_2]$ *holds.*

In the previous definition, it should be noted that v_2 results from resetting variables on a transition in case of the discrete transition rule, while it results from the continuous evolution of the variables in case of the continuous transition rule. Intuitively, an execution of a hybrid automaton corresponds to a sequence of transitions from a state to another. Therefore we define the valid run as follows.

Definition 2.4 (Run: micro level). *A run of a hybrid automaton* $\sum = \sigma_0\sigma_1\sigma_2, \ldots,$ *is a finite or infinite sequence of admissible states, where the transition from a state* σ_i *to a state* σ_{i+1} *is related by either a discrete or a continuous transition and* σ_0 *is the initial state.*

It should be noted that the continuous change in the run may generate an infinite number of reachable states. It follows that state-space exploration techniques require a symbolic representation way in order to represent the set of states in an appropriate way. A good way is to use mathematical intervals. This interval captures all possible states. We call this interval *region*, which is defined as follows:

Definition 2.5 (Region). *given a run* $\sum$, *a sub-sequence of states* $\Gamma = (\sigma_{i+1} \cdots \sigma_{i+m}) \subseteq \sum$ *is called a region, if for all states* σ_{i+j} *with* $1 \leq j \leq m$, *it holds* $q_{i+j} = q$ *and if there exist a state* σ_i *and a state* σ_{i+m+1} *with respective locations* q_1 *and* q_2, *then it must hold* $q_1 \neq q$ *and* $q_2 \neq q$. *Conventionally, a region* Γ *is written as* $\Gamma = \langle q, V, T \rangle$, *where* $t_{i+1} \leq T \leq t_{i+m}$ *is the interval of continuous time, and* V *is the set of intervals* V_k *of the interval defined by the values of* $x_k \in X$ *in the time interval* T. *A region* Γ *is called admissible if each state* $\sigma \in \Gamma$ *is admissible.*

The previous definition reveals that a region captures the possible states that can be reached using continuous transitions in each location $q \in Q$. Therefore, T represents the continuous reached time. Additionally, a region captures the continuous values for each variable $x_i \in X$. These continuous values can be represented as an interval V_i of real values. Therefore, $V = \{V_1, V_1, ..., V_n\}$ represents a set of intervals of the variables in X. Now, the run of hybrid automata can be rephrased in terms of reached regions, where the change from one region to another is fired using a discrete step.

Definition 2.6 (Run: macro level). *A run of hybrid automaton* H *is* $\sum_H = \Gamma_0\Gamma_1, ...,$ *a sequence of (possibly infinite) admissible regions, where a transition from a region* Γ_i *to a region* Γ_{i+1} *is enabled (written as* $\Gamma_i \xrightarrow{a} \Gamma_{i+1}$), *if there is* $\sigma_i \xrightarrow{a} \sigma_{i+1}$, *where* $\sigma_i \in \Gamma_i$, $\sigma_{i+1} \in \Gamma_{i+1}$ *and* $a \in Event$ *is the generated event before the control goes to the region* Γ_{i+1}. Γ_0 *is the initial region obtained from a start state* σ_0 *by means of continuous transitions.*

The operational semantics is the basis for verification of a hybrid automaton. In particular, model checking of a hybrid automaton is defined in terms of the reachability analysis of its underlying transition system. The most useful question to ask about hybrid automata is the reachability of a given state. Thus, we define the reachability of states as follows.

Definition 2.7 (Reachability). *A region* Γ_i *is called reachable in* $\sum_H$, *if* $\Gamma_i \subseteq \sum_H$. *Consequently, a state* σ_j *is called reachable, if there is a reached region* Γ_i *such that* $\sigma_j \in \Gamma_i$

The classical method to compute the reachable states consists of performing a state space exploration of a system, starting from a set containing only the initial state and spreading the reachability information along control locations and transitions until fixed regions are obtained. Stabilization of a region is detected by testing, whether the current region is included in the union of the reached regions obtained in previous steps. It is worth mentioning that checking reachability for hybrid automata is generally undecidable. However, under various constraints, reachability is decidable for certain classes of hybrid automata including timed and initialized rectangular automata (Henzinger et al., 1998). A rectangular automaton is initialized if each continuous variable is reset every time a discrete transition is taken.

2.4 State Machine Composition

For the specification of complex systems, we extend hybrid automata by parallel composition. Basically, the parallel composition of hybrid automata can be used for specifying larger systems (multi-agent systems), where a hybrid automaton is given for each part of the system, and communication between the different parts may occur via shared variables and synchronization labels. Technically, the parallel composition of hybrid automata is obtained from the different parts using a product construction of the participating automata. The transitions from the different automata are interleaved, unless they share the same synchronization label. In this case, they are synchronized on transitions. As a result of the parallel composition a new automaton called composed automaton, is created, which captures the behavior of the entire system. In turn, the composed automata are given to a model checker that checks the reachability of a certain state.

It is of advantage to do this during the verification process, instead of constructing the parallel composition before involving in the verification phase. Intuitively, the composition of hybrid automata H_1 and H_2 can be defined in terms of synchronized or interleaved regions of the regions produced from run of both H_1 and H_2. As a result from the composition procedure, compound regions are constructed, which consists of a conjunction of a region $\Gamma_1 = \langle q_1, V_1, T \rangle$ from H_1 and another region $\Gamma_2 = \langle q_2, V_2, T \rangle$ from H_2. Therefore, each compound region takes the form $\Lambda = \langle (q_1, V_1), (q_2, V_2), T \rangle$ (shortly written as $\Lambda = \langle \Gamma_1, \Gamma_2, T \rangle$), which represents the reached region at both control locations q_1 and q_2 the during a time interval T. Now the run of composed automata can be defined as the sequence $\sum_{H_1 \circ H_2} = \Lambda_0, \Lambda_1, ...$ of compound regions, where a transition between compound regions $\Lambda_1 = \langle \Gamma_1, \gamma_1, T_1 \rangle$ and $\Lambda_2 = \langle \Gamma_2, \gamma_2, T_2 \rangle$ (written as $\Lambda_1 \xrightarrow{a} \Lambda_2$) is enabled, if one of the following holds:

- $a \in Event_{H_1} \cap Event_{H_2}$ is a joint event, $\Gamma_1 \xrightarrow{a} \Gamma_2$, and $\gamma_1 \xrightarrow{a} \gamma_2$. In this case , we say that the region Γ_1 is synchronized with the region γ_1.
- $a \in Event_{H_1} \setminus Event_{H_2}$ (respectively $a \in Event_{H_2} \setminus Event_{H_1}$), $\Gamma_1 \xrightarrow{a} \Gamma_2$ and $\gamma_1 \rightarrow \gamma_2$, such that both γ_1 and γ_2 have the same control location (i.e., they relate to each other using a continuous transition).

The previous procedures give the possibility to construct the composition dynamically during the run/verification phase. Obviously, as it has been said, computing the composition in such a way is advantageous. This is because only the active parts of the state space will be taken into consideration during the run instead of producing the composition procedure prior to the verification phase. This can relieve the state space problem raised from modeling multi-agent systems.

In the following, we show how the previous procedure can be performed with the help of constraint logic programming.

2.5 Constraint-Based Modeling

In Mohammed & Furbach (2009) we showed how to encode the syntax and semantics of hybrid automata, described previously, as a constraint logic program (CLP) (Jaffar & Lassez, 1987). A primary version of this model was given by Mohammed & Furbach (2008b). There are diverse motivations beyond choosing *CLP* as a modeling prototype to implement the framework. Firstly, hybrid automata can be described as a constraint system, where the constraints represent the possible flows, invariants, and transitions. Secondly, constraints can be used to characterize certain parts of the state space (e.g., the initial states or a set of unsafe states). Further, there are close similarities in operational semantics between *CLP* and hybrid automata. Ideally, state transition systems can be represented as a logic program, where the set of reachable states can be computed. Moreover, constraints enable us to represent infinite states symbolically as a finite interval. For instance, the infinite states can be handled efficiently as an interval constraint that bounds the set of infinite reachable state as a finite interval (i.e., $0 \leq X \leq 250$). Hence, a constraint solver can be used to reason about the reachability of a particular state inside this interval. A further motivation to choose *CLP* is its enrichment with many efficient constraint solvers of various domains. For example, *CLP* contains a constraint solver over real interval constraints, which can be used to represent the continuous flows as constraint relations to the time, as well as to reason about a particular valuation. On the other hand *CLP* contains a constraint solver over symbolic domains, which are appropriate to represent the synchronization events (communication messages) among agents. Last but not least, by employing *CLP* the automata composition can be constructed on the fly (during models checking). This can be done by investigating the constraints appeared during running models. In turn, the previous can relieve the state space problem raised from specifying MAS.

Our implementation prototype was built using ECLiPSe Prolog (Apt & Wallace, 2007). The prototype follows the definitions of both the formal syntax and semantics of hybrid automata, which are defined in the previous section. To start implementing a hybrid state machine, we primarily begin with modeling the locations and their constraints (e.g. flows, invariants), which are modeled as the predicate *automaton* as follows:

```
%%% automaton(+Location,?Vars,+Vars0,+T0,?Time)
%%% models invariant and flow inside location
automaton(Location,Vars,Vars0,T0,Time):-
        Flow(Vars),
        Inv(Vars),Time $>=T0.
```

Here, *automaton* is the name of the automaton itself, and *Location* represents the actual name of the current locations of the automaton. *Vars* is a list of real variables participating in the automaton, whereas *Vars0* is a list of the corresponding initial values. $Inv(Vars)$ is the list of invariant constraint on *Vars* inside the location. The constraint predicate $Flow(vars)$ models the continuous flows of the variables *Vars* with respect to time *T0* and *Time*, given initial values *Vars0* of the variables *Vars* at the start of the flow. *T0* is the initial time at the start of the continuous flow. As it has been described in Subsection 2.2, a hybrid automaton is classified according to the constraints on the continuous flow. Therefore, $Flow(Vars)$ is represented in terms of constraints as $Vars = Var0 + c \cdot (Time - T0)$ in case of a linear hybrid automaton, as $Var0 + c \cdot (Time - T0) \leq Vars \leq Var0 + c \cdot (Time - T0)$ in case of a rectangular hybrid automaton, and as $Vars = Var0 - c2/c1 + c2/c1 \cdot \exp(c1 \cdot (Time - T0))$ in case of a non-linear hybrid automaton. Here, $(Time - T0)$ models the delay inside the location. It should be noted that after executing the predicate *automaton*, *Vars* and *Time* holds the reached valuations of the variables together with the reached time respectively. The following is an example showing the concrete implementation of the location *injured* in the automaton *Civilians* Fig. 2. The $ symbol in front of the (in)equalities is the constraint relation for interval arithmetic constraints (library *ic* in ECLiPSe Prolog).

```
civilians(injured,[W],[W0],T0,Time):-
     W $= W0-(Time-T0),
     W $>=0, Time $>=T0.
```

According to operational semantics defined in Def. 2.3, a hybrid automaton has two kinds of transitions: *continuous* transitions, capturing the continuous evolution of variables, and *discrete* transitions, capturing the changes of location. For this purpose, we encode transition systems into the predicate *evolve,* which alternates the automaton between a discrete and a continuous transition. The automaton evolves with either discrete or continuous transitions according to the constraints appearing during the run.

```
%%% evolve(+Automaton,+State,-Nextstate,+T0,+Time,?Event)
evolve(Automaton,(L1,Var1),(L2,Var2),T0,Time,Event) :-
     continuous(Automaton,(L1,Var1),(L1,Var2),T0,Time,Event);
     discrete(Automaton,(L1,Var1),(L2,Var2),T0,Time,Event).
```

When a *discrete* transition occurs, it gives rise to updating the initial variables from $Var1$ into $Var2$, where $Var1$ and $Var2$ are the initial variables of locations $L1$ and $L2$ respectively. Otherwise, a delay transition is taken using the predicate *continuous*. It is worth noting that there are infinite states due to the continuous progress. However, this can be handled efficiently as an interval constraint that bounds the set of infinite reachable state as a finite interval (i.e., $0 \leq X \leq 250$).

In addition to the variables, each automaton is augmented with a set events called $Event_{Automaton}$. An example of this set of events of the automaton *FirebrigadeMain* is denoted as $\{reported, emergency\}$. For this reason, each transition is augmented with the variable *Event*, which is used to define the parallel composition from the automata individuals sharing the same event. The variable *Event* ranges over symbolic domains and guarantees that whenever an automaton generates an event, the corresponding synchronized automata have to be taken into consideration simultaneously. It should be mentioned that the declaration of automata events must be provided in the modeling example. For instance, the declaration of the possible events domain of Fig. 2. is coded as follows :

```
:- local domain(events(emergency,reported,hlep,burn)).
```

This means that the domains of events are declared symbolically to capture the set of all possible events applicable to the underlying modeled system. The appropriate solver of a symbolic domain deals with any defined constraints in terms of the declared domains. Now after defining the domains of events, a variable of type *events* can be declared as follow:

```
Event &:: events, Event &= domain_value.
```

The variable *Event* is declared with domain values defined by *events,* and is initialized with a specific value from its domain. The & symbol is a constraint relation for symbolic domains (library *sd* in ECLiPSe Prolog).

The following is the general implementation of the predicate *discrete,* which defines transitions between locations.

```
%%% driver(+State1,+State2,...,+Staten,+T0,-Regions,+PastRegion).
%%% perform composition and reachability
driver((L1,Var01),(L2,Var02),...,(Ln,Var0n),T0,[Reg|NxtReg],PastReg) :-

     automaton1(L1,Var1,Var01,T0,Time),
     automaton2(L2,Var2,Var02,T0,Time),
     ... ,
     automatonn(Ln,Varn,Var0n,T0,Time),

     evolve(automaton1,(L1,Var01),(NxtL1,Nvar01),T0,Time,T,Event),
     evolve(automaton2,(L2,Var02),(NxtL2,Nvar02),T0,Time,T,Event),
     ... ,
     evolve(automatonn,(Ln,Var0n),(NxtLn,Nvar0n),T0,Time,T,Event),

    \+ member((L1,L2,..,Ln,Var1,Var2,..,Varn,_,Event), PastReg),
    Reg = (L1,L2,..,Ln,Var1,Var2,..,Varn,Time,Event),
    NpastReg =[Reg|PastReg],

driver((NxtL1,Nvar01),(NxtL2,Nvar02),...,(NxtLn,Nvar0n),T,NxtReg,NpastReg).
```

Fig. 3. A state machine to drive the execution of automata.

```
%%% discrete(+Automaton,+State1,-State2,+IntTime,-Time,-Event)
discrete(Automaton,(Loc1,Var1),(Loc2,Var2),T0,Time,Event):-
         automaton,(Loc1,Var1,Var,T0,Time),
         jump(Var), reset(Var2),
         Event &::events,Event &=domain_value.
```

In the previous predicate, *domain_value* must be a member in $Event_{Automaton}$. Here, when the *discrete* predicate is fired, the automaton generates an event by constraining the variable *Event* to the suitable value from its domain.

The following is an instance showing the concrete implementation of the *discrete* predicate between locations *no fire* and *burning* in automaton *fire*.

```
discrete(fire,(no_fire,[B0,N0]),(burning,[BB0,NN0]),T0,Time,Event):-
         fire(no_fire,[B0,N0],[BB0,NN0],T0,Time),
         BB0 $=3, NN0 $=120,
         Event &::events, Event &=burn.
```

Once the locations and transition rules have been modeled, a state machine needs to be implemented in order to execute the model. For this purpose, a driver program is implemented as shown in Fig. 3.

The *driver* is a state machine that is responsible to generate and control the behaviors of the concurrent hybrid automata, as well as to provide the reachable regions symbolically. The *driver* takes the starting state for each participating automaton (i.e. a control location as input argument as well as the list of initial valuations of the variables). In addition, it takes the starting time *T0* as begin of the execution, followed by the list of reached regions, which is needed for the purpose of the verification. It should be noted that during the course of the execution of the driver, there is a symbolic domain variable *Event* shared among automata, which is used by the appropriate solver to ensure that only one event is generated at a time. Precisely when an automaton generates an event, due to a discrete transition of one of the predicates *evolve* of the concurrent automata, the symbolic domain solver will exclude all the

domain of values of the other automata that are not coincident with the generated event. This means that only one event is generated at a time. If more than one automaton generates different events at the same point of time, then the symbolic domain solver will handle only one of them at a time, but the other events will be handled using backtracking.
Since each automaton generates an event by a discrete step at the end of its continuous evolution, then the precedence of events that appear during the run is important to both composition and the verification process. For this reason, an obvious way to deal with this precedence is to use constraints on the time of the generated events. To accomplish this, we constraint the execution of each automaton with a shared variable *Time*. The constraint solver, in turn, binds this variable with the minimum execution time among the automata. It follows that this variable *Time* eventually holds the minimum time needed to generated an event. The previous computation partitions the state space into regions, where the transition from one region to another depends on the minimum time needed to generate an event. Consequently, this shows how the automata composition can be implicitly constructed efficiently on the fly (i.e. during the computation).
It has been said that we are not only concerned with running and composing the automata, but also with the their verification. For this purpose, the *driver* is augmented with the list of reached compound regions. At each step of the execution of the *driver* execution, a compound region, of the form $\langle locations, Variables, Time, Event \rangle$ is added to the list of reached regions. This region symbolically represents the set of reached states and times to each control location as mathematical constrains. Additionally, each region contains the generated event before the control goes to another region using a discrete step. Technically, the *driver* computes the set of reached regions until fixed regions are obtained. This is computed by checking, in each iteration of *driver*, if the reached region is not contained in the list of the previously reached regions. For this purpose, the last argument of the *driver* holds for the list of these regions. Due to the undecidability of hybrid automata (Henzinger et al., 1998), the termination of the driver to reach to a fixed regions is not guaranteed generally. To overcome the non termination problem, we augment the predicate *driver* with a depth limit, by which the driver is enforced to stop upon reaching a given depth.
Reachable regions should contain only those variables, which are important for the verification of a given property. Therefore, the last argument list of the predicate *driver* can be expanded or shrunk as needed to contain the significant variables.
As soon as the *driver* has been built, the complete model should be invoked for the purpose of execution and hence verification. For this reason, the predicate *runmodel* is implemented to invoke the *driver* with the initial states of the hybrid automata. An example showing how to query the driver on the running scenario (see Fig. 2) takes the form:

```
runmodel(Reached) :-
      driver((idle,[wlMax,0,0]),(injured,[10]),(no_fire,0),0,Reached,[]).
```

2.5.1 Verification as Reachability Analysis

Now we have an executable constraint-based specification, which can be used to verify properties of a multi-agent system. In particular, one can check properties on states using reachability analysis. For this we have two basic steps. Firstly, we compute the state space of the automaton under consideration by using the predicate *driver*. Secondly, we search for states

that satisfy or contradict given properties. This is done with the help of standard Prolog predicates like *member*/2 and *append*/3. Thus it is possible to implement our entire framework by some very simple Prolog rules.
In terms of *CLP*, a state is reached iff the constraint solver succeeds in finding a satisfiable solution for the constraints representing the intended state. In other words, assuming that *Reached* represents the set of all reachable states computed by the *CLP* model from an initial state, then the reachability analysis can be generally specified, using *CLP*, by checking whether *Reached* $\models \Psi$ holds, where Ψ is the constraint predicate that describes a property of interest. In practice, many problems to be analyzed can be formulated as a reachability problem. For example, a safety requirement can be checked as a reachability problem, where Ψ is the constraint predicate that describes forbidden states, and then the satisfiability of Ψ wrt. *Reached* is checked. An example would be to check that the state where the fire can be put out is reached. The following very simple *CLP* query gives us the answer *yes*:

```
?- runmodel(L),
member((_firebrigade,_civilian,Fire,_var1,_var2,_var3,_time,_event),Reached),
Fire $=put_out.
```

Other properties concerning the reachability of certain states can be verified similarly. Additionally, constraint solvers can be used to reason about the reachability of interesting properties within a region, like properties of the variables that model the continuous dynamics of a model. For example, we can reason about the water level of the firebrigade after putting out the fire.
Mohammed & Furbach (2009) provide various verification rules based on reachability analysis. For example, finding the time delay between events is possible within the framework. This is because both the events and time are recorded at reached regions. Another example is to find a condition on a certain variable, which is necessary to to reach a particular state. We also did some experiments on a set of benchmarks taken from the domain of hybrid automata. The experiments have been compared with HyTech (Henzinger et al., 1995a). HyTech was chosen as a reference tool, because it is one of the most well-known tools for the verification of hybrid automata, and it tackles verification similarly based on reachability analysis. In HyTech, however, the automata working in parallel are composed before they are involved in the verification phase.
The experimental results revealed that our framework has a slight advantage wrt. In terms of the run-time of checking the properties of the benchmarks. With respect to the expressiveness, our approach is more powerful, because HyTech can not deal directly with non-linear hybrid automata. The continuous dynamics of non-linear hybrid automata have to be approximated in a linear form, before applying the model checking. Additionally, HyTech cannot verify simple properties that depend on the occurrence of events – i.e. checking the reachability of the event *help* –, despite of the fact that events are used to synchronize the automata. HyTech is able to verify time properties of events; however, this can be checked only after augmenting the original automata with an extra automaton. Its functionality is to observe the model without changing its behavior and to record the time of occurring events. In contrast to our framework, verifying this type of properties can be checked without any extra automaton, since the events and time are recorded in the reached regions. For further details about the experimental results, the reader is referred to Mohammed & Furbach (2009).

3. Hybrid Statecharts

So far, we have used hybrid Finite State Machines (FSMs) to specify and verify a group of agents. Unfortunately, classical FSMs lack support for modularity, which is very important when modeling complex systems that contain similar subsystems. All states are equally visible and are considered to be at the same level of abstraction, which makes modeling cluttered and unreadable. In practice, to describe complex systems using FSMs, several extensions are useful. Statecharts have been introduced by Harel (1987) to overcome the limitations of traditional FSM. The most important extension is hierarchy, or what is called *hierarchical (nested) FSM*. Such a hierarchy has descriptive advantages over ordinary FSM in a sense that hierarchy of states offers a convenient structuring mechanism that allows us to specify systems with different levels of view. For their expressiveness, statecharts have become part of the Unified modeling language (UML) (UML, 2009).
The main purpose of statecharts has been the description of complex reactive systems. However, in order to cope with those reactive systems that exhibit continuous timed behaviors, it seems to be advantageous to extend statecharts with continuous actions inside states. This extension allows complex/multi-agent systems to be modeled with different levels of abstraction and provides a formal way to analyze the dynamical behavior of the modeled systems. There are two possibilities of combination, namely combining statecharts with differential equations or extending hybrid automata with hierarchy. Therefore, both terms hierarchical hybrid automata (HHA) and hybrid statecharts can be used interchangeably.
Basically, the straightforward way to analyze hierarchical machines is to flatten them and apply model checking tools on the resulting ordinary FSM. For example, Möller et al. (2003) have presented hierarchical specification of timed automata. In order to verify a hierarchical model, it has to be transformed first to flat timed automata, which in turn can be used as input for the model checker tool UPPAAL (Behrmann et al., 2004). Similarly, Ruh (2007) has presented a translator tool that automatically converts hybrid hierarchical statecharts, defined as an ASCII-formatted specification, into an input format for a model checker of simple hybrid automata (Henzinger et al., 1995a). In this section, we show, how hierarchical hybrid automata can be analyzed without getting involved in the flattening process.

Let us come back to the illustrative RoboCup rescue scenario given in Sec. 2.1. Suppose that the specification of the fire brigade agent consists of the main control structure (*FirebrigadeMain*) and a rescue sub-system (*FirebrigadeRSS*) which are supposed to run in parallel. The latter just records the detected civilians. In addition to the fire brigade, the model should include a fire station, whose responsibility to inform and assign a fire brigade to a fire as soon as a fire alarm received. Now let us describe the scenario in a hierarchical way. At the top level is the rescue scenario, which in turn comprises at the lower level *Fire, Civilians, Firestation,* and *Firebrigade*. The latter can be described in a further lower level, which is *FirebrigadeMain*, and *FirebrigadeRSS*. The specification of this hierarchical structure is shown in Fig. 4. In the following, the hierarchical specification will be described in a formal flavor.

3.1 Formal Hierarchy

As illustrated by Fig. 4, for hierarchical hybrid automata (HHA), locations are generalized into a set Q of locations, which is partitioned into three disjoint sets: Q_{simple}, Q_{comp}, and Q_{conc} — called *simple*, *composite* and *concurrent* locations. There is one designated start state, which is the topmost location in the hierarchy. In essence, the locations of plain hybrid automata correspond to simple location in the hybrid FSM. Composite and concurrent locations belong

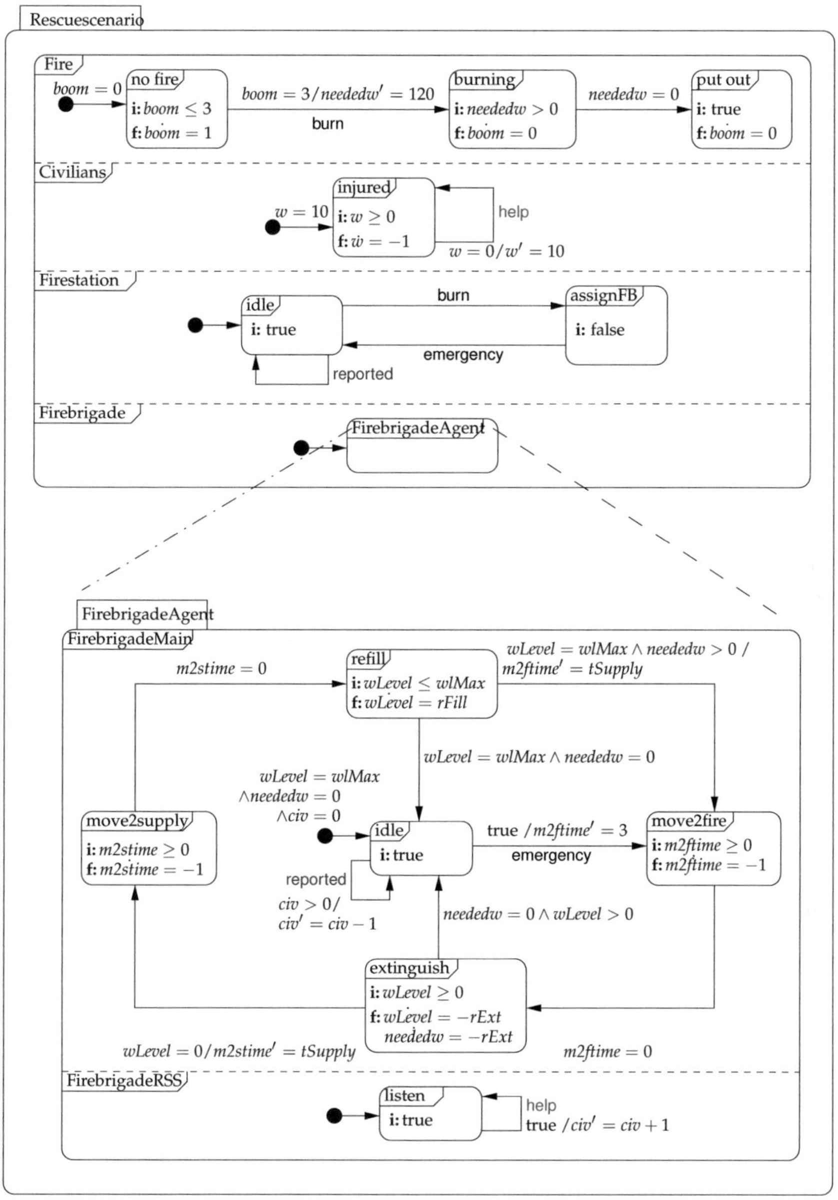

Fig. 4. A hierarchical hybrid state machine for a RoboCup rescue scenario.

to the definition of statecharts (Harel, 1987) and have become part of UML (UML, 2009). They are useful for expressing the overall system on several levels of abstraction and multi-agent aspects, respectively. Events are treated as global variables in this context. Based on this, we will now introduce the concepts of HHA more formally (Furbach et al., 2008).

Definition 3.1 (Hierarchy components)**.** *The basic components of* HHA *are the following disjoint sets:*

Q : *a finite set of locations, which is partitioned into three disjoint sets: Q_{simple}, Q_{comp}, and Q_{conc} — called simple, composite and concurrent locations, containing one designated* start state $q_0 \in Q_{comp} \cup Q_{conc}$.

In the rescue example (Fig. 4), *idle*, *extinguish* or *listen* are simple locations, and *FirebrigadeAgent* is a concurrent location and *FirebrigadeMain* and *FirebrigadeRSS* are composite locations, which are separated by a dashed line. *m2ftime* and *wLevel* are examples for real valued variables.

Definition 3.2 (Location hierarchy). *Each location q is associated with zero, one or more* initial locations $\alpha(q)$: *a simple location has zero, a composite location exactly one, and a concurrent location more than one initial location. Moreover, each location $q \in Q \setminus \{q_0\}$ is associated to exactly one superior state $\beta(q)$. Therefore, it must hold $\beta(q) \in Q_{conc} \cup Q_{comp}$. A concurrent state must not directly contain other concurrent ones and all transitions (q_1, q_2) must keep to the hierarchy, i. e. $\beta(q_1) = \beta(q_2)$.*

For the example in Fig. 4, according to the previous Def. 3.2, it holds e.g.:

α(*Civilian*) =injured.
α(*FirebrigadeAgent*)={ FirebrigadeMain,FirebrigadeRSS}.
α(*Rescuescenario*)={Fire,Civilian,Firestation,Firebrigade}
β(*burning*) = Fire.
α(*Firebrigade*) = FirebrigadeAgent.
α(*Fire*) = no_fire.
α(*Firestantion*) = idle.
β(*Fire*) = Rescuescenario.

The function β from the previous definition naturally induces a location tree with q_0 as root. This is shown for the running example in Fig. 5. While processing, each composite location of the state machine contains only one active location. These locations also form a tree, called configuration. A configuration of the given state machine, is indicated by the thick lines in Fig. 5. Let us now define the notion configuration more formally.
As shown in Def. 2.3, a hybrid automaton may change in two ways: *discretely*, from location q_1 to another location q_2, when the transition $e \in E$ between the two locations is enabled (i.e., the jump condition holds) and *continuously* within a control location $q \in Q$, by means of a finite (positive) time delay t. The semantics of our automata can now be defined by alternating sequences of discrete and continuous steps between configurations.

Definition 3.3 (Semantics). *The state machine starts with the* initial configuration, *i.e. the completed topmost initial state s_0 of the overall state machine. In addition, an initial condition must be given as a predicate with free variables from X. The current* situation *of the whole system can be characterized by a triple (c, v, t) where c is a configuration, v a valuation (i. e. a mapping $v : X \to \mathbb{R}^n$), and t the current time. The* initial situation *is a situation (c, v, t) where c is the initial configuration, v satisfies the initial condition, and $t = 0$. The following steps are possible in the situation (c, v, t):*

discrete step: *a discrete/micro-step from one configuration c of a state machine to a configuration (c', v', t) by means of a transition $(q, q') \in E$ with some jump condition in the current situation (written $c \to c'$) is possible iff:*

1. *c contains a node labeled with q;*
2. *the jump condition of the given transition holds in the current situation (c, v, t);*
3. *c' is identical with c except that q together with its sub tree in c is replaced by the completion of q';*
4. *the variables in X are set by executing specific assignments.*

```
complete(T,Rest,State,[State:Var|Complete]) :-
        init(T,State,[Var|Rest],Init,_),
        maplist(complete(T,[Var|Rest]),Init,Complete).

discrete(T,Rest1,Rest2,[State1:Var1|_],[State2:Var2|Conf]) :-
        trans(T,State1,[Var1|Rest1],State2,[Var2|Rest2]),
        complete(T,Rest2,State2,[State2:Var2|Conf]).
discrete(T,Rest1,Rest2,[Top:Var1|Sub],[Top:Var2|Tree]) :-
        Sub \= [],
        maplist(discrete(T,[Var1|Rest1],[Var2|Rest2]),Sub,Tree).

continuous(T1,T2,Rest1,Rest2,[State:Var1|Sub],[State:Var2|Tree]) :-
        flow(T1,T2,State,[Var1|Rest1],[Var2|Rest2]),
        maplist(continuous(T1,T2,[Var1|Rest1],[Var2|Rest2]),Sub,Tree).
```

Fig. 6. Code for the abstract state machine for HHA in CLP. The `Rest` variables host nested lists of the variables declared in the states superior to the current state. The built-in predicate `maplist` is a macro for applying a predicate call (first argument of `maplist`) to a list of arguments (second and third argument) one by one.

continuous step: *a continuous step/flow within the actual configuration to the situation (c, v', t') requires the computation of all $x \in X$ that are valid in c at the time t' according to the conjunction of all state conditions (i.e. flow conditions plus invariants) of the active locations $q \in c$, where it must hold $t' > t$.*

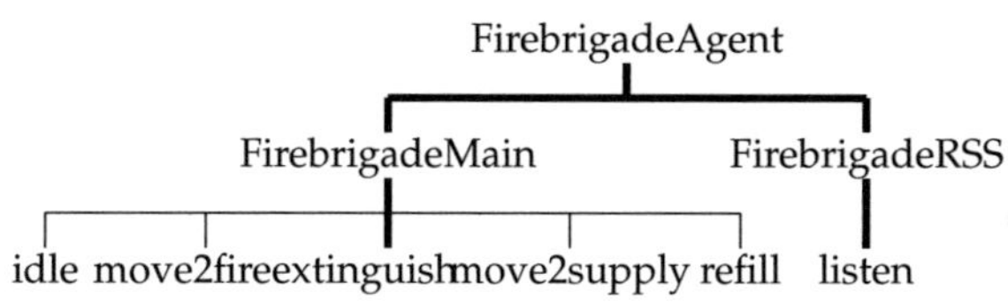

Fig. 5. Location hierarchy and configuration tree (thick lines).

Definition 3.4 (Configuration and Completion). *A* configuration *c is a rooted tree of locations where the root node is the topmost initial location q_0 of the overall state machine. Whenever a location q is an immediate predecessor of q' in c, it must hold $\beta(q') = q$. A configuration is* completed *by applying the following procedure recursively as long as possible to leaf nodes: if there is a leaf node in c labeled with a location q, then introduce all $\alpha(q)$ as immediate successors of q.*

3.2 Hierarchy Implementation with CLP

In Sec. 2.5, a CLP implementation of concurrent hybrid automata was given which implements hybrid finite state machine. Now we will show how to implement an abstract state machine for HHA, treating hierarchies and concurrency more explicitly (Mohammed & Stolzenburg, 2008; 2009). This leads to a lean implementation of hybrid automata, where efficient CLP solvers are employed for performing complex analyses.

Fig. 6 shows parts of the abstract state machine in Prolog, namely the code for completion and for performing discrete and continuous steps according to Def. 3.3 and 3.4. Discrete steps take zero time; continuous steps remain within the same configuration, but the variable values may differ. The flow conditions of active locations (in the configuration) must be applied, as

time passes by. In this context, configurations are encoded in Prolog lists, where the head of a list corresponds to the root of the respective configuration tree. In addition, each location is conjoined by a colon (`:`) with its list of local variables. Thus, according to Def. 3.4, the completed start configuration will be represented as shown below.
The use of lists is straightforward and allows us to implement the abstract state machine for HHA (see Fig. 6) within only a dozen lines of CLP/Prolog code. By this technique, explicit composition of automata is avoided. For each location, its initial states have to be declared together with their continuous flow conditions. For all discrete transitions, the jump conditions have to be stated. Local variables are expressed by nested lists of variables valid in the respective state. Since the abstract state machine is of constant size and the abstract machine computes complex configurations only on demand, there is a one-to-one correspondence between the elements of the HHA and its CLP/Prolog implementation. Thus, the program size is linear to the size of the HHA.
In the concrete implementation of the rescue example, the overall start location q_0 is indicated by the predicate `start`, while `init` defines the initial states for each state (α values according to Def. 3.2). The flow and the jump conditions have to be expressed by means of the predicates `flow` and `trans`. The reader can easily see from Fig. 7 that the size of the CLP program is only directly proportional to the size of the given HHA, because there is a one-to-one correspondence between the graphical specification and its encoding in Prolog, whereas computing the composition of concurrent automata explicitly leads to an exponential increase. Furthermore, since the overall system behavior is given by the abstract state machine (Fig. 6), this approach is completely declarative and concise.
Similarly, in the CLP model of hybrid FSM, reachability analysis is performed by computing the state space of HHA under consideration starting from the initial configuration. For details as well as experiments on benchmarks, the reader is referred to (Mohammed & Stolzenburg, 2009).

4. A Tool: Automatic Design and Verification

In the previous sections, we have shown a framework to specify and verify multi-agent system by means of hybrid automata. Traditionally, in order to verify a certain model with any hybrid automata model-checking tool, one has to specify such model textually with a suitable description language of a model checker. In our framework, one has to specify a multi-agent system in a constraint logic approach. However, in order to textually specify a certain scenario, generally two alternatives can be used: either designing a scenario prior to put it conveniently in a textual specification format to a model checker, or starting to specify the scenario directly with the suitable description languages, which is definitely a tedious and undesirable work, particularly when specifying safety critical systems. From this we may conclude, that it is favorable to graphically specify and automatically verify a certain scenario. For this, a combination of the graphical notations from software engineering with the formal methods realm is necessary.
Generally, the graphical notation is becoming more and more accepted, as it is expected that designers will be more familiar with graphical notation. Therefore, several researchers have approached specifying the behaviors of multi-agent systems using graphical notations, namely UML statechart. For instance, Murray (2004) presents the statechart editor *StatEdit* that is used to graphically specify behaviors multi-agent systems with a layered structured. He has used *StatEdit* to design agents for the RoboCup simulation league. However, neither model checking, nor timed notation are allowed in the tool. In order to combine the formal

```
%%% rescue scenario
start(rescuescenario).
init(T,rescuescenario,[[Event]],
     [fire,civilians,firestation,firebrigade],_) :-
     Event = none.
flow(T1,T2,rescuescenario,[[Event]],[[Event]]).

%%% fire
init(T,fire,[[Boom,Neededw]|_],[no_fire],rescuescenario) :-
    Boom $= 0.
flow(T1,T2,fire,_,_).

init(T,no_fire,[[]|_],[],fire).
flow(T1,T2,no_fire,[[],[Boom1,Neededw]|_],[[],[Boom2,Neededw]|_]) :-
        Boom2 $=< 3,
        Boom2 $>= Boom+(T2-T1).

trans(T,no_fire,[[],[Boom,Neededw],[Event1]],burning,[[],
        [Boom,Neededw],[Event2]]) :-
        Event2 = burn,
        Neededw $= 120.
```

Fig. 7. A part of the HHA implementation of the rescue example.

verification with graphical models, there already exist a number of tools proposed for validation of UML statecharts by translating models into the input language of existing model checkers. For example, Lilius & Porres (1999) have presented the tool *vUML* for model checking systems, which have been modeled by UML statecharts. They have used *SPIN* model checker (Holzmann, 1997) as the underlying verification engine in their tool. On the other hand, in order to graphically specify real time software using UML models, several researchers have extended the standard UML with time notation (Graf et al., 2006). For this purpose, several tools have been developed in order verify the timed UML models by mapping them to input languages of timed automata, which in turn are verified using existing timed automata model checkers. For example, (Del Bianco et al., 2002) have used Kronos (Yovine, 1997) as a model checker to verify their system, whereas (Knappi et al., 2002) have used UPPAAL (Behrmann et al., 2004) as a model checker for the purpose .
Stemming from the previous discussion, we find that it seems advantageous to implement a tool (see Fig.8) that combines both design and verification in the same framework, instead of generating an intermediate specification, which in turn is given to a model checkers. To our knowledge, there is no tool that supports the integration of graphical notations and formal verification of hybrid automata with two different views of multi-agent systems, namely the concurrent and the hierarchical view. For this aim, the tool *HieroMate* (Mohammed & Schwarz, 2009) has been presented, which aims to simplify the process of specification and verification of MAS by combining the advantages of both graphical notations of software engineering together with formal methods. In the tool, the specification of MAS together with their requirements are graphically specified, then the process of verification is achieved automatically.

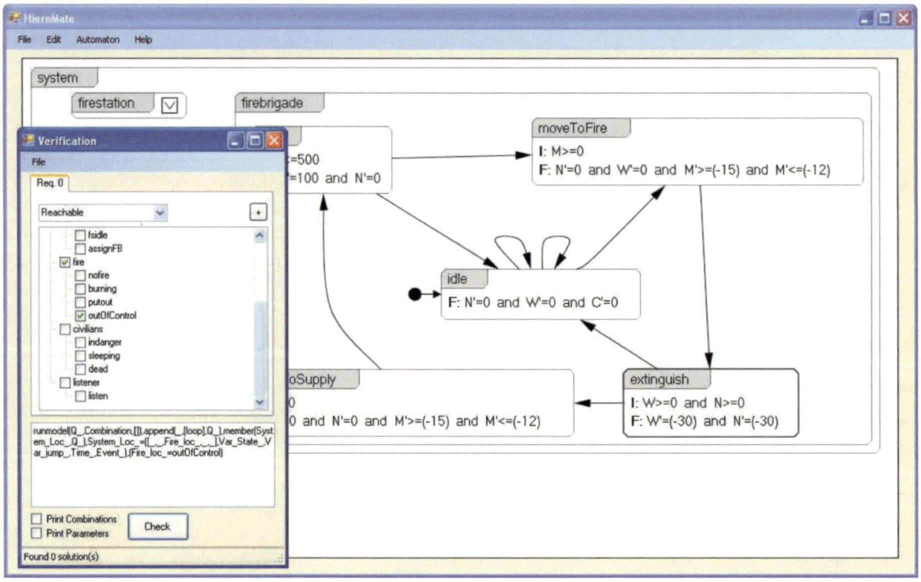

Fig. 8. A tool for modeling and verification based on CLP.

A designer interacts with *HieroMate* using context sensitive menus that allows only meaningful actions. For example, the user is able to add a location to an automaton by right clicking onto the automaton and selecting *Add location* from a context menu. After a model has been built, the specification should be given for formal verification. Actually, a user can either specify queries manually using CLP Prolog, use the tool to generate simple queries automatically, or combine both methods.

4.1 Examples with Model checking

As we already mentioned, the formal semantics of our framework gives the possibility to apply formal methods in order to prove certain properties of specified systems, e.g. by model checking. However, in the context of hybrid automata the term *model checking* usually refers to *reachability* testing, i.e. the question whether some (unwanted) state is reachable from the initial configuration of the specified system. For this purpose, some exemplary model checking tasks for the rescue scenario can be investigated.
For the behavior specification shown in Fig. 4 we conducted several experiments with *HieroMate*. The tool performs reachability tests on the state space of the model. This is done by first computing all reachable states from the initial state/configuration, and then checking the resulting set for the needed properties. In the following, we present some exemplary model checking tasks for the rescue scenario.

Is it possible to extinguish the fire? When the state of the automaton modeling the fire changes from *no fire* to *burning*, the variable *neededw* stores the amount of water needed for putting out the fire ($neededw = 120$ in the beginning). When the fire is put out, i.e. $neededw = 0$, the automaton enters the state *put out*. Thus the fire can be extinguished, iff there is a reachable configuration c_{out} where fire is in the state *put out*. It is easy to see from the specification, that this is indeed the case, as *neededw* is only decreased after the initial setting, and so the transition from *burning* to *put out* is eventually forced.

Does the agent try to extinguish with an empty water tank? To answer this question, we should check the reachability of certain intervals in the continuous valuation of the automaton. The fact that the fire brigade agent tries to put out the fire without water corresponds to the simple state *extinguish* being active while $wLevel < 0$. Note that we must not test for $wLevel \leq 0$, as the state *extinguish* is only left when the water level is zero, so including a check for equality leads to false results.

Won't the fire brigade move to the fire if it is not burning? This is a kind of questions that needs to check the reachability of composed locations in the same time. This can be checked by investigating that no location where *firebrigade* is in location *move2fire* and *fire* is in location *nofire*, or *putout* is reachable

Does the agent report all discovered civilians? We can check properties about the history of a certain state and the reachable states from a given state, this allows more complex questions like this question. Actually, this question contains two properties to be checked:

(a) all discovered civilians are reported eventually, and

(b) the agent does not report more civilians than it found.

The property (a) corresponds to the fact that from every reachable state there is a state reachable where all discovered civilians have been reported. This again means that the number of transitions labeled with *help* equals the number of transitions labeled with *reported*. Property (b) holds if in the history of each reachable state the number of transitions labeled with *help* is always greater or equal to the number of transitions that are labeled with *reported*.
All properties described above could be successfully proven using our framework.

5. Related Works

Hybrid automata have not only been used in the context of robot soccer, but also in many other applications of multi-agent and multi-robot systems. Therefore, we will give a brief overview on related works on modeling, specification, and model checking such systems with focus on approaches that employ CLP.
Using hybrid automata (Henzinger, 1996) is a well accepted method to model and analyze (mobile) multi-agent systems (Alur et al., 1999; 1996). Hierarchical hybrid automata (HHA) can be used for building up and describing multi-layer control architectures based on physical motion dynamics of moving agents (Borges de Sousa et al., 2007; Furbach et al., 2008). In many applications they form a link between multi-robot systems and theories of hybrid systems as in Zelinski et al. (2003). CLP as a programming paradigm has already been applied to modeling hybrid systems including solving differential equations (Hickey & Wittenberg, 2004b). Several authors propose the explicit composition of different concurrent automata by hand leading to one single automaton, before a CLP implementation is applied. This is a tedious work, especially when the number of automata increases. The latter case is exemplified in Urbina (1996) and Jaffar et al. (2004), where approaches to model and analyze hybrid systems using CLP(R) (Jaffar et al., 1992) are introduced.
In Banda & Gallagher (2008), it is shown how reachability analysis for linear hybrid automata can be done by means of CLP, again by computing compositions of (simple) hybrid automata. Events are handled as constraints, which avoids some of the effort for computing composition, which leads to an exponential increase in the number of clauses in general. In our approach, however, we compute configurations of the overall system only if required.

In contrast to our approach, some authors approached modeling the behavior of hybrid systems as an automaton using CLP, but they do not handle a hybrid system consisting of different interacting hybrid automata. For example, Hickey & Wittenberg (2004a) present a hybrid system modeled as an automaton using CLP(F) (Hickey & Wittenberg, 2004b), but neither handling concurrency nor hierarchies. Other authors employ CLP for implementing hybrid automata (Ciarlini & Frühwirth, 2000; Delzanno & Podelski, 1999; Gupta & Pontelli, 1997), but restrict attention to a simple class of hybrid systems (e.g. timed systems). They do not construct the overall behavior prior to modeling, but model each automaton separately. However, the run of the model takes all possible paths into consideration, resulting from the product of each component, which leads to unnecessary computation.
Another interesting approach on model checking hybrid systems is presented in Gulwani & Tiwari (2008). There, an analysis technique is proposed that is able to derive verification conditions, i.e. constraints that hold in reachable states. These conditions are universally quantified and transformed into purely existentially quantified conditions, which is more suitable for constraint solving. For this, an implementation in Lisp is available employing a satisfiability modulo theories (SMT) solver, whereas the Prolog implementation presented in this chapter, allows to express discrete transitions explicitly and allows the use of several constraint solvers.
Another approach for verification of hybrid systems is presented in Fränzle & Herde (2007). In particular, the authors apply so-called bounded model checking (BMC) (Biere et al., 1999) to linear hybrid automata, by encoding them into predicative formulae suitable for BMC. For this reason, they developed a tool called HySAT that combines a SAT solver with linear programming, where the Boolean variables are used for encoding the discrete components, while real variables represent the continuous component. The linear programming routine is used to solve large conjunctive systems of linear inequalities over reals, whereas the SAT solver is used to handle disjunctions. Similar to this approach, our approach presented in this chapter has the essence of BMC. However, instead of checking the satisfiability of formulae to some given finite depth k, we find the set of reachable states and verify various properties on this set. In Biere et al. (1999), neither concurrency nor hierarchy of hybrid automata is taken into consideration.
Differently to this chapter, Jha et al. (2007) introduce symbolic reachability analysis of lazy linear hybrid automata. They provide a verification technique based on bounded model checking and k-induction for reachability analysis. In their technique, SAT-based decision procedures are used to perform a symbolic analysis instead of an enumerative analysis. However, they did not show how the interacting concurrent components can be handled in their approach.

6. Conclusion

In this chapter, we have shown a framework to formally specify and verify physical multi-agent systems by means of hybrid automata, especially for those agents that are defined through their capability to continuously react to a physical environment, while respecting some time constraints. The framework provided two different views of behaviors' specifications, namely, the concurrent and the hierarchical view. In the concurrent view, it has been demonstrated how to avoid the composition of the agents before getting involved into the verification phase,which, in turn can relieve the state explosion problem that may raise as the result of specifying multi-agent systems. On the other hand, in the hierarchical view, we show how multi-agent systems can be hierarchically specified and formally verified without flattening the hierarchy, as it is commonly done. We have shown the implementations of both views by means of constraint logic programming, which forms the specification and the verification engine of the framework. In addition, we have presented a tool that graphically specifies both views, in order to combine the powerful of software engineering into our framework. A case study taken from RoboCup rescue simulation has been depicted to show applicability of our approach.

7. References

Alur, R. & Dill, D. (1994). A Theory of Timed Automata, *Theoretical Computer Science* **126**(2): 183–235.

Alur, R., Esposito, J. M., Kim, M., Kumar, V. & Lee, I. (1999). Formal modeling and analysis of hybrid systems: A case study in multi-robot coordination, *World Congress on Formal Methods*, pp. 212–232.
URL: *citeseer.ist.psu.edu/article/alur99formal.html*

Alur, R., Henzinger, T. A. & Ho, P.-H. (1996). Automatic symbolic verification of embedded systems., *IEEE Transactions on Software Engineering* **22**(3): 181–201.

Apt, K. R. & Wallace, M. (2007). *Constraint Logic Programming Using Eclipse*, Cambridge University Press, Cambridge, UK.

Arai, T. & Stolzenburg, F. (2002). Multiagent systems specification by uml statecharts aiming at intelligent manufacturing, *AAMAS '02: Proceedings of the first international joint conference on Autonomous agents and multiagent systems*, ACM, New York, NY, USA, pp. 11–18.

Banda, G. & Gallagher, J. P. (2008). Analysis of linear hybrid systems in CLP, *in* M. Hanus (ed.), *Pre-Proceedings of LOPSTR 2008 – 18th International Symposium on Logic-Based Program Synthesis and Transformation*, Technical University of Valencia, Spain, pp. 58–72.

Behrmann, G., David, A. & Larsen, K. G. (2004). A tutorial on Uppaal, *in* M. Bernardo & F. Corradini (eds), *Proceedings of 4th International School on Formal Methods for the Design of Computer, Communication, and Software Systems – Formal Methods for the Design of Real-Time Systems (SFM-RT)*, LNCS 3185, Springer, Berlin, Heidelberg, New York, pp. 200–236.

Biere, A., Cimatti, A., Clarke, E. M. & Zhu, Y. (1999). Symbolic model checking without BDDs, *Proceedings of 5th International Conference on Tools and Algorithms for Construction and Analysis of Systems (TACAS)*, LNCS 1579, Springer, Berlin, Heidelberg, New York, pp. 193–207.

Borges de Sousa, J., Johansson, K. H., Silva, J. & Speranzon, A. (2007). A verified hierarchical control architecture for coordinated multi-vehicle operations, *International Journal of Adaptive Control and Signal Processing* **21**(2-3): 159–188. Special issue on autonomous adaptive control of vehicles.

Ciarlini, A. & Frühwirth, T. (2000). Automatic derivation of meaningful experiments for hybrid systems, *Proceeding of ACM SIGSIM Conf. on Artificial Intelligence, Simulation, and Planning (AIS'00)* .

Clarke, E., Grumberg, O. & Peled, D. (1999). *Model checking*, Springer.

da Silva, V., Choren, R. & de Lucena, C. (2004). A UML Based Approach for Modeling and Implementing Multi-Agent Systems, *Proceedings of the Third International Joint Conference on Autonomous Agents and Multiagent Systems-Volume 2*, IEEE Computer Society Washington, DC, USA, pp. 914–921.

Del Bianco, V., Lavazza, L. & Mauri, M. (2002). Model checking uml specifications of real time software, p. 203.

Delzanno, G. & Podelski, A. (1999). Model checking in CLP, *Proceedings of 5th International Conference on Tools and Algorithms for Construction and Analysis of Systems (TACAS)*, LNCS 1579, Springer, Berlin, Heidelberg, New York, pp. 223–239.

Egerstedt, M. (2000). Behavior Based Robotics Using Hybrid Automata, *Proceedings of the Third International Workshop on Hybrid Systems: Computation and Control*, Springer, pp. 103–116.

Fränzle, M. & Herde, C. (2007). HySAT: An efficient proof engine for bounded model checking of hybrid systems, *Formal Methods in System Design* **30**(3): 179–198.

Frehse, G. (2005). PHAVer: Algorithmic verification of hybrid systems past HyTech, *in* M. Morari & L. Thiele (eds), *Hybrid Systems: Computation and Control, 8th International Workshop, Proceedings*, LNCS 3414, Springer, Berlin, Heidelberg, New York, pp. 258–273.

Furbach, U., Murray, J., Schmidsberger, F. & Stolzenburg, F. (2008). Hybrid multiagent systems with timed synchronization – specification and model checking, *in* M. Dastani, A. El Fallah Seghrouchni, A. Ricci & M. Winikoff (eds), *Post-Proceedings of 5th International Workshop on Programming Multi-Agent Systems at 6th International Joint Conference on Autonomous Agents & Multi-Agent Systems*, LNAI 4908, Springer, pp. 205–220.

Graf, S., Ober, I. & Ober, I. (2006). A real-time profile for UML, *International Journal on Software Tools for Technology Transfer (STTT)* **8**(2): 113–127.

Gulwani, S. & Tiwari, A. (2008). Constraint-based approach for analysis of hybrid systems, *in* J.-F. Raskin & P. S. Thiagarajan (eds), *Proceedings of 20th International Conference on Computer Aided Verification (CAV 2008)*, LNCS 5123, Springer, Berlin, Heidelberg, New York, Princeton, NJ, pp. 190–203.

Gupta, G. & Pontelli, E. (1997). A constraint-based approach for specification and verification of real-time systems, *Proceedings of IEEE Real-time Symposium* pp. 230–239.

Harel, D. (1987). Statecharts: A visual formalism for complex systems, *Science of Computer Programming* **8**: 231–274.

Henzinger, T. (1996). The theory of hybrid automata, *Proceedings of the 11th Annual Symposium on Logic in Computer Science*, IEEE Computer Society Press, New Brunswick, NJ, pp. 278–292.

Henzinger, T. A., Ho, P.-H. & Wong-Toi, H. (1995a). HyTech: The Next Generation, *IEEE Real-Time Systems Symposium*, pp. 56–65.

Henzinger, T., Ho, P.-H. & Wong-Toi, H. (1995b). A user guide to HyTech, *Proceedings of International Conference on Tools and Algorithms for the Construction and Analysis of Systems (TACAS)*, LNCS 1019, Springer, Berlin, Heidelberg, New York, pp. 41–71.

Henzinger, T., Kopke, P., Puri, A. & Varaiya, P. (1998). What's Decidable about Hybrid Automata?, *Journal of Computer and System Sciences* **57**(1): 94–124.

Hickey, T. J. & Wittenberg, D. K. (2004a). Rigorous modeling of hybrid systems using interval arithmetic constraints, *in* R. Alur & G. J. Pappas (eds), *Proceedings of 7th International Workshop on Hybrid Systems: Computation and Control (HSCC 2004)*, LNCS 2993, Springer, Berlin Heidelberg, New York, Philadelphia, PA, USA, pp. 402–416.

Hickey, T. J. & Wittenberg, D. K. (2004b). Using analytic CLP to model and analyze hybrid systems, *in* V. Barr & Z. Markov (eds), *Proceedings of the 17th International Florida Artificial Intelligence Research Society Conference*, AAAI Press.

Holzmann, G. (1997). The model checker SPIN, *IEEE Transactions on software engineering* **23**(5): 279–295.

Jaffar, J. & Lassez, J. (1987). Constraint logic programming, *Proceedings of the 14th ACM SIGACT-SIGPLAN symposium on Principles of programming languages*, ACM New York, NY, USA, pp. 111–119.

Jaffar, J., Michaylov, S., Stuckey, P. & Yap, R. (1992). The CLP(R) language and system, *ACM Transactions on Programming Languages and Systems* **14**(3): 339–395.

Jaffar, J., Santosa, A. & Voicu, R. (2004). A clp proof method for timed automata, *Real-Time Systems Symposium, IEEE International* **0**: 175–186.

Jha, S., Brady, B. A. & Seshia, S. A. (2007). Symbolic reachability analysis of lazy linear hybrid automata, *in* J.-F. Raskin & P. S. Thiagarajan (eds), *Proceedings of 5th International Conference on Formal Modeling and Analysis of Timed Systems (FORMATS 2007)*, LNCS 4763, Springer, Berlin, Heidelberg, New York, Salzburg, Austria, pp. 241–256.

Knappi, A., Merzi, S. & Rauh, C. (2002). Model Checking Timed UML State Machines and Collaborations, *Proceedings of the 7th International Symposium on Formal Techniques in Real-Time and Fault-Tolerant Systems: Co-sponsored by IFIP WG 2.2*, Springer, p. 395.

Lilius, J. & Porres, I. (1999). Formalising UML state machines for model checking, *The Unified Modeling Language: UML'99: Beyond the Standard: Second International Workshop, Fort Collins*, Springer, p. 430.

Mohammed, A. & Furbach, U. (2008a). Modeling multi-agent logistic process system using hybrid automata, *Modelling, Simulation, Verification and Validation of Enterprise Information Systems, Proceedings of the 6th International Workshop on Modelling, Simulation, Verification and Validation of Enterprise Information Systems, MSVVEIS-2008*, INSTICC PRESS, pp. 141–149.

Mohammed, A. & Furbach, U. (2008b). Using CLP to model hybrid systems, *Proceedings of Annual ERCIM Workshop on Constraint Solving Programming (CSCLP2008)*.
URL: *http://pst.istc.cnr.it/CSCLP08/program*

Mohammed, A. & Furbach, U. (2009). Multi-agent systems: Modeling and verification using hybrid automata, *Proceedings of the 7th International Workshop on Programming Multi-Agent Systems (ProMAS 2009), May 10-15, 2009, Budapest, Hungary*. Extended version available as Technical Report 8/2009, Department of Computer Sceince,University of Koblenz-landau.

Mohammed, A. & Schwarz, C. (2009). HieroMate: A graphical tool for specification and verification of hierarchical hybrid automata, *in* B. Mertsching, M. Hund & Z. Aziz (eds), *KI 2009: Advances in Artificial Intelligence, Proceedings of 32nd Annual German Conference on Artificial Intelligence*, LNAI 5803, Springer, Berlin, Heidelberg, New York, Paderborn, pp. 695–702.

Mohammed, A. & Stolzenburg, F. (2008). Implementing hierarchical hybrid automata using constraint logic programming, *in* S. Schwarz (ed.), *Proceedings of 22nd Workshop on (Constraint) Logic Programming*, University Halle Wittenberg, Institute of Computer Science, Dresden, pp. 60–71. Technical Report 2008/08.

Mohammed, A. & Stolzenburg, F. (2009). Using constraint logic programming for modeling and verifying hierarchical hybrid automata, *Technical Report 6/2009*, Department of Computer Science, Universität Koblenz–Landau.

Möller, O., David, A. & Yi, W. (2003). Verification of uml statechart with real-time extensions, *Fundamental Approaches to Software Engineering (FASE'2002), LNCS 2306*, Springer-Verlag, pp. 218–232.

Murray, J. (2004). Specifying agents with UML statecharts and StatEdit, *in* A. Bonarini, B. Browning, D. Polani & K. Yoshida (eds), *RoboCup 2003: Robot Soccer World Cup VII*, Vol. 3020 of *Lecture Notes in Artificial Intelligence*, Springer, pp. 145–156.

Murray, J., Obst, O. & Stolzenburg, F. (2002). RoboLog Koblenz 2001, *in* A. Birk, S. Coradeschi & S. Tadokoro (eds), *RoboCup 2001: Robot Soccer World Cup V*, LNAI 2377, Springer, Berlin, Heidelberg, New York, pp. 526–530. Team description.
URL: *http://link.springer-ny.com/link/service/series/0558/bibs/2377/23770526.htm*

Ruh, F. (2007). *A translator for cooperative strategies of mobile agents for four-legged robots*, Master thesis, Hochschule Harz, Wernigerode.

Ruh, F. & Stolzenburg, F. (2008). Translating cooperative strategies for robot behavior, *in* G. J. Nalepa & J. Baumeister (eds), *Proceedings of 4th Workshop on Knowledge Engineering and Software Engineering at 31st German Conference on Artificial Intelligence*, Kaiserslautern, pp. 85–96. CEUR Workshop Proceedings 425.
URL: *http://sunsite.informatik.rwth-aachen.de/Publications/CEUR-WS/Vol-425/paper9.pdf*

Tadokoro, S. et al. (2000). The RoboCup-Rescue project: A robotic approach to the disaster mitigation problem, *Proceedings of IEEE International Conference on Robotics and Automation (ICRA 2000)*, pp. 4089–4104.

UML (2009). *OMG Unified Modeling Language (OMG UML): Infrastructure; Superstructure.*

Urbina, L. (1996). Analysis of hybrid systems in CLP(R), *Proceedings of 2nd International Conference on Principles and Practice of Constraint Programming (CP'96)*, LNAI 1118, pp. 451–467.

Yovine, S. (1997). Kronos: A verification tool for real-time systems, *International Journal on Software Tools for Technology Transfer (STTT)* **1**(1): 123–133.

Zelinski, S., Koo, T. J. & Sastry, S. (2003). Hybrid system design for formations of autonomous vehicles, *Proceedings of 42nd IEEE Conference on Decision and Control*, Vol. 1, pp. 1–6.

12

RFuzzy: an easy and expressive tool for modelling the cognitive layer in RoboCupSoccer

Susana Muñoz Hernández
Technical University of Madrid
Spain

1. Introduction

The idea of robot playing soccer has been developed since early 90s Chen et al. (2003). Soccer environment is a dynamically changing environment which requires individual skill as well as team skill and therefore is an interesting research field on Artificial Intelligence and robotics. Prolog is a programming language that represent logic reasoning. Is is a perfect tool to represent human reasoning, so it seems to be a good choice for implementing the cognitive layer of soccer players that is a simulation of human behaviour related to this game. For example, applying the rule "if the goal keeper is not at the goal then kick to ball". But many of the most important decisions that are made by soccer players deal with non-crisp issues. They are related to fuzziness (e.g. "if other player of my team is FAR from me then don't pass him/her the ball"), uncertainty (e.g. "if I CAN get the goal then kick the ball"), or incompleteness (e.g. "if I cannot see the position of a player, by default I'm not going to pass him the ball").
In this work we are going to provide a programming framework to Robot Soccer programmers to model robot control in an expressive but simple way. We propose the possibility of using fuzzy concepts for this modelization and we are going to provide some conclusions about this tool based on a bench of practical experiments that we have done and that we describe here.
In the rest of this section we introduce RoboCupSoccer field (section 1.1) and we discuss some previous fuzzy approaches in logic programming (sections 1.2 and 1.3. In section 2 we describe our framework, *RFuzzy* enumerating the features that characterize its expressivity for modelling problems in general. From the following section we focus on the Robot Soccer use of our tool. Section 3 describes the environment for our experimentation (the general arquitecture at section 3.1 and the particular Prolog code arquitecture at section 3.2). We have described in detail our experiments in section 4. We provide information about the decision making analysis that we have done (section 4.1.1) and the action execution analysis (section 4.1.2). We finally conclude at section 5.

1.1 RoboCupSoccer

RoboCup is an international annual event promoting research on Artificial Intelligence, robotics, and related field. The original motivation of RoboCup is RoboCupSoccer. As the nature of soccer game, autonomous robots participating in RoboCupSoccer should have individual ability such as moving and kicking the ball, cooperative ability such as coordinating with team mates, and of course, the ability to deal with dynamic environment.

RoboCupSoccer consists of several leagues, providing test beds for various research scale: Simulation League, Small Size Robot League (F-180), Middle Size Robot League (f-2000), Four-Legged Robot League, Humanoid League, E-League and RoboCup Commentator Exhibition. The first E-League was held at RoboCup 2004.This league is a simplified version of Small Size Robot League, where vision processing and communications are factored out, thus provided by the league. Each team in this league consists of four small sized autonomous robots, one of whom can be a goalkeeper. The match lasts for two equal periods of 10 minutes.
We employ RoboCupSoccer Simulation League for the sake of simplicity because we are just focused on the robot control layer.

1.2 Fuzzy Approaches in Logic Programming

Introducing Fuzzy Logic into Logic Programming has provided the development of several fuzzy systems over Prolog. These systems replace its inference mechanism, SLD-resolution, with a fuzzy variant that is able to handle partial truth. Most of these systems implement the fuzzy resolution introduced by Lee in Lee (1972), as the Prolog-Elf system Ishizuka & Kanai (1985), the FRIL Prolog system Baldwin et al. (1995) and the F-Prolog language Li & Liu (1990). However, there is no common method for fuzzifying Prolog, as noted in Shen et al. (1989).
Some of these Fuzzy Prolog systems only consider fuzziness on predicates whereas other systems consider fuzzy facts or fuzzy rules. There is no agreement about which fuzzy logic should be used. Most of them use min-max logic (for modelling the conjunction and disjunction operations) but other systems just use Łukasiewicz logic Klawonn & Kruse (1994).
There is also an extension of constraint logic programming Bistarelli et al. (2001), which can model logics based on semiring structures. This framework can model min-max fuzzy logic, which is the only logic with semiring structure. Another theoretical model for fuzzy logic programming without negation has been proposed by Vojtáš in Vojtas (2001), which deals with many-valued implications.

1.3 Fuzzy Prolog

One of the most promising fuzzy tools for Prolog was the "Fuzzy Prolog" system Vaucheret et al. (2002); Guadarrama, Munoz-Hernandez & Vaucheret (2004). The most important advantages against the other approaches are:

1. A truth value is represented as a finite union of sub-intervals on $[0,1]$. An interval is a particular case of union of one element, and a unique truth value (a real number) is a particular case of having an interval with only one element.
2. A truth value is propagated through the rules by means of an *aggregation operator*. The definition of this *aggregation operator* is general and it subsumes conjunctive operators (triangular norms Klement et al. (n.d.) like min, prod, etc.), disjunctive operators Trillas et al. (1995) (triangular co-norms, like max, sum, etc.), average operators (averages as arithmetic average, quasi-linear average, etc) and hybrid operators (combinations of the above operators) Pradera et al. (2002)).
3. Crisp and fuzzy reasoning are consistently combined Munoz-Hernandez et al. (2002).

Fuzzy Prolog adds fuzziness to a Prolog compiler using CLP($\mathcal{R}$) instead of implementing a new fuzzy resolution method, as other former fuzzy Prologs do. It represents intervals as constraints over real numbers and *aggregation operators* as operations with these constraints, so it uses Prolog built-in inference mechanism to handle the concept of partial truth. From the implementation point of view, Fuzzy Prolog is implemented over Ciao

Fig. 1. *RFuzzy* architecture.

Prolog CLIP Lab (n.d.). The Ciao Prolog System offers a complete Prolog system supporting ISO-Prolog. Its modular design allows restriction and extension of the language both syntactically and semantically. The Ciao Prolog Development System provides many libraries including a constraint logic programming system and interfaces to some programming languages. In Ciao Prolog terminology, a library is implemented as either a module or a package. Fuzzy Prolog described in Guadarrama, S.Muñoz & C.Vaucheret (2004) and *Extending Prolog with Incomplete Fuzzy Information* (2005);
Default values to handel Incomplete Fuzzy Information (2006) is implemented as the package "fuzzy.pl", a syntactic extension of the CLP($\mathcal{R}$) system in the Ciao Prolog System.

2. *RFuzzy* tool expressiveness

Besides the advantages of Fuzzy Prolog, it truth value representatio based on constraints is too general that it is complex to interpret for regular users. That was the reason for implementing a simpler variant that we called *RFuzzy*. In *RFuzzy* the truth value is represented by a simple real number.
RFuzzy is implemented as a Ciao Prolog CLIP Lab (n.d.) package because Ciao Prolog offers the possibility of dealing with a higher order compilation through the implementation of Ciao packages.
The compilation process of a *RFuzzy* program has two pre-compilation steps: (1) the *RFuzzy* program is translated into CLP($\mathcal{R}$) constraints by means of the *RFuzzy* package and (2) the program with constraints is translated into ISO Prolog by using the CLP($\mathcal{R}$) package. Fig. 1 shows the whole process.
As the motivation of *RFuzzy* was providing a tool for practical application, it was loaded with many nice features that represent an adventage with respect to previous fuzzy tools to model real problems. In this section we enumerate and describe some of the most interesting characteristics of *RFuzzy* expressiveness through its syntax (to show its simplicity that is the other advantage of *RFuzzy*). For the examples we are going to use intuitive concepts related to soccer vocabulary although many of ther are not use for the simulator because it use just simple variables of position and speed but they are more ilustrative in the interest of concepts understanding.

2.1 Types definition

Prolog does not have types. The problem of not having types is that it is impossible to return constructive answers but using constraints. *RFuzzy* does not use constrains because they are not friendly to return constructive results and that is the reason for having types instead.
In *RFuzzy* types are defined according to (1) syntax.

$$\textbf{:- set\_prop}\ pred/ar\ \texttt{=>}\ type\_pred\_1/1\ [,\ type\_pred\_n/1\ ]^*\ . \qquad (1)$$

where *set_prop* is a reserved word, *pred* is the name of the typed predicate, *ar* is its arity and *type_pred_1*, *type_pred_n* ($n \in 2,3,\ldots,ar$) are predicates used to define types for each argument of *pred*. They must have arity 1. The definition is constraining the values of the n_th argument

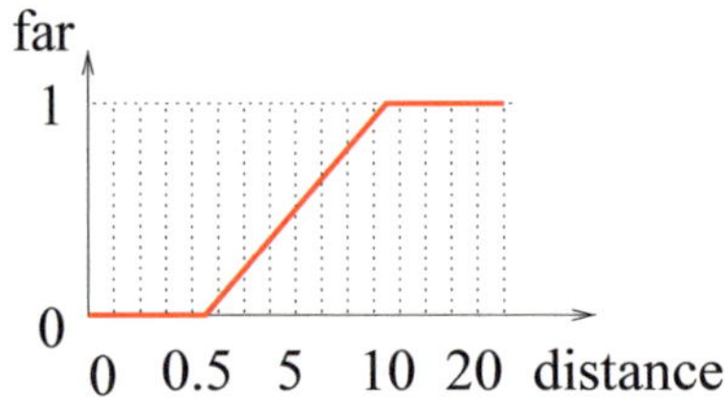

Fig. 2. Far truth value continuous representation

of *pred* to the values of the type *type_pred_n*. This definition of types ensures that the values assigned to the arguments of *pred* are correctly typed.
The example below shows that the arguments of predicates *is_striker/1* and *is_faster_than/2* have to be of type *player/1*. The domain of type *player* is enumerated.

$$:-set\_prop\ is\_striker/1\ =>\ player/1.$$
$$:-set\_prop\ is\_faster\_than/2\ =>\ player/1,\ player/1.$$
$$player(robot1). \quad player(robot2). \quad player(robot3).$$
$$player(robot4). \quad player(robot5).$$

2.2 Simple truth value assignment

It is possible to assign a truth value to an individual using fuzzy facts. Their syntax, that we can see in (2), is different than regular Prolog facts syntax.

$$pred(args)\ \mathbf{value}\ truth\_val. \tag{2}$$

Arguments, *args*, should be ground and the truth value, *truth_val*, must be a real number between 0 and 1. The example below defines that the player *robot3* is a *fast_player* with a truth value 0.9.

$$fast\_player(robot3)\ value\ 0.9.$$

2.3 Continuous function to represent truth values

Facts definition (see subsection 2.2) is worth for a finite (and relative small) number of individuals. Nevertheless, it is very common to represent fuzzy truth using continuous functions. Fig. 2 shows an example in which the continuous function assigns the truth value of being *far* to a distance.
Functions used to define the truth value of some group of individuals are usually continuous and linear over intervals. To define those functions there is no necessity to write down the value assigned to each element in their domains. We have to take into account that the domain can be infinite.
RFuzzy provides the syntax for defining functions by stretches. This syntax is shown in (3). External brackets represent the Prolog list symbols and internal brackets represent cardinality in the formula notation. Predicate *pred* has arity 1, *val1, ..., valN* should be ground terms representing numbers of the domain (they are possible values of the argument of *pred*) and *truth_val1, ..., truth_valN* should be the truth values associated to these numbers. The truth value of the rest of the elements is obtained by interpolation.

$$pred\ \mathbf{:\#}\ ([(val1, truth\_val1),\ (val2, truth\_val2)\ [,\ (valn, truth\_valn)\ ]^{*}])\ . \tag{3}$$

The *RFuzzy* syntax for the predicate *far/1* (represented in Fig.2) is:

$$teenager : \#([(0.5,0),(10,1)]).$$

2.4 Rule definition with truth values and credibility

A tool which only allows the user to define truth values through functions and facts lacks on allowing him to combine those truth values for representing more complex situations. A rule is the tool to combine the truth values of facts, functions, and other rules.

Rules allow the user to combine truth values in the correct way (by means of aggregation operators, like *minimum, maximum, product,* etc.). The aggregation operator combines the truth values of the subgoals of the body of the rule to obtain the truth value of the head of the rule. Appart from this, rules are assigned a credibility value to obtain the final truth value for the head of the clause. Credibility is used to express how much we trust a rule. It is used another opperator to aggregate the truth value obtained (from the aggregation of the subgoals of the body) with the rule's credibility.

RFuzzy offers a simple syntax for representing these rules, defined in (5). There are two aggregation operators, *op2* for combining the truth values of the subgoals of the rule body and *op1* for combining the previous result with the rule's credibility. The user can choose for any of them an aggregation operator from the list of the available ones[1] or define his/her own aggregation operator.

$$pred(arg1\ [,\ argn]^*\ )\ [\ \mathbf{cred}\ \mathbf{(}op1,value1\mathbf{)}\ ] : \sim op2 \quad pred1(args\_pred\_1)\ [,\ predm(args\_pred\_m)]\ . \tag{4}$$

The following example uses the operator *prod* for aggregating truth values of the subgoals of the body and *min* to aggregate the result with the credibility of the rule (which is 0.8). "**cred** ($op1,value1$)" can only appear 0 or 1 times.

$$good\_player(J)\ cred(min,0.8) : \sim prod\ swift(J),\ agile(J),\ has\_experience(J).$$

2.5 General and Conditioned Default Truth Values

Unfortunately, information provided by the user is not complete in general. So there are many cases in which we have no information about the truth value for a fuzzy predicate of an individual or a set of them. This happend many times in Robot soccer (not in the simulator but in games with real robots) when the camara does not detect correctly any player of the ball position. Nevertheless, it is interesting not to stop a complex query evaluation just because we have no information about one or more subgoals if we can use a reasonable approximation. A solution to this problem is using default truth values for these cases. The *RFuzzy* extension to define a default truth value for a predicate when applied to individuals for which the user has not defined an explicit truth value is named *general default truth value*. The syntax for defining a general default truth value is shown in (5).

Conditioned default truth value is used when the default truth value only applies to a subset of the domain. This subset is defined by a membership predicate which is true only when an individual belongs to the subset. The membership predicate (*membership_predicate/ar*) and the

[1] Aggregation operators available are: *min* for minimum, *max* for maximum, *prod* for the product, *luka* for the Łukasiewicz operator, *dprod* for the inverse product, *dluka* for the inverse Łukasiewicz operator and *complement*.

predicate to which it is applied (*pred/ar*) need to have the same arity (*ar*). The syntax is shown in (6).

$$\text{:- } \textbf{default}(pred/ar,\ truth\_value)\ . \tag{5}$$

$$\text{:- } \textbf{default}(pred/ar,\ truth\_value) => membership\_predicate/ar. \tag{6}$$

pred/ar is in both cases the predicate to which we are defining default values. As expected, when defining the three cases (explicit, conditioned and default truth value) only one will be given back when doing a query. The precedence when looking for the truth value goes from the most concrete to the least one.
The code from the example below joint with the code from examples in subsections 2.1 and 2.2 assigns to the predicate *fast_player* a truth value of 0.8 for *robot2* (default truth value), 0.6 when it is *robot1* (conditioned default truth value for the goal keeper) and 0.9 when it is *robot3* (explicit truth value).

$$:-default(fast\_player/1, 0.6) => goal\_keeper/1.$$
$$:-default(fast\_player/1, 0.8).$$
$$goal\_keeper(robot1).$$

2.6 Constructive Answers

A very interesting characteristic for a fuzzy tool is being able to provide constructive answers for queries. The regular (easy) questions ask for the truth value of an element. For example, how fast is *robot3*? (See left hand side example below)

$?-fast\_player(robot3, V).$	$?-fast\_player(X, V), V > 0.7.$
$V = 0.9?;$	$V = 0.9, X = robot3?;$
no	$V = 0.8, X = robot2?;$
	$V = 0.8, X = robot4?;$
	$V = 0.8, X = robot5?;$
	no

But the really interesting queries are the ones that ask for values that satisfy constraints over the truth value. For example, which players are very fast? (See right hand side example above). *RFuzzy* provides this constructive functionality.

3. Environment

In this work we have prepared a complete framework with all the interfaces necessary for someone interested in defining strategies for robot control. In this section we are going to describe all the components of the environment that we have used for our experiments and we provide them for their free use.

3.1 Architecture and Implementation Details

Based on agent system architecture proposed by García et al. (2004), in Hernndez & Wiguna (2007) we proposed a generic system architecture for RoboCup offering flexibility on choice of programming language and minimal modification to switch between leagues. This architecture is shown in figure 3. Prolog is proposed for cognitive layer, and in Hernndez & Wiguna (2007) we use Fuzzy Prolog Guadarrama, Munoz-Hernandez & Vaucheret (2004) for implementing the cognitive layer and we use the Atan library and a RoboCupSoccer Simulator.

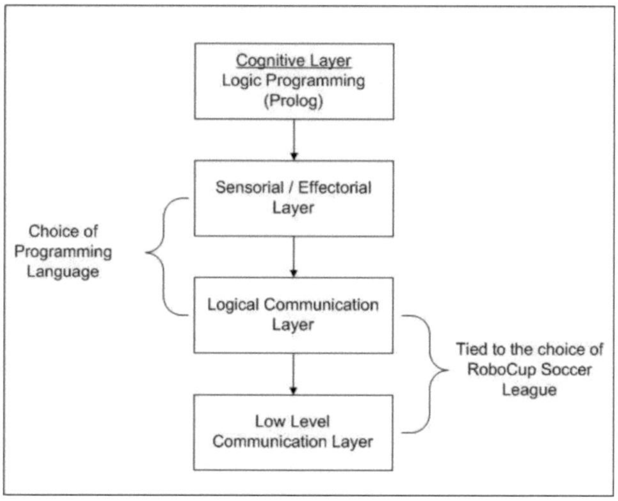

Fig. 3. Generic System Architecture

3.1.1 Low Level Communication Layer

As the name suggests, this is the lowest layer of our architecture. This layer includes all hardwares and softwares provided by the league. The robots, infrared transmitter, video camera, communication network, and vision systems belong to this layer. Different leagues in RoboCupSoccer are represented by different Low Level Communication Layer. E-League has the robots, Doraemon vision package, and communication server as part of this layer, whereas Simulation League has only The RoboCup Soccer Simulator as part of this layer.

3.1.2 Logical Communication Layer

This layer acts as the interface between low level communication layer and the upper layers. It is intended to hide physical structure of the environment from the upper layer. As long as the interface of the services offered by this layer remain unchanged, then the rest of the upper layer can also remain unchanged García et al. (2004). Basic services that should be offered for E-league are :

- Reading the packets generated by video server.
- Establishing communication with the communication server.
- Continuous sensing for the referee decision.

We have used the Simuro environment FIRA (n.d.). SimuroSot consists of a server which has the soccer game environments (playground, robots, score board, etc.) and two client programs with the game strategies. A 3D color graphic screen displays the match. Teams can make their own strategies and compete with each other without hardware.

3.1.3 Sensorial/Effectorial Layer

This layer serves as a bridging layer between the logical communication layer and the cognitive layer. It translates visual information into the representation needed by cognitive layer, and also translates output from cognitive layer into basic action to be performed by the robots. In our implementation for Simulation League which use Prolog programs as cognitive layer

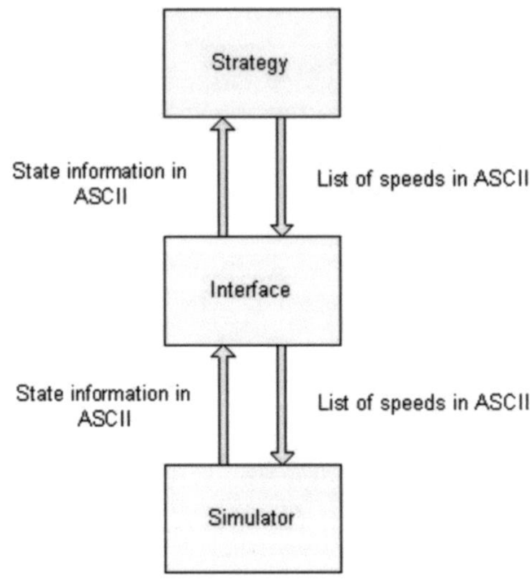

Fig. 4. Environment Architecture

and Java library as logical communication layer, this means translating visual information into prolog predicates and interpreting prolog query result. We use the Rakiduam UNCOMA (2006) interface with the dll files ("tcpb.dll" and "tcpy.dll") that are necessary in between the simulator and the Prolog compiler.

3.1.4 Cognitive Layer

Cognitive layer is where the strategy is implemented. It is the highest level layer. Our work is focused in this layer where we employ The Ciao Prolog System Hermenegildo et al. (1999), and in particulas the RFuzzy Prolog library, to reason over the provided information. Our approach is providing the capability of handling fuzzy, uncertain and incomplete information at the cognitive layer. This information is very close to the human reasoning, so this framework is improving the human-like control of this layer. A strategy can be easily implemented on this layer without having to put effort on low level technical details more related to the machine than to the human mind.
In this contribution we have changed (with respect to Hernndez & Wiguna (2007)) the fuzzy library and the RoboCupSoccer simulator to obtain more precise results related the use of fuzzy and crisp rules in the control of the robots. The fuzzy library that we use here is RFuzzy (that is decribed in detail in section 2), and for the simulation we use Rakiduam UNCOMA (2006). We can see in figure 4 the environment architecture.

3.2 Prolog Code Arquitecture

For this comparative study we have implemented two main modules and a set of auxiliary modules. We have also used a couple of communication modules from UNCOMA (2006).
The complete modules arquitecture is represented in figure 5 whose internal running is described below.

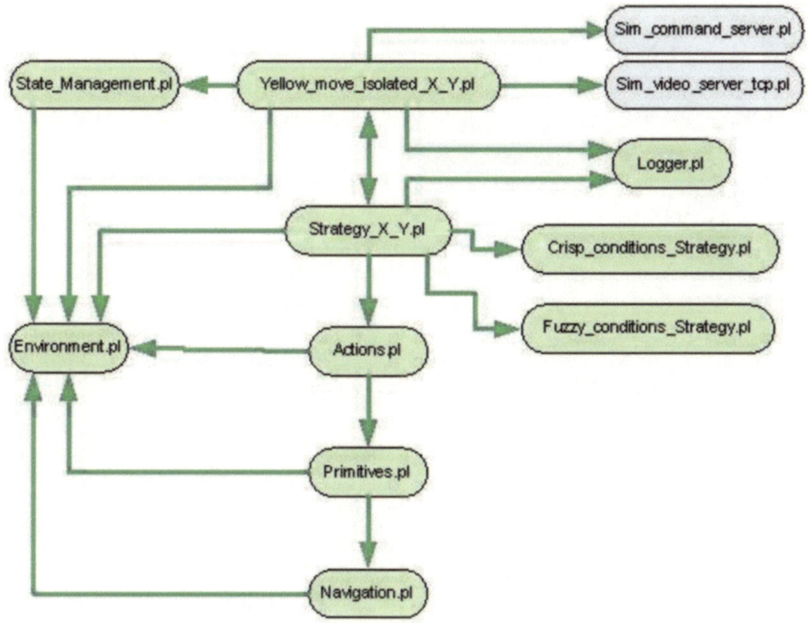

Fig. 5. Prolog code Architecture

3.2.1 Main Modules

- The module **yellow_move_isolated_X_Y.pl** starts the comunication in between the interface and the team strategy. After stablishing the connection goes into a loop for deciding (with the help of module strategy_X_Y.pl) an action for each robot. These actions should be transmited to the simulator server as a list of speeds of the set of robots. During the loop a trace of the game is generated also.
- The module **strategy_X_Y.pl** provide the strategy information to the above module to take a decision about the robots action.

3.2.2 Auxiliary Modules

- In the module **stage_management.pl** the value of the stage variable (that contains the data of the environment as for example the position of the players and the ball.
- All predicates related with loading stored data from the stage variable and creating new variables for the strategy are in the module **environment.pl**. For example getting the ball position or calculating its speed from its two last positions.
- The module **crisp_conditions_strategy.pl** contains the predicates that evaluate the conditions of the environment in each cycle and, acording to them, determine (using crisp rules) the action to develope by each robot.
- The module **fuzzy_conditions_strategy.pl** contains the predicates that evaluate the conditions of the environment in each cycle and, acording to them, determine (using fuzzy rules) the action to develope by each robot.
- The set of high level actions (e.g. shoot, pass, etc.) that the robots can developed are implemente in the module **actions.pl**.

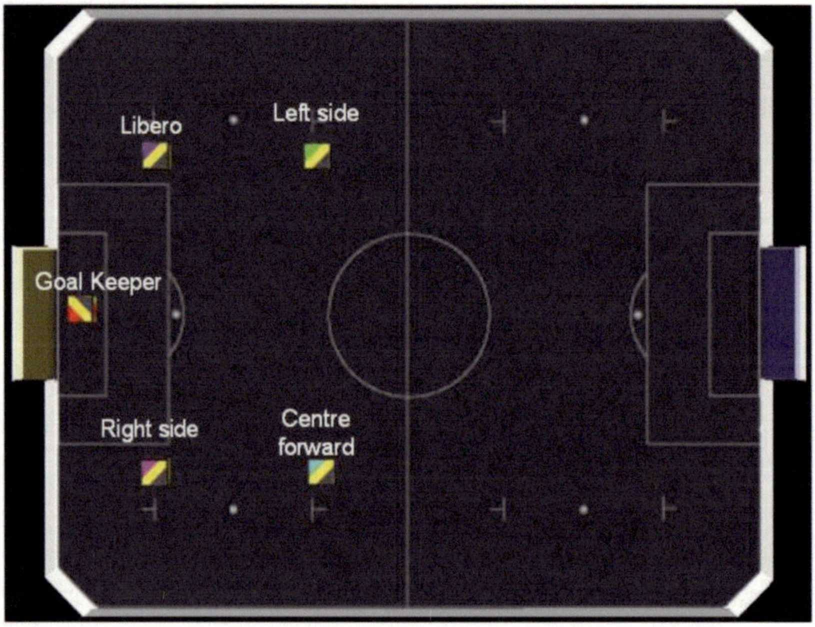

Fig. 6. Robots positions

- The set of medium level actions (e.g. moving the robot to a concrete position) that the robots can developed are implemente in the module **primitives.pl**.
- The set of medium level actions (e.g. moving the robot to a concrete position) that the robots can developed are implemente in the module **navegation.pl**.
- The module **logger.pl** provide the trace of the the game

3.2.3 Communication Modules

- The module **sim_video_server_tcp.pl** abstracts the strategy programming of the communication of the video server. It receives the ambient data from the video server and it transforms them into logic rules in Prolog.
- The module **sim_command_server.pl** abstract the logic programming to the communication with the interface. It maintains the communication with the command server and decodify the Prolog logic rules for being understandable by the command server.

4. Comparative Study

For testing the use of different logics we have define the structure of the comparative study. We have design and implement all modules (described in section 3.2) that are necessary to model the basic strategy of a math, the decission making, the actions execution and the test cases.

Figure 6 identifies the name of the position of each robot to reference them in the rest of the paper.

In this section we are going to describe some test cases. Some of them are simple moves and others are strategies in matches.

4.1 Simple Movements

For a set of simple movements we have study the behavior of the players in two aspects: decision making and action execution. We compare the control of a player implemented using crisp rules and the control of a player implemented using fuzzy rules for both aspects:

- **Decision Making**: Which one is the best action to chose. The crisp and the fuzzy variants will take different decisions sometimes. This is analyzed in the comparative study. When the decision is the same, the execution of the action will be the same because this experiments just use basic actions that are not taking into account the environment conditions. The code for this part is in the module **conditions_X_strategy.pl** where X can be crisp or fuzzy.
- **Action Execution**: How the action is executed. We have chosen some actions that depend on some factors as speed or direction of the ball. The execution of the action is examined but not the previous decision that have taken us to do it. The code for defining the execution of the basic actions (executed after the decision making process of the first experiments) and the actions that are programmed using crisp and fuzzy logic (for their comparison) is in module **actions.pl**.

The tests are clasified attending to the action that is expected to do the robot (shoot, clear or pass). We have used for the bench of tests that we have made the same structure for an action X:

1. Action X: description of the players that participate and showing (through an image) the initial position of the robots.
2. Analysis of the crisp logic in Action X
3. Analysis of the fuzzy logic in Action C
4. Crisp and Fuzzy logic comparison

4.1.1 Decision Making Analysis

The action that is made for a robot is chosen attending to a set of variables. These variables describe the environment of the game. The variables used for decision making using crip logic are different from the set of variables that are used for decision making through fuzzy logic. In figure 7 there are some of these variables that are used by module **conditions_X_strategy.pl** to decide which action to perform by the robot.
Close to the definition of values for the variables is the distribution of areas in which we have divided the game field. It is represented in figure 8.
We are going to provide a brief description of these variables to understand their relevance in the comparative study:

- **Relative position** is the position of the ball in the field attending to the position of the players with respect the bal. It is the same in fuzzy and crisp logic. There are five possible values (ofensive left side area, ofensive right side area, ofensive centre, ofensive closure, defensive area).
- **Crisp ball position** is the area from 8 where the ball is. It has twelve possible values.
- **Crisp goal direction** is the angle (in grades and always positive) that if defined in between the trayectory of the ball and the segment that joins the ball with the centre of the opposite goal area. This variable provides information about the direction of the ball with respect to the opposite goal area. The possibles values are in the following ranges: $[0^o..45^o]$ or $[135^o..180^o]$ if it is in the direction of the goal area, and $[45^o..135^o]$ if it is not.

Crisp logic variables	Fuzzy logic variables
Relative position	Relative position
Crisp ball position	Fuzzy ball position X
	Fuzzy ball position Y
Crisp goal direction	Fuzzy ball direction
Crisp ball speed	Fuzzy ball speed

Fig. 7. Decision making variables

- **Crisp ball speed** is the speed of the ball that is calculated as the distance that the ball is able to cover during a server cycle in the simulator. The possible values are: slow (less than 0.5), regular (into 0.5 and 1) and fast (greater than 1).
- **Fuzzy ball position X** is the distance from the ball to the back line of the own field. The values are in the range [0..86].
- **Fuzzy ball position Y**is the distance from the ball to the left line of the own field. The values are in the range [0..70].
- **Fuzzy ball direction** is the same concept that the crisp goal direction. The values belong the range $[0^o..180^o]$
- **Fuzzy ball speed** is the same concept that crisp ball speed but the domain is continue ($\text{speed} \in R^+$).

For the comparison of the decision making we have considered a set of crisp rules and a set of fuzzy rules to determine the best action for the robots in each situation. The rules are different for each robot depending on its position (goal keeper, striker, midfielder, etc.) Let's see a couple of examples of rules.
Crisp rules (implemented in Prolog) are of the form:

$$shoot\_striker \leftarrow ofensive\_area \wedge opposite\_area\_ball$$

where it is said that the striker should shoot if it is in the ofensive area and the ball is in the opposite area.
The list of crisp rules should be ordered according to priority because in each situation it is executed the first one that is satisfied. If any of them is satisfied, then the robot should mantain the base position that is calculated as the average point into the position of the ball and the own goal place. But to observed better the cases in which the player is not deciding to do any action we have change this base position to the same position. So, the robot that is not making any action is going to maintain its position.
Fuzzy rules (implemented in RFuzzy) are of the form:
$shoot\_striker(V) \leftarrow_{prod} relative\_position\_good\_shoot(V1), ball\_position\_X\_good\_shoot(V2),$
$ball\_position\_Y\_good\_shoot(V3), ball\_direction\_good\_shoot(V4), ball\_speed\_good\_shoot(V5)$

Fig. 8. Areas of the field

where it is said that it is going to be calculated the truth value for the action of shooting for a striker. The truth value will be a value in the range [0..1](0 means that it is a bad action for the player in its situation and 1 means that it is the perfect action for the player). Intermediate values have intermediate meanings. In this case the truth value of a set of concepts will be calculated first (if the relative position of the player is good for shooting, if the position of the ball is good for shooting, if the direcction of the ball is good for shooting and if the speed of the ball is godd for shooting) and them the truth value for shooting, V, will be calculated as the aggregation (using product operation in this case) of all the truth values previously calculated (V1, V2, V3, V4 and V5).
In figure 9 we can see the fuzzy functions that represent the concepts that are involved in the definition of the rule that we have defined above. They represent the truth value for each value of the environment.
In the case of fuzzy rules, all rules will be executed and the best action for the player situation (the one with higher truth value) will be executed. The option by default of keeping the base position has a default value of 0.2.
In a particular case, we can have all players and the ball in a particular position (as in figure 10) and then try to find out the best action to perform by a particular robot. For our experiments we use crisp and fuzzy batery of rules and we compare the results.
For the rest of the study we will analyse always a robot of the left team of the screen, so the yellow team.
We have studied different starting positions of the game that lead the robots to do the actions of shooting, passing and clearing. The results of the comparison in between the crisp and the fuzzy control for these tests are detailed in section 5.

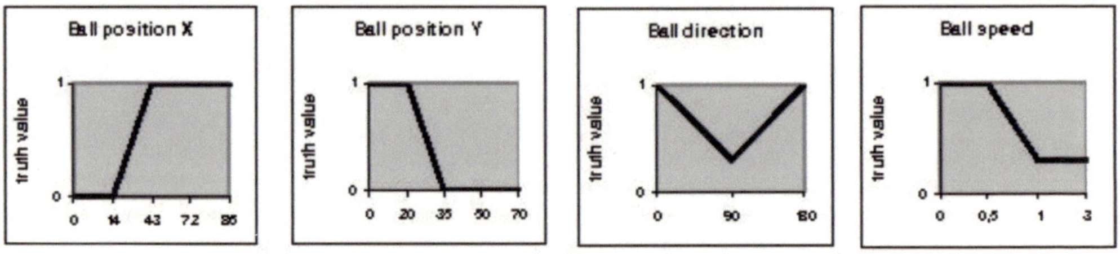

Fig. 9. Fuzzy functions to represent concepts related shooting

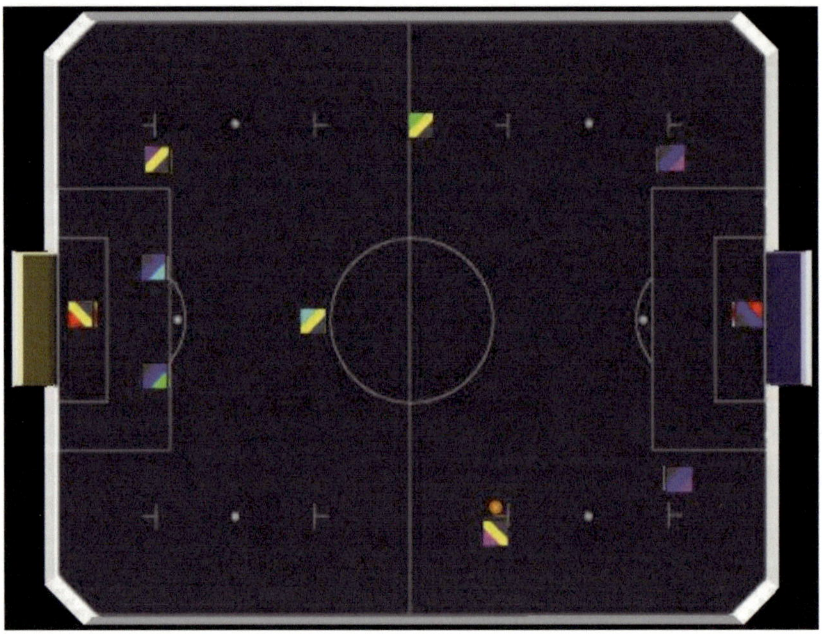

Fig. 10. Starting positions for shooting

4.1.2 Action Execution Analysis

In the analysis of the action execution we have studied starting situations in which the robot has already decided to make a particular action. Then we have observed the result when it do it using crisp and fuzzy execution rules.

The variables that we have used for shooting are listed in figure 11. Ball distance is the same in the crisp and in the fuzzy logic. The variables that we have used for passing are shown in figure 12.

The speed and direction variables are the same that the ones used in section 4.1.1 for decision making analysis. The rest of variables are defined as follow:

- **Ball distance** is the distance in between the robot that is shooting and the ball (distance $\in R^{+}$).

Crisp logic variables	Fuzzy logic variables
Crisp goal direction	Fuzzy ball direction
Crisp ball speed	Fuzzy ball speed
Ball distance	Ball distance

Fig. 11. Shoot action variables

Crisp logic variables	Fuzzy logic variables
Crisp centre forward position	Fuzzy centre forward position

Fig. 12. Pass action variables

- **Crisp centre forward position** is the distance from the striker that is passing the ball and the opposite passing point. It is calculated as the difference (always positive) of the striker X coordinate and the X coordinate of the opposite passing point (that is aproximately in the penalty point).
- **Fuzzy centre forward position** is the same concept that the crisp centre forward position but the values are in the range [0..75].

We have just study the most representative starting situations that lead the robot to shoot and pass. In section 5 are discussed the results of the comparison in between the crisp and the fuzzy control for these actions.

4.2 Matches

Besides studying single movements we have uses proof tests for complete matches. We have program in a different way each team. The crisp team uses crisp logic for decision making and also crips logic for action execution. The fuzzy team uses fuzzy logic for decision making and also crips logic for action execution. We will use also the Lingo team and the Rakiduam team. In the tests the yellow team is team 1 and the blue team is team 2.

For each scenario the test is a 30 seconds match where it is compared the number of gols of each team and the percentage of time that each team control the ball.

We have evaluated the mathes: crisp team vs Lingo team, crisp team vs Lingo team, crisp logic vs Rakiduam07, fuzzy logic vs Rakiduam07, and crisp team vs fuzzy team.

5. Conclusions

RFuzzy has many advantages related its expressivity and some advanced characteristics of *RFuzzy* are missing in other similar tools as FLOPERMoreno (2006); Morcillo & Moreno (2008). *RFuzzy* has been designed with a simple syntax to facilitate programmers from other fields (as

in this case from Robot Soccer programming) to model their problems in a simple way. This is the reason why *RFuzzy* is much more convenient that *Fuzzy Prolog* (that use constrains that are much more difficult to handle and understand that real numbers that are used in *RFuzzy*). Extensions added to *Prolog* by *RFuzzy* are: types (subsection 2.1), default truth values conditioned or general (subsection 2.5), assignment of truth values to individuals by means of facts (subsection 2.2), functions (subsection 2.3) or rules with credibility (subsection 2.4).
One of the most important consequences of these extensions is the constructivity of the answers with the possibility of constraining the truth value in the queries as we describe in section 2.6.
There are countless applications and research lines which can benefit from the advantages of using the fuzzy representations offered by *RFuzzy*. Some examples are: Search Engines, Knowledge Extraction (from databases, ontologies, etc.), Semantic Web, Business Rules, Coding Rules, etc.
In particular in this work we have studied the possibilites of this tool for modelling the robot control in Robot Soccer.
It is well known that logic programming is a perfect environment for dealing with the cognitive layer at RoboCupSoccer league as it is in general to implement cognitive and control issues in robotics.
Our goal is to provide a programming framework to Robot Soccer programmers to model robot control in an expressive but simple way.
After some preliminary groundwork Hernndez & Wiguna (2007) we have developed a better engine for rules execution (*RFuzzy* instead of the discrete constraint variant used in Hernndez & Wiguna (2007), called *dfuzzy*) and we have designed and provided a set of unitary test to compare the behaviour of a crisp and a fuzzy strategy for simple movements and for complete matches. Our goal is to provide this framework and some practical results (based in our experimentation) for its use in the strategy programming at Robot Soccer.
After evaluating some study tests we can provide the following conclusions:

- Using fuzzy logic we can model some concepts that are impossible to represent in an adequate way using crisp logic or other representation (i.e. fast, slow, close, far, etc.) Due to this the rules to define robot control are much more expressive and alike to human reasoning.
- *RFuzzy* lets us define continue functions over real numbers using syntactic sugar. Other tools requires to provide values for all elements of the domain. This is impossible for an infinite domain (that is the general case). So, a simple syntax is available.
- Using fuzzy logic we can distinguish the level of satisfaction of a rule. In crisp logic, rules can be satified or not. In *RFuzzy* we can obtain different truth vales of satisfaction for the set of rules that can be applied in a particular situation. So the robot can choose at any time the best rule (the one with highest truth value) that is suppose to provide it the best decision about which action to make.
- The tests comparing single movements show similar efectivenes in fuzzy logic and crisp logic. When both take the same decision, fuzzy logic is faster because it has not to wait till any limit to asign a truth value while crisp logic depend on when the ball across limit areas.
- Atending to decision making, fuzzy control is much better and the disadvantage comes from the speed. In this case it does not affect the experiments because it depends on the cycle and it comes determined by the simulator.

- The tests comparing complete match strategies show that fuzzy control is much better in taking decisions. Due to the importance of speed in this kind of game, an ofensive strategy can obtain better results even if it fails frecuently in the decisions.

Despite the results are good for our experiments in Robot Soccer, they are much better for scenarios in which it is more important to take the right decision that to decide fast. Do not fail in the decision is important in some parts of the Robot Soccer strategy but not in all of it because in much parts the speed is the decisive parameter.

6. References

Baldwin, J. F., Martin, T. P. & Pilsworth, B. W. (1995). *Fril: Fuzzy and Evidential Reasoning in Artificial Intelligence*, John Wiley & Sons.

Bistarelli, S., Montanari, U. & Rossi, F. (2001). Semiring-based constraint Logic Programming: syntax and semantics, *ACM TOPLAS*, Vol. 23, pp. 1–29.

Chen, M., K.Dorer & E.Foroughi (2003). *Users Manual RoboCup Soccer Server.*

CLIP Lab (n.d.). The ciao prolog development system www site.
URL: *http://www.clip.dia.fi.upm.es/Software/Ciao/*

Default values to handel Incomplete Fuzzy Information (2006). Vol. 14 of *IEEE Computational Intelligence Society Electronic Letter, ISSN 0-7803-9489-5*, IEEE.

Extending Prolog with Incomplete Fuzzy Information (2005). Proceedings of the 15th International Workshop on Logic Programming Environments.

FIRA (n.d.). Simurosot environment for soccer game.
URL: *http://www.fira.net/soccer/simurosot/overview.html*

García, A., G.I.Simari & T.Delladio (2004). Designing an Agent System for Controlling a Robotic Soccer Team. Argentine Conference on Computer Science (CACIC 2004).
URL: *http://www.cs.umd.edu/ gisimari/publications/cacic2004GarciaSimariDelladio.pdf*

Guadarrama, S., Munoz-Hernandez, S. & Vaucheret, C. (2004). Fuzzy Prolog: A new approach using soft constraints propagation, *Fuzzy Sets and Systems* **144**(1): 127–150. ISSN 0165-0114.

Guadarrama, S., S.Muñoz & C.Vaucheret (2004). Fuzzy prolog: A new approach using soft constraints propagation, *Fuzzy Sets and Systems* **144**(1): 127–150.

Hermenegildo, M., Bueno, F., Cabeza, D., García de la Banda, M., López, P. & Puebla, G. (1999). The CIAO Multi-Dialect Compiler and System: An Experimentation Workbench for Future (C)LP Systems, *Parallelism and Implementation of Logic and Constraint Logic Programming*, Nova Science, Commack, NY, USA.

Hernndez, S. M. & Wiguna, W. S. (2007). Fuzzy cognitive layer in robocupsoccer, *Proceedings of the 12th International Fuzzy Systems Association World Congress (IFSA 2007). Foundations of Fuzzy Logic and Soft Computing*, Springer, Cancn, Mxico, pp. 635–645.

Ishizuka, M. & Kanai, N. (1985). Prolog-ELF incorporating fuzzy Logic, *International Joint Conference on Artificial Intelligence*, pp. 701–703.

Klawonn, F. & Kruse, R. (1994). A Łukasiewicz logic based Prolog, *Mathware & Soft Computing* **1**(1): 5–29.
URL: *citeseer.nj.nec.com/227289.html*

Klement, E., Mesiar, R. & Pap, E. (n.d.). Triangular norms, Kluwer Academic Publishers.

Lee, R. C. T. (1972). Fuzzy Logic and the resolution principle, *Journal of the Association for Computing Machinery* **19**(1): 119–129.

Li, D. & Liu, D. (1990). *A Fuzzy Prolog Database System*, John Wiley & Sons, New York.

Morcillo, P. & Moreno, G. (2008). Floper, a fuzzy logic programming environment for research, *Proceedings of the Spanish Conference on Programming and Computer Languages, PROLE 2008*, Gijón, Spain.

Moreno, G. (2006). Building a fuzzy transformation system., *SOFtware SEMinar 2006: Theory and Practice of Computer Science*, pp. 409–418.

Munoz-Hernandez, S., Vaucheret, C. & Guadarrama, S. (2002). Combining crisp and fuzzy Logic in a prolog compiler, *in* J. J. Moreno-Navarro & J. Mariño (eds), *Joint Conf. on Declarative Programming: APPIA-GULP-PRODE 2002*, Madrid, Spain, pp. 23–38.

Pradera, A., Trillas, E. & Calvo, T. (2002). A general class of triangular norm-based aggregation operators: quasi-linear t-s operators, *International Journal of Approximate Reasoning* **30**(1): 57–72.

Shen, Z., Ding, L. & Mukaidono, M. (1989). Fuzzy resolution principle, *Proc. of 18th International Symposium on Multiple-valued Logic*, Vol. 5.

Trillas, E., Cubillo, S. & Castro, J. L. (1995). Conjunction and disjunction on ([0,1],<=), *Fuzzy Sets and Systems* **72**: 155–165.

UNCOMA (2006). Diseño e implementación de un sistema multiagente: Un equipo de fútbol con robots.
URL: *http://code.google.com/p/rakiduam*

Vaucheret, C., Guadarrama, S. & Munoz-Hernandez, S. (2002). Fuzzy prolog: A simple general implementation using clp(r), *in* M. Baaz & A. Voronkov (eds), *Logic for Programming, Artificial Intelligence, and Reasoning, LPAR 2002*, number 2514 in *LNAI*, Springer-Verlag, Tbilisi, Georgia, pp. 450–463.

Vojtas, P. (2001). Fuzzy logic programming, *Fuzzy Sets and Systems* **124**(1): 361–370.

13

Soccer at the Microscale: Small Robots with Big Impact

S. L. Firebaugh[1], J. A. Piepmeier[1] and C. D. McGray[2]
[1]*United States Naval Academy*
[2]*Semiconductor Electronics Division, National Institute of Standards and Technology*
USA

1. Introduction

The robots in the RoboCup Nanogram demonstration events are the size of dust, and compete on a millimeter-scale playing field under a microscope (Fig. 1). It seems unlikely that the robots of the Nanogram events will meet the RoboCup goal of beating the human World Cup champion team, so what is the point? The potential for these robots in applications beyond soccer is vast, but so are the engineering challenges. The Nanogram challenge provides an intermediate target for microroboticists, and the innovation that has resulted has been astounding. Some day heart surgery may be as simple as swallowing a pill of microrobots only a few generations evolved from a microscale soccer team.

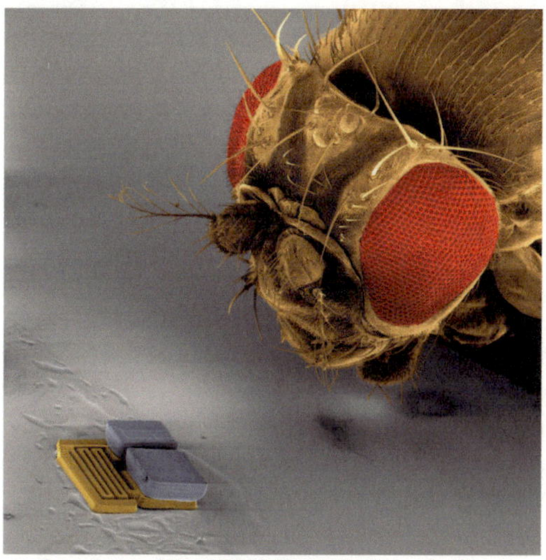

Fig. 1. The Magmite robot, developed by ETH Zurich, in relation to a fruit fly, as captured in a scanning electron micrograph (Vollmers et al., 2008; Kratochvil et al., 2009; Frutiger et al., 2008; Frutiger et al., 2009). The Magmite has participated in the RoboCup Nanogram events since their inception. Figure used with permission from the authors.

The field of microrobotics has been defined "to include the robotic manipulation of objects with characteristic dimensions in the millimeter to micrometer range (micromanipulation) as well as the design and fabrication of robotic agents in a smilar size range (microrobots)" (Abbott et al., 2007). This distinguishes the robots being discussed in this chapter from robots on the millimeter to centimeter scale that are sometimes called "microrobots," but would more accurately be termed "miniature robots" (Dario et al., 1992).

The fundamentals of traditional robotics - including kinematics, dynamics, motion planning, computer vision, and control (Spong et al., 2006) -- apply soundly to the field of microrobotics. However, the nature of the actuators that function best within the micro-domain often dominates how these fundamental concepts are applied. Microrobots competing in the Nanogram events have utilized electrostatic and magnetic actuation methods. Thermal, vibratory, piezoelectric, and biological actuation have also been used for microrobotics, and so these technologies are also reviewed. While microscale end effectors or grippers have been made, Nanogram participants have only utilized passive designs to aid in ball handling. To date, sensors and controllers are all off-board for the Nanogram League, and the control strategies used with the various microrobots are discussed below.

1.1 Microrobot Applications

The classic example of robot applications is the set of line assembly tasks performed in the manufacture of large durable goods, such as automobiles, where robots have been used commercially for decades. Yet these days, robots are rapidly moving into other niche markets such as surgery, security, and entertainment at a pace that has raised the eyebrows of many industry leaders and technology visionaries. As robots become smaller, cheaper, more versatile, and more reliable, some analysts have forecast the coming of a robotics industry of large enough scope to mirror in the next thirty years the economic and societal impact of the computer revolution thirty years ago (Gates, 2007). In response, Microsoft has established a robotics product line, the EU is doubling its investment in robotics research, and U.S. lawmakers have formed a new congressional caucus on robotics.

Some of the most exciting future applications of robotics will require systems that are many orders of magnitude smaller than today's commercial robots. These micro-robots would have masses and characteristic dimensions that are best measured in nanograms and micrometers, respectively. Their tiny size would allow them to penetrate small pores and cavities and to explore microscopic environments, such as the vascular system of the human body. As with today's transistors, the material costs for such small robots are insignificant, so building a million need not be much more expensive than building a dozen. This opens up the possibility of disposable, massively parallel robotic systems for search and rescue in disaster sites, exploration of planetary surfaces or military encampments, and other as yet unimagined tasks.

1.2 Challenges of the Microscale

Realization of commercially viable microrobots to meet these application requirements will require solutions to numerous engineering challenges. At the heart of these challenges lies the fact that the physical forces that dominate locomotion and manipulation tasks in micro-

and nano-scale size regimes are different from those most relevant at larger scales. For objects much below a millimeter in their characteristic dimensions, surface phenomena such as friction, adhesion, heat transfer, and electrostatic attraction become significantly more important, while inertia, weight, heat capacity, and other body phenomena become comparatively less important (Madou, 1997). Furthermore, these surface phenomena can change whenever two micro-scale parts come into contact as a result of triboelectric charging and other factors. These phenomena can be difficult to measure, but must be tracked and controlled to prevent failure modes such as altered transfer functions and irreversible adhesion.

With surface forces dominating motion at these scales, adsorption of contaminants - even water - can lead to devastating results. Proper environmental control can be critical, especially for electrostatic devices (Liu, 2006). Devices and operating environments must be kept meticulously clean, and humidity should be regulated. In contrast to larger robots, once a microrobot is damaged it can rarely be fixed and must instead be discarded. These devices must therefore be manufactured in quantity, at comparatively low cost. While the very small size of these devices makes low-marginal-cost manufacturing practical, it also makes it very difficult to identify damaged devices prior to critical failure. The ability to identify a damaged device in time to replace it before it fails can be a crucial difference between a winning team and a losing one.

Even the very smallest of batteries available today remain far too big to be mounted on a microrobot. As a result, providing a source of power to these devices is a significant problem. One promising approach is to provide a source of wireless power that well-designed microrobots can receive from their environment. This wireless power can be electrostatic (Donald et al., 2003), optical (Sul et al., 2006), magnetic (Vollmers et al., 2008), vibrational (Yasuda et al., 1994), or alternative modes yet to be developed. When multiple (competing) teams are involved, this ambient power must be provided in a way that will not unintentionally (or intentionally!) interfere with the operation of opposing teams. On-board power systems remain a challenging possibility, with thin-film batteries becoming increasingly energetic (Patil et al., 2008) and nuclear power presenting a feasible and exciting approach (Lal et al., 2005).

Control strategies familiar to robot coordination must be carefully reconsidered for applicability to the new field of microrobotics. For example, the closed-loop feedback approach to systems engineering is widely considered a necessity in micro-scale and smaller robots, but the optical systems often employed for this feedback at the macro scale can be inappropriate to micro-environments (Abbot et al., 2007). The depth and breadth of field under the microscope can be limiting factors for vision systems; it is often not possible to employ stereo cameras; and the pin-hole camera models widely used at macro-scales break down under the microscope (Abbot et al., 2007). Other feedback mechanisms must be similarly adapted. For example, in macro-scale actuators, servoing systems are often implemented with optical feedback from diodes or lasers, but capacitive feedback may be more effective at the micro- and nano-scale (Yamahata et al., 2008).

Communication poses particular difficulties in micro-soccer competitions, where individual robots may be smaller than the wavelength of the radio signals typically used to communicate with their larger brethren. If an ambient power source is used, communication signals can be embedded in the power waveform, to be mechanically decoded by the micro-robots' physical structure (Donald et al., 2006; Donald et al., 2008). New communication methods must be devised, and standards must be established to divide the available bandwidth between competing teams of multiple microrobots.

1.3 RoboCup Nanogram Events

Microrobotic soccer competitions have been used to drive innovation in micro-electromechanical systems (MEMS) and microrobotic technologies and to provide educational opportunities. Games were held annually from 2007 through 2009 in association with the RoboCup Federation. Teams competed in each of three events:

- The Two-Millimeter Dash
- The Slalom Drill
- The Ball Handling Drill

In the two-millimeter dash, microrobots must demonstrate maximum speed and the ability to start and stop on command. The microrobots are placed onto a field of play that contains two open goals that are 500 micrometers deep, with two millimeters of space between opposite goal lines. A microrobot is placed inside one of the goals so that the entire body of the microrobot is behind the goal line. Upon a signal from the event timing system, the microrobot must sprint to the opposite goal line, stop inside the goal, and signal completion to the timing system.

The slalom drill proceeds in the same fashion as the two-millimeter dash, except that the area between the goal lines is obstructed by obstacles formed from thick-film photoresist. The microrobots must be maneuverable enough to navigate a path between the obstacles into the opposite goal.

To compete in the ball handling drill, microrobots must be capable of controlled planar pushing manipulation of small disks whose diameter is only 100 micrometers. The microrobot begins inside one of the goals as in the case of the other two events, and the space between the goals is obstructed as in the slalom drill. Tiny "micro-soccer balls," which consist of a 100 μm diameter disk of silicon, are placed amongst the obstacles. Upon a signal from the event timing system, the microrobot moves out onto the playing field to push the micro-soccer balls into the opposing goal.

2. Electrostatic Microrobots

In the field of micro-electromechanical systems (MEMS), electrostatic transducers have long been a favored class of actuator (Kovacs, 1998). Forces are generated between insulated bodies having opposite net charge, resulting in motion. Electrostatic actuators can be comparatively easy to build using established microfabrication techniques, and they can generate considerable forces on micro-scale components due to large surface area to volume ratios. Classic electrostatic actuators from the field of MEMS include comb drive actuators

(Tang et al., 1990), rotary side-drive micro-motors (Fan et al., 1988), and scratch drive actuators (Akiyama et al., 1993). Several teams have used robots that rely on electrostatic actuation in RoboCup Nanogram events, including the United States Naval Academy, Simon Fraser University, and Carnegie Mellon University, whose robots from the 2007 competition in Atlanta are shown in Fig. 2 below.

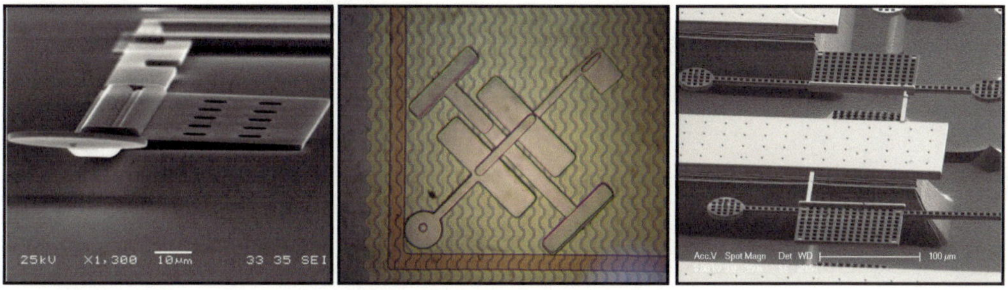

Fig. 2. Electrostatic actuators from the 2007 RoboCup Nanogram event. **Left:** The microrobot from the U.S. Naval Academy team (Firebaugh and Piepmeier, 2009), based on work by Donald et al. (Donald et al., 2006). **Center:** A polymer micromachined electrostatic microrobot from Simon Fraser University. Photo courtesy of Dan Sameoto. **Right:** CMOS-MEMS electrostatic microrobots from Carnegie Mellon University. Photo courtesy of Fernando Alfaro, Chiung Lo and Professor Gary Fedder.

2.1 Electrostatic Actuation

The forces generated by electrostatic actuators can be predicted using simple capacitive circuit models. The energy stored in a capacitor of capacitance C at voltage V is given by:

$$U_c = \frac{1}{2}CV^2 \tag{1}$$

Therefore, the force generated in a given direction, r, is therefore given by:

$$F = \left(\frac{V^2}{2}\right)\frac{\partial C}{\partial r} \tag{2}$$

Knowing the geometry of an electrostatic actuator allows us to approximate its capacitance, either analytically or using finite element methods. For example, in the simplest case of a parallel plate capacitor, the capacitance is $\varepsilon wL/g$, where L is the length of the overlapping area of the two plates, w is the width of the overlap, g is the gap between the plates, and ε is the permittivity of the material in the gap. If the plates are misaligned so that L would be increased by shifting the top plate along the x axis, then the x-component of the electrostatic force on the top plate would be $F_x = \varepsilon w V^2/2g$. Similarly, the force of attraction between the two plates is $F_z = \varepsilon wLV^2/2g^2$.

The force of attraction between the capacitor plates is a non-linear function of the plate separation. This leads to the useful phenomenon of snap-down voltages (Nathanson et al.,

1978). Imagine that the top plate were suspended over the bottom plate by a linear spring of constant K. For a given extension, z, the energy stored in the spring is:

$$U_S = K\frac{z^2}{2} \tag{3}$$

The energy stored in the capacitor, U_c, is:

$$U_C = \frac{\varepsilon L w V^2}{2(g_0 - z)} \tag{4}$$

where g_0 is the plate separation at zero spring extension. Minimizing the energy of the system, the relationship between the plate separation and the applied voltage can be calculated as follows:

$$V = \frac{g\sqrt{2\varepsilon LwK(g_0 - g)}}{\varepsilon Lw} \tag{5}$$

This function defines the curve shown in Fig. 3. To the right of the maximum voltage, the system is statically stable. To the left, the system is unstable and will collapse, extending the spring until the plates come into contact or encounter a hard stop. Solving for V and maximizing with respect to g produces the voltage and gap at which this snap-down event occurs (Nathanson et al., 1978):

$$\begin{aligned} V_{SD} &= \sqrt{\frac{8Kg_0^3}{27\varepsilon Lw}} \\ g_{SD} &= \frac{2g_0}{3} \end{aligned} \tag{6}$$

If the plates are prevented from electrical contact by, for example, a layer of insulation, they will remain snapped down until the voltage is reduced below the release voltage. The release voltage for a given gap can be determined from the left-hand side of the graph in Fig. 3. Each hysteresis band between a snap-down voltage and a release voltage can be used as a mechanical memory bit for electrostatic microrobotic systems (Donald et al., 2006), allowing the robots to be configured as simple state machines without any on-board transistors.

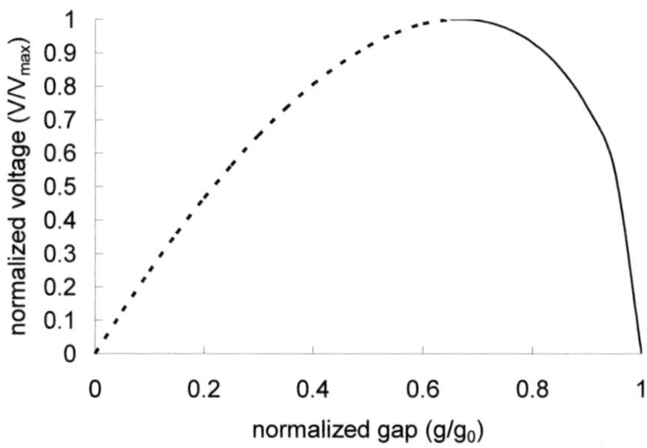

Fig. 3. Graph of capacitor plate separation versus applied voltage. To the right of the maximum, the system is statically stable. To the left, the system is unstable, and will collapse, extending the spring until the plates come into contact or encounter a hard stop. After Nathanson (Nathanson et al., 1978).

For microrobotic applications, one of the more common electrostatic actuators is the scratch drive. A scratch drive actuator (Akiyama et al., 1993) is composed of a thin polysilicon plate with a support at the front end. The plate is typically in the range of 60 µm to 80 µm on a side, and 1 µm to 2 µm thick. The support height is typically in the 1 µm to 2 µm range. The operation of the scratch drive is shown in Fig. 4. When a voltage is applied between the polysilicon plate and the substrate beneath it, the plate is drawn down into flat contact with the dielectric layer. Since the front of the plate is supported by the bushing, strain energy is stored in the plate, and the edge of the bushing is pushed forward. When the voltage is removed, the strain is released, and the scratch drive plate snaps back to its original shape, slightly in front of where it began. When a periodic pulse is applied, this cycle is continuously repeated, and the scratch drive moves forward in a step-wise manner, achieving maximum speeds on the order of millimeters per second.

Steering can be accomplished by exploiting snap-down and release voltages of cantilever beams mounted on the scratch drive body. Fig. 2 (left) shows one such steering cantilever with a two-micron-diameter stylus at the tip. If the periodic drive pulse of the scratch drive actuator nests within the hysteresis band of these cantilevers, the cantilevers' position (raised or lowered) can be controlled independently from the operation of the actuator (Donald et al., 2006).

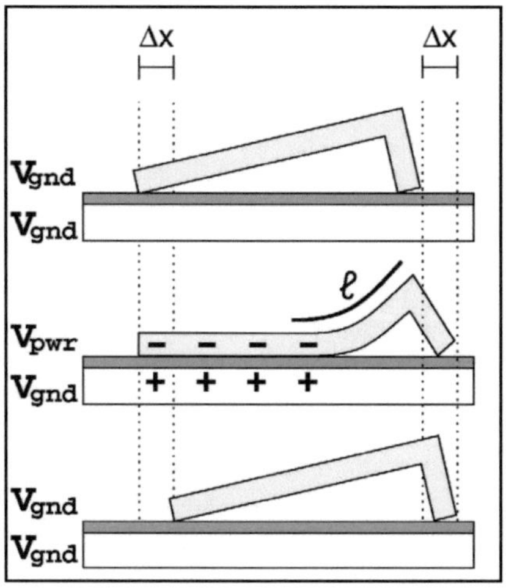

Fig. 4. A schematic of the operation of a scratch drive actuator (Akiyama et al., 1993). The length of the curved region of the plate, ℓ, and the step size, Δx, are determined by the voltage.

2.2 Electrostatic Microrobot Fabrication

Fabrication of thin-film electrostatic actuators like the scratch drive is performed through photolithographic techniques that were originally developed for the integrated circuit industry. Each device is composed of multiple layers, each of which is defined by a lithographic mask. Typically, a material such as silicon, glass, or metal is deposited by chemical or physical vapor deposition and is then coated with a film of photoresist. The photoresist is exposed through a lithographic mask and developed to remove the unwanted portion of the film. The pattern defined by the remaining film is transferred into the material layer through an etching process, and then the residual photoresist is removed. As this process is repeated, the device is built up layer by layer. The equipment required for these processes is available at microfabrication laboratories in many universities. In addition, many standard thin-film microfabrication processes are available commercially as multi-project wafer services, where each wafer is divided between many researchers, producing significant cost decreases (Markus et al., 1995; Sniegowski et al., 1996; Tea et al., 1997). Participants in these processes receive a set of die containing devices built from their own supplied designs. The die can then be post-processed for additional customization if desired (Donald et al., 2006; Huikai et al., 2002).

2.3 Electrostatic Microrobot Control

The work by Donald (Donald et al., 2006) uses the pre-image motion planning strategy developed by Lozano-Perez, Mason, and Taylor (Lozano-Perez, et al., 1984). The planner starts with the goal position and computes backwards the sequence of single velocity motion primitives that leads to the initial position of each robot. This method assumes perfect robot motion. Since robots rarely live up to this assumption, a closed-loop error correction method is employed. The trajectory is recalculated at periodic intervals, making on-the-fly adjustments for observed error. In the case of (Donald et al., 2006) the feedback is provided by a digital microscope camera, and trajectory adjustments are made by changing the relative percentages of the interleaved control primitives.

The microrobotic devices may by modeled as nonholonomic mobile robots similar to a Dubins vehicle (Dubin, 1957) limited to forward motion and left turns. The state of the robot q is given by:

$$q = \begin{bmatrix} x \\ y \\ \theta \end{bmatrix} \tag{7}$$

where (x,y) is the location of the robot, and θ is the orientation of the device. The kinematics of the device are given by:

$$\dot{q} = \begin{bmatrix} v\cos\theta \\ v\sin\theta \\ \sigma\omega\theta \end{bmatrix} \tag{8}$$

where v is a positive velocity with an expected constant value that is related to the stepping frequency in the forward motion wave form shown in Figure 7. The angular velocity w is also a fixed value, and $\sigma \in \{0,1\}$ is the state of the steering arm (0= up, 1=down). It should be noted that w may be positive or negative depending on the construction of the arm. Without loss of generality, we will assume the value is positive for the remainder of this discussion. The robot is globally controllable, but not small time locally controllable (STLC).

As discussed in (G. Lionis and K.J. Kyriakopoulos, 2007) the radius of curvature can be varied by alternating the relative percentage of the two voltage waveforms that moves the robot either forward or in a turn within a short time period. The device kinematics of the ideal system can be decomposed using control vectors that characterize all of the possible motions of the scratch drive actuators. For scratch drive actuators, these vectors are given by the following

$$g_1 = \begin{bmatrix} v_1\cos\theta \\ v_1\sin\theta \\ 0 \end{bmatrix}, g_2 = \begin{bmatrix} v_2\cos\theta \\ v_2\sin\theta \\ \omega_2 \end{bmatrix} \tag{9}$$

A key departure here from the work (G. Lionis and K.J. Kyriakopoulos, 2007) is the absence of the ability to achieve a $-w$ or $-v$. The work in (G. Lionis and K.J. Kyriakopoulos, 2007) considered a larger microrobot with piezo-actuated legs similar to that discussed in (Martel, 2005). That system had the ability to go forwards and backwards as well as turn both left and right with fixed radius turns. Despite these differences, the large portions of the discussion of (G. Lionis and K.J. Kyriakopoulos, 2007) are still applicable. Of note is the analysis concerning a pulse width modulated (PWM) control strategy. The approach is to alternate between two of the permissible vectors. Fig. 5 shows the two waveforms that would be sent to the scratch drive actuator in order to switch between forward motion and turning motion.

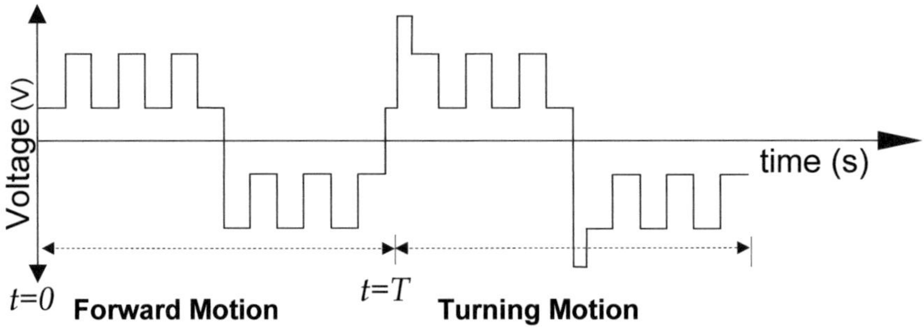

Fig. 5. Pulse width modulated (PWM) control scheme that alternates between forward motion for time T and turning motion with time aT.

To create a PWM control scheme, a switching function is defined

$$\sigma(t,a,T)=\begin{cases} 1 & 0\le t<T \\ 0 & T\le t<(a+1)T \\ \sigma(t-(a+1)T,a,T) & t>(a+1)T \end{cases} \tag{10}$$

Physically, the switching is achieved by lowering and raising the steering arm. The control input is defined as $C^{a}_{g_2-g_1}(q_0)$, and is constructed by selecting a, T, g_1, and g_2 such that

$$\dot{q}=\sigma(t,a,T)g_1+(1-\sigma(t,a,T))g_2 \tag{11}$$

Note that T defines the time for the motion primitive g_1 and a controls the length of time for the second motion primitive g_j relative to g_i.

It was shown in (G. Lionis and K.J. Kyriakopoulos, 2007) that as $T \to 0$, the microrobot with a fixed turning radius will move as a unicycle; a device capable of arbitrary curvature. For example, switching between g_1 and g_2 will result in

$$g_{12a}=\begin{bmatrix} v\cos\theta \\ v\sin\theta \\ \dfrac{a}{a+1}\omega_2 \end{bmatrix} \tag{12}$$

While these results hold in the ideal case, implementation issues for scratch drive actuators violate two of the assumptions made in the analysis. First, the motion of the robots can be inconsistent. Devices often exhibit mild turning characteristics even when they are controlled by the straight motion wave function (Donald et al., 2006). Secondly, the practicalities of waveform generation dictate that T is small, but does not necessarily approach zero. In (Donald et al., 2006) the length of T was set at 0.25 ms. In (Piepmeier, 2010) it was shown that the following model gives a better approximation of the motion of the device for larger control periods, T. The following vector field provides a closer approximation of the motion produced by a PWM-controlled scratch drive device.

$$\tilde{g}_{12a} = \begin{bmatrix} \left(\frac{v_1}{a+1} + \frac{v_2 a}{a+1} \right) \cos\theta \\ \left(\frac{v_1}{a+1} + \frac{v_2 a}{a+1} \right) \sin\theta \\ \frac{1}{a+1}\omega_1 + \frac{a}{a+1}\omega_2 \end{bmatrix} \tag{13}$$

3. Magnetic Microrobots

Two groups participating in the RoboCup Nanogram events, one from Carnegie Mellon University, and the other from ETH Zurich University, have demonstrated magnetic microrobots. These systems have shown less sensitivity to environmental variation than microrobots based on scratch drive actuators. The two magnetic microrobotic systems operate on different principles, as described below.

The Carnegie Mellon microrobot, developed by Steven Floyd, Chytra Pawashe, and Metin Sitti, is simply a laser machined slug of neodymium-iron-boron, a hard magnetic material, which is manipulated through externally applied magnetic fields (Floyd et al., 2008 Pawashe, 2008; Pawashe, 2009, a; Pawashe, 2009, b). This robot is similar to the microrobots developed by a number of groups for biomedical applications (Gauthier & Piat, 2004; Yesin et al., 2006; Tamaz et al., 2008; Yamazaki et al., 2004). Robot technologies such as this, where the robot itself is a simple shape that is translated through externally applied magnetic fields without internally moving parts, will be termed "ferromagnetic-core-based robots." Most of these microrobot technologies are targeted towards operation in a three-dimensional fluid environment, such as the human body. The distinction for the Carnegie Mellon microrobot is that the control system has been tailored to allow the robot to operate on a surface.

The ETH Zurich microrobot (Vollmers et al., 2008; Kratochvil et al., 2009; Frutiger et al., 2008, a; Frutiger et al., 2008, b; Frutiger et al., 2009) utilizes an alternating magnetic field to excite mechanical resonances within the structure, allowing the robot to move through a stick-slip method. Bidirectional motion is facilitated by a z-axis clamping force. Robots of this type will be termed "resonant magnetic robots." Before describing these technologies in more detail, however, it is useful to review the physics of magnetics.

Fig. 6. Electromagnetic cage used by Floyd et al. for microrobot actuation (Floyd, 2008; Pawashe, 2008; Pawashe, 2009, a; Pawashe, 2009, b). Figure used with permission from the authors.

3.1 Magnetic Force Generation

For both of these technologies, the playing field area is encased inside of a cage of independent electromagnet coils, as is illustrated in Fig. 6 for the Carnegie Mellon microrobot. The Biot-Savart law determines the relationship between the current, I, through a coil, C, consisting of N turns, and the magnetic field density vector, $\mathbf{B}$, at a distant position (Hayt & Buck, 2006):

$$\vec{B} = \frac{\mu_0 I}{4\pi} \oint_C \frac{d\vec{L} \times \vec{a}_R}{R^2} . \tag{14}$$

where μ_0 is the permeability of free space, $\mathbf{dL}$ is an infinitesimal line segment along the direction of integration, $\mathbf{a_R}$ is the unit vector pointing from $\mathbf{dL}$ to the point of interest, and R is the distance from the line segment to the point of interest. Of course, in a case where multiple independent coils are used, the net magnetic field at the microrobot location must be found by adding the contributions from each coil at the location of the microrobot. The important thing to note about this equation is that *the field intensity is inversely proportional to the square of the distance between the coil and the robot.* This is an important consideration for the control of the robot, as the resulting force is highly non-linear with position.

It's important to distinguish between magnetic field intensity vector, $\mathbf{H}$ (units: A/m), and the magnetic field density vector, $\mathbf{B}$ (units: T or Wb/m^2), which is the induced total magnetic field within a given material (Liu, 2006). The relationship between the two quantities is given by:

$$\vec{B} = \mu_0\left(\vec{H} + \vec{M}\right) = \mu_0\left(\vec{H} + \chi\vec{H}\right) = \mu_0\mu_r\vec{H} \tag{15}$$

where μ_0 is the magnetic permeability of space, μ_r is the relative permeability of the magnetic material, χ is the susceptibility, and **M** the internal magnetization. The relative permeability is large for ferromagnetic materials such as iron and nickel, and these are the materials most commonly used in magnetic microactuators.

There are two classes of ferromagnets: hard magnets, which retain some of their magnetic polarization even after an applied external magnetic field is removed, and soft magnets, which exhibit internal magnetization only in the presence of an external magnetic field. In general, higher magnetization levels are available in hard magnetic materials. Shape anisotropy, more so than the orientation to the induction field, determines the direction of polarization in the material. A rod-shaped piece of material, for example, will usually have an internal field pointing in the longitudinal direction, and a thin plate will usually exhibit magnetization in the plane of the plate, even when the induction field is oriented in another direction. This is a particular consideration for soft-magnetic materials-- hard magnetic materials can be magnetized before being cut into the desired shape, allowing the designer control over the orientation of the magnetization relative to the shape.

In the presence of an external magnetic field, both hard and soft magnets will rotate until their internal magnetization is parallel to the local external magnetic field lines. A net force for translational motion, however, is only exerted in the case of non-uniform magnetic field. The equations for torque and force are given by (Yesin et al., 2006):

$$\begin{aligned} \vec{T} &= V\vec{M} \times \vec{B} \\ \vec{F} &= V\left(\vec{M} \bullet \nabla\right)\vec{B} \end{aligned} \tag{16}$$

where V is the volume of the slug of ferromagnetic material. The force will draw the slug in the direction of increasing field intensity, or towards the magnet.

The challenge for magnetic actuation is not getting the robot to move, but rather getting it to move in a controllable fashion. Combining the distance dependencies of the magnetization and of the gradient of the magnetic field, the force for a soft magnet is inversely proportional to the fifth power of the distance between the robot and the electromagnet coil (Yesin et al., 2006). For a hard magnet, it depends only on the inverse of the cube. Once the robot begins to move, it quickly accelerates towards the electromagnet, and without careful control it will snap to the electromagnet. This problem is addressed in part by sizing and placing the electromagnets so that their interior volume is much greater than the volume in which the robot is meant to operate. This reduces the variation in the field strength over the playing field area and allows the opposing electromagnet to counter one electromagnet's force at a reasonable current level. The non-linearity is also addressed by using oscillatory motion or multiple coils operating simultaneously to build a linear field, such as the Helmholtz or Maxwell coil configurations, which are common in the magnetic resonance imaging (MRI) industry (Yesin et al., 2006).

3.2 Ferromagnetic-Core-Based Microrobots

Floyd et al. use a hard magnet as their robotic element (Floyd et al., 2008; Pawashe, 2008; Pawashe, 2009, a; Pawashe, 2009, b). The robot is cut from a magnetized sheet of NdFeB using laser machining. The robot, shown in motion in Fig. 7, is chevron-shaped, 100 μm thick, and approximately 250 μm by 130 μm in the *xy*-plane. It has a mass of 25.6 μg and a magnetization of 500 kA/m, and it operates within a cubic workspace approximately 2 cm on a side with gradient fields as strong as 149 mT/m. They use short duration, periodic fluctuations in the magnetic field to control robot motion.

The robot body is controlled using five independent electromagnetic coils. Four of these are in the plane of the robot's playing field and the fifth is below the playing field. Two actuation techniques can be employed, and both of them take advantage of stick-slip motion of the robotic mass induced by pulsed magnetic fields.

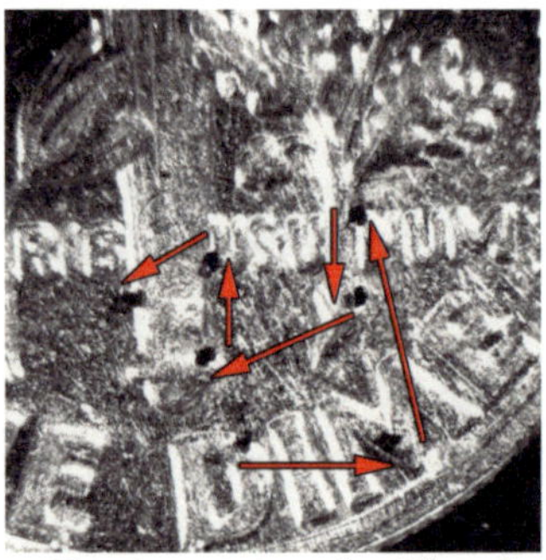

Fig. 7. The Carnegie Mellon Mag-μBot in motion, by Floyd and Pawashe at Carnegie Mellon University. The robot is approximately 300 μm x 300 μm x 100 μm(Floyd, 2008; Pawashe, 2008; Pawashe, 2009, a; Pawashe, 2009, b). Figure used with permission from the authors.

The first technique, In-Plane Pulsing (IPP), first uses the fifth coil to clamp the robot to the playing field and then orients the robot using the other four coils. Then, with the clamping coil still in effect, translation is effected by applying a saw tooth magnetic field waveform with the four in plane coils. The second technique, Out-of-Plane Pulsing (OPP), reverses the role of the clamping coil and the four in-plane coils. The coil beneath the work surface is pulsed to create a saw tooth magnetic field, and the in-plane coils are held constant with similar stick-slip translational motion as a result. Experimental results show that robot velocities increase with increased pulsing frequencies for both IPP and OPP methods; however IPP produces more consistent results fitting an exponential attack curve. OPP produces higher velocities, on the order of 2.8 mm/s, where the maximum velocity for IPP was only 700 μm/s. The authors suggest OPP for coarser motion and IPP for finer control.

While this robot is tailored to surface motion, several groups have pursued similar strategies for microrobots tailored to biomedical applications. The group of S. Martel at Ecole Polytechnique de Montreal has adapted a clinical magnetic resonance imaging (MRI) platform for control of a ferromagnetic bead within a fluid environment, and have demonstrated their work in vivo in swine (Mathieu et al., 2006; Martel, 2007; Tamaz et al., 2008). Yesin et al. at ETH Zurich have developed a nickel, football-shaped microrobot that could be used for microsurgery in the human eye as well as cardiovascular applications (Yesin et al., 2006).

ETH Zurich has also developed "nanohelices" or tiny spiral structures (about 60 µm in length) that have a soft magnetic head which allows them to be manipulated by a magnetic field (Kratochvil et al., 2008). With these structures they have demonstrated controlled conversion between rotary and linear motion, which finds application both as a biomimetic swimming microrobot actuator (like a bacteria flagellum) and as an instrumentation tool for rotating structures within a scanning electron microscope. Similar work at a larger size-scale (about 1 mm in length) using a spiral of tungsten wire has been demonstrated by Yamazaki et al. (Yamazaki et al., 2004). Biomimetic, bacteria-flagella-type motion has also been demonstrated by Dreyfus et al. (Dreyfus et al., 2005), who formed their actuator from a string of magnetic beads held together by DNA and attached to a red blood cell.

3.3 Resonant Magnetic Robots

The resonant magnetic robots, or "Magmites" developed by the group at ETH Zurich (Vollmers et al., 2008; Kratochvil et al., 2009; Frutiger et al., 2008; Frutiger et al., 2009) consist of a base frame that contains a spring and two ferromagnetic masses. One mass is attached to the frame and the other to the spring. The entire robot covers an area of 300 µm x 300 µm and is about 150 µm thick. In a spatially uniform magnetic field, the two adjacent pieces of ferromagnetic material magnetize and experience interaction forces but no net force. When the field is oscillated at the appropriate frequency, it excites the mechanical resonance of the structure, swinging the mass attached to the spring (the hammer) within the plane of the frame. By applying a clamping field at the right moment in the cycle, the momentum of this hammer action is transformed into translational motion. The direction of motion depends on the phase difference between the magnetic excitation field and the clamping field. Clamping is accomplished electrostatically using the playing field electrode array. Unidirectional motion is even possible in the absence of the clamping field due to the non-uniformity of the friction forces during the resonator cycle.

Magmites are fabricated through a surface micromachining process that utilizes copper or gold for the frame and spring structure and electroplated nickel for the soft magnetic masses, as is illustrated in Fig. 8. Dimples on the bottom of the frame are used to reduce the contact area with the substrate. Magmites have been shown to operate in an environment of up to 60% relative humidity. They use fields as low as 2 mT operating in the frequency range of 1 kHz to 5 kHz. In the 2009 competition in Graz, Austria, they set the record for the 2 mm dash with a time of 0.326 s (a speed of 6.1 m/s, faster speeds were demonstrated informally). Multirobot control is demonstrated by using robots engineered for different resonance frequencies (Kratochvil et al., 2009).

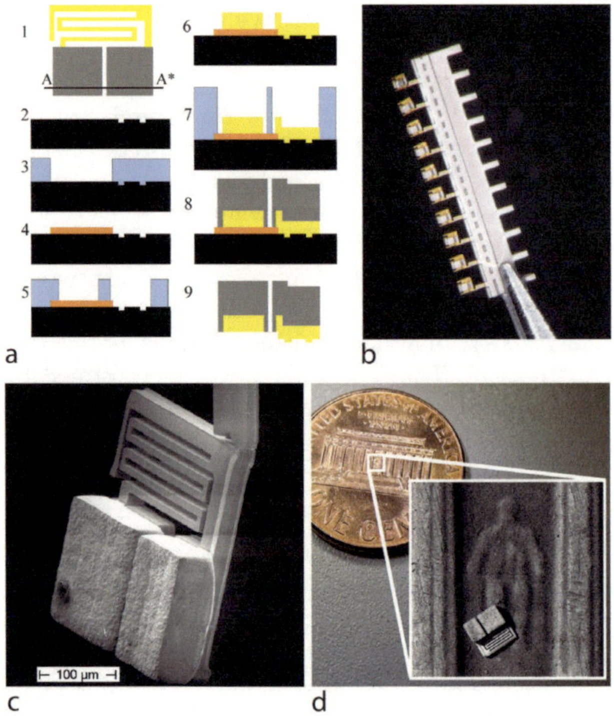

Fig. 8. a) The fabrication procedure for first-generation magmites uses surface micromachining with electroplated gold and nickel layers; b) A strip of released Magmites still attached to a tether for handling; c) a scanning electron micrograph of a Magmite; d) the Magmite on an American penny, reprinted from (Frutiger, 2009). Figures by Frutiger, Vollmers and Kratochvil and used with permission by the authors.

3.4 Magnetic Microrobot Challenges

Fabrication limitations pose a significant challenge for this class of microrobots. Traditional silicon micromachining incorporates few materials which are ferromagnetic. Only Nickel is available in a commercial microfabrication process, and, with a relative permeability of only 600, it is far less ferromagnetic than materials such as permalloy used in common macroscale magnetic systems. Custom electroplating can be used to integrate a handful of other materials, but the best magnetic materials are still not available through these methods (Cugat et al., 2003). Techniques such as laser cutting allow for greater material breadth, but it is difficult to then integrate the magnetic material with other microstructures.

Multirobot cooperation also presents a challenge, particularly for ferromagnetic-core-based robots, as there is no way to selectively address a particular robot with the magnetic field, alone. However, Pawashe et al. (Pawashe et al., 2009) have used an underlying electrode array that uses localized electrostatic clamping to differentiate between robots. For resonant microrobots, individual robots can be targeted by their resonant frequencies.

Even with these challenges, magnetic microrobots have outperformed the scratch drive actuator-based robots thus far in the Nanogram soccer competitions. It seems likely that fabrication methods will expand to incorporate more magnetic material variety if there is sufficient need. Furthermore, given the prevalence of magnetic resonance imaging (MRI) systems in medicine, magnetic microrobots are an enticing candidate for medical applications.

4. Other Microrobot Actuation Methods

While only electrostatic and magnetic actuators have been used by teams competing in the Nanogram events, there are a variety of other actuation methods that have been used for microrobotics. In particular, thermally actuated microrobotics has been demonstrated by several groups (Ohmichi et al., 1997; Ebefors et al., 2000; Kladitis et al., 2000; Baglio et al., 2002; Sul et al., 2006; Brown et al., 2007). Several groups have also investigated robots powered by external vibrations (Yasuda et al., 1994; Yasuda et al., 1995; Saitou et al., 2000), and piezoelectric actuators (Smits, 1989; Wood et al., 2003; Nguyen and Martel, 2006; Edqvist, 2009). Some of these systems are not microrobot systems in the sense of a microscale independent actor, but rather micro- or nano-positioning systems where the actuators are fixed to a substrate and are used to manipulate micro-sized components, like a microscale conveyer belt. In the fluid domain, a few groups have even been working with bacteria flagella-based propulsion (Behkam and Sitti, 2007; Martel et al., 2008) and semiconductor diode propulsion (Chang et al., 2007). These technologies are not as easily adapted to a surface walking application, but are briefly reviewed as they relate to the larger nanorobotic goals of minimally invasive surgery, microassembly, and micropositioning.

4.1 Thermal Actuation

Thermal actuation is prevalent in microsystems, and so, while there has as yet been no thermally actuated competitor in the Nanogram events, the future appearance of a robot of this type seems likely. These robots rely on thermal expansion for motion. There are two large classes of thermally actuated robots: "inchworm" drives and "impulse" or "impact" drives. In an inchworm drive cyclic deformation results in forward motion in a manner that is relatively independent of the time scale of the deformation. In such devices the speed should simply be linearly dependent on the excitation frequency. In contrast, for an "impulse" drive momentum generation is essential to the forward motion, making the actuation method dependent on the time scale of the excitation. In other words, the inchworm drive is a sort of shuffle step, while the impulse drive requires an initial sharp kick. (It's the Charleston of actuation mechanisms.)

4.1.1 The physics of thermal actuation

Thermal actuators are based on thermal expansion. Most materials expand when heated, including most of the materials common in micromachining. Thermal actuators can be divided into three broad classes: thermal bimorphs, single material structures, and multiphase devices. The final category, which includes ink jet heads where the expansion of a gas bubble is used to force a fixed volume of ink out of a small cavity, has not been applied to microrobotics.

Thermal bimorphs consist of layered structures containing materials with different thermal expansion properties. When heated, the structures will bend away from the side of the structure containing the material with the higher expansion coefficient. Because of fabrication limitations, thermal bimorphs are mostly used for out-of-plane motion. In contrast, single-material thermal actuators are more commonly used for in-plane motion, although they can be used for both in-plane and out-of-plane motion. Single-material actuators sometimes rely on localized heating to create motion. This is certainly the case for laser-heated structures, but also can occur in Joule-heated devices if the device geometry confines the heating to a small area by creating an area that is thermally confined. An example of this is shown on the left of Fig. 9.

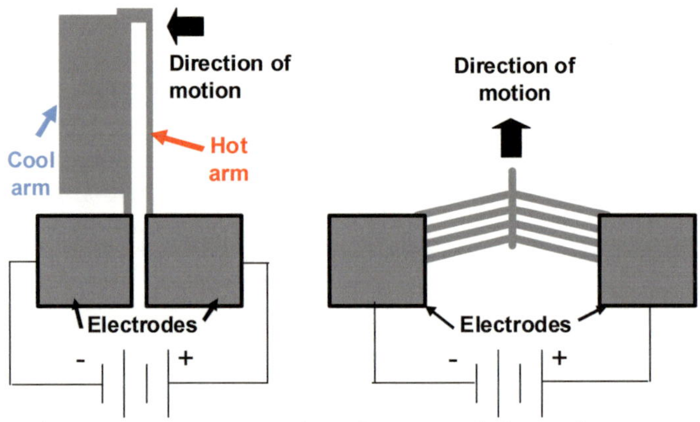

Fig. 9. Illustration of two common types of single-material thermal actuators. On the left, the geometry of the structure results in differential heating causing the thin arm to expand and bend the structure towards the thick arm. In a bent beam actuator like on the right, the beams are oriented so that thermal expansion leads to forward motion of the shuttle structure in the middle.

Another common type of thermal actuator is a "bent-beam" actuator, illustrated on the right of Fig. 9. In these chevron-shaped structures, current is passed through the beam causing heating. Because of the bend in the center of the beam, expansion leads to translational motion in-plane in the direction of the chevron's base.

While much of the interest in thermally actuated robotic systems has been directed at miniature robots (Kladitis and Bright, 2000, Ebefors et al., 2000), a number of groups have demonstrated impulse-drive microrobots that are actuated through local heating by lasers—an idea that was introduced by (Baglio et al., 2002). Sul et al. at UNC Chapel Hill have illustrated an elegant tripod shaped robot that is actuated by local heating (Sul et al., 2006). The three legs of the robot are metal-film bimorphs. At equilibrium the legs arch down to the substrate because of residual stress from the fabrication process. When heated, the legs further deflect due to differential thermal expansion coefficients in the two materials. The rapid heating of one leg leads to a stepwise transition on a low-friction surface. There are two phases to the motion, the contraction phase, in which there is rapid motion of the device which breaks the adhesive contact and overcomes sliding friction for the contacts, and the relaxation phase where the device returns to its original shape. In the contraction phase for the UNC microrobot, only one leg is heated, but, in the relaxation phase, the heat has spread throughout the device and dissipates in a more uniform fashion, causing the device to move away from the leg that was heated (Sul et al., 2006). Thus the device can be steered by focusing the laser on the appropriate leg.

A similar micro-impact drive mechanism was described by Ohmichi et al. on the millimeter scale (Ohmichi et al., 1997). These robots were fabricated from aluminum alloy by precision cutting techniques. Like the UNC microrobot, these structures relied on a fast initial stage where the motion is generated, followed by a slow relaxation period. The Ohmichi robot consisted of a main body that is separated from a weight block by a narrow neck. A laser was directed onto the neck causing rapid thermal expansion which pushed the weight away from the body, resulting in an impulsive force that exceeded the static friction force. The heat then dissipated throughout the structure causing a slower relaxation to the original shape that does not counteract the initial motion. With repeated optical pulsing (up to 5 kHz) the team achieved speeds of up to 31 mm/s. The robot was approximately 1.7 mm x 0.6 mm x 0.4 mm.

One challenge for these types of robots is the requirement for optical access and fine control of the optical system. A further challenge for these systems is that the complexity of the control system would significantly increase for multirobot cooperation. Finally, these systems have relied on special low-friction surfaces.

Friction forces are exploited by Brown et al. at Dalhousie University (Brown et al., 2007). Their "frictional crawler" consists of 3 "feet" linked by two actuators. The locomotion mechanism uses a shuffle step that takes advantage of the differences in contact area that result from coupling one or two feet to achieve net translational motion. The team used bent-beam actuators, of the type illustrated in Fig. 9 (right). Electrical connection to the actuators was accomplished through rails on the substrate. The structures were fabricated using a silicon-on-insulator process by the Micragem process[1] (CMC Microsystems, 2009) that results in a single-crystal-silicon actuators. The overall device dimensions were 1400 μm X 525 μm X 10 μm. The actuators required 2.75 V for operation. At this voltage, each drew about 69 mA. The device traveled at 0.7 mm/s at its maximum frequency of 300 Hz (the limit for the thermal actuators) and was observed to develop a horizontal force greater than 130 μN. Significantly, it operated reliably even when carrying a load over 100 times its own weight of 1 μN. Good contact between the device and the rails was essential to operation, and the team used a thin film of electrically conductive grease on the rails to maintain that contact in the face of device imperfections. The rails, of course, limit the range and turning capability of this robot, but the overall locomotion method is quite interesting.

[1] Certain commercial equipment, instruments, or materials are identified in this review to adequately specify the experimental procedure. Such identification does not imply recommendation or endorsement by the National Institute of Standards and Technology, nor does it imply that the materials or equipment identified are necessarily the best available for the purpose.

4.2 Vibration Actuation

Vibration was used for microrobot locomotion by Yasuda et al., who have built a millimeter-scale walking robot (Yasuda et al., 1994; Yasuda et al., 1995). Their robot had six legs: four of which supported the body and transmitted the vibration energy to the robot, and two of which were free to "kick" the base to direct its motion. The kicking legs were cantilevered mass-spring units designed for particular resonant frequencies. One of the kicking legs caused right turns and the other left turns. The resonant frequency for the kicking legs was well below that of the four support legs. By changing the spectra of the vibration field, one, both, or neither of the legs can be activated allowing for turning, straight motion, and a full stop. The microrobot measured 1.5 mm by 0.7 mm. It was fabricated by surface micromachining with a sacrificial layer of phosphosilicate glass (PSG) and structural layers of polysilicon and polyimide (used to create hinges). Following fabrication, the final structure was created by folding using micro probes. The kicking legs were tailored to resonance frequencies in the 100 Hz to 1000 Hz range, and motion of up to 7 mm/s was observed.

Saitou et al. use microcantilever impactors designed for specific mechanical resonance frequencies in their micropositioning system (Saitou, 2000). The actuators are anchored to the substrate, and they control a shuttle intended for fine positioning of a generic microcomponent. The actuators are suspended masses that resonate in the plane of the substrate. As they resonate they impact against the side of the shuttle structure moving it forward or backward depending on the relative orientation of the actuator and the shuttle. Two actuators with one resonance frequency are positioned to cause forward motion, and another two actuators with a distinct resonance frequency are positioned to cause reverse motion. When the entire system is exposed to a vibration field, the frequency of that vibration determines the direction of motion of the shuttle. The system was fabricated using the PolyMUMPS process, which is a multi-user, polysilicon-based, surface micromachining process (MEMSCAP, 2009). The resonance frequencies used for the structure are in the range of 1 kHz to 10 kHz, and result in motion at the speed of 2 mm/s to 5 mm/s.

While robots with this vibration actuator scheme have not yet been used in the Nanogram events, the vibration actuation mechanism offers many of the same advantages as the Magmite resonant magnetic actuation method. The use of resonance allows for multirobot cooperation by allowing individual robots to be targeted with a global energy field through frequency control.

4.3 Piezoelectric Actuation

The presence of an electric field induces stress in piezoelectric materials. This property has been exploited to create a variety of electrically powered actuators. A few groups have used piezoelectric actuators for millimeter-to-centimeter scale robots, including flying insect-modeled miniature robots (Wood et al., 2003) and vibratory walking miniature robots (Nguyen and Martel, 2006, Edqvist et al., 2009). While these existing systems are a one to two orders of magnitude larger than microrobots, the scaling limits of the piezoelectric actuation methodhave not yet been fully explored.

Piezoelectricity was first identified in quartz crystals and is widely applied in sonar systems and in quartz crystal oscillators. Material processing presents a significant challenge to the development of piezoelectric MEMS. Techniques that are commonly used for preparing bulk piezoelectric materials are not suited for microfabrication, and the piezoelectric coupling coefficients for thin-film materials are often significantly lower than their bulk counterparts (Liu, 2006). However, recent advances in deposition techniques for PZT (lead zirconate titanate) (Wang, 2003) and PVDF (polyvinylidenfluoride, a piezoelectric polymer) (Manohara et al., 1999) have increased the prevalence of this transduction mechanism.

4.4 Biological and Other Actuation Methods

The work of several groups in which ferromagnetic core based robots were used to form flagella-like propulsion structures was reviewed in the magnetic actuation section (Kratochvil et al., 2008; Yamazaki et al., 2004; Dreyfus et al, 2005). Other groups, however, work directly with biological structures for micron-scale aqueous propulsion. Martel et al. have worked with "Magnetotactic Bacteria (MTB)," which is bacteria that contains magnetosomes—membrane-based nanoparticles of a magnetic iron that respond to magnetic fields (Martel et al., 2008). They have attached these flagellated bacteria to microbeads and demonstrated controlled motion using a magnetic resonance imaging (MRI) system. Behkam and Sitti use chemicals instead of magnetic fields to control their bacterial flagella actuators, which use Serratia marcescens bacteria attached to polystyrene beads (Behkam & Sitti, 2007). Other mechanisms for micro- and nano-scale actuation in fluids have been described in the literature that are beyond the scope of this work. Many are reviewed by Chang et al., who also describe a novel diode-based actuation method for motion in fluids (Chang et al., 2007). These systems rely on a fluid environment and are not suitable for the surface crawling application of microsoccer. However, these systems may some day find a use in medical applications.

5. Multirobot Cooperation

Recent work by Bretl (Bretl, 2007), Donald et al. (Donald et al., 2008), and Kratochovil et al. (Kratochvil et al., 2009) has also investigated control of multiple scratch drive actuators moving on the same substrate. In the case of (Donald et al., 2008) the authors take advantage of a group of individual scratch drive actuators with slightly different properties (arm lengths, etc.) that respond to various motion primitives (waveforms generated by altering the period and amplitude) differently. The preimage planning mentioned earlier takes this into account.

Bretl's work (Bretl, 2007) explores the possible control of multiple identical robots that are exposed to the same voltage control sequence. He shows how two robots with unicycle dynamics and a bounded turning radius can be controlled to an arbitrary point from arbitrary positions.

Multi-robot control for magnetic devices has also been demonstrated by Floyd et al. (Floyd et al., 2008), by utilizing electrostatic clamping of the playing field in addition to the pulsed magnetic field actuation. In subsequent work by Pawashe (Paswashe, 2009), the playing field is composed of an array of independent interdigitated electrodes. In other words, the playing field is composed of m×n subfields similar to those used for actuation of scratch drive actuators. For purpose of demonstration, a two by two array was created. By applying a voltage field to an individual field, any robotic device on that subfield is held static by electrostatic clamping. In this manner, a particular device can be held still while other devices are controlled by the pulsing magnetic fields.

Kratochvil et al. (Kratochvil et al., 2009) demonstrated multirobot movement by building two different microrobot devices that are sensitive to different resonant frequencies. By varying the frequency of the oscillating magnetic field, either one or both of the devices will move. Their work is illustrated in Fig. 10.

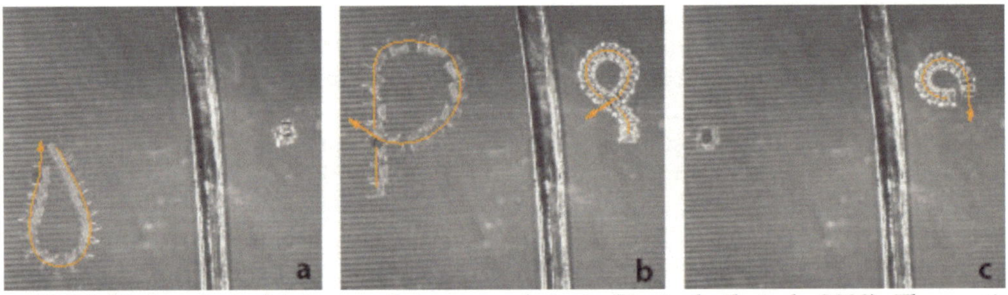

Fig. 10. Multiple robots driving on the same substrate (Kratochvil et al., 2009). The vertical line is a physical barrier to help prevent robot collisions and expedite experiments. In a and c, one robot is moving while the other is stationary. In b, the two robots are moving in different patterns. Figures are by Frutiger, Vollmers and Kratochvil and are used with permission by the authors.

6. Conclusion

After only three years, the RoboCup Nanogram events have produced a variety of creative microrobot systems. Competitors so far only used magnetic and electrostatic actuation, but other actuation technologies may emerge. Thermal, vibratory, and piezoelectric actuation all offer potential performance advantages. As a step down the road to a revolution in minimally invasive surgery and micromanufacturing, these small robots might indeed have a big impact.

7. References

Abbott, J ; Nagy, Z. ; Beyeler, F & Nelson, B. (2007). Robotics in the small. *IEEE Robotics and Automation Magazine,* Vol. 14, No. 2, pp. 92-103.

Akiyama, T. & Shono, K. (1993). Controlled stepwise motion in polysilicon microstructures. *Journal of Microelectromechanical Systems,* Vol. 2, pp. 106-110.

Baglio, S. ; Castorina, S. ; Fortuna, L. & Savalli, N. (2002). Novel microactuators based on photo-thermo-mechanical actuation strategy. Proceedings of IEEE Sensors 2002, Orlando, Florida, pp. 192-197.

Behkam, B. & Sitti, M. (2007). Bacterial flagella-based propulsion and on/off motion control of microscale objects. *Applied Physics Letters,* Vol. 90, 023902.

Bretl, T. (2007). Control of many objects using few instructions. *Proceedings of Robotics : Science and Systems,* Atlanta, Georgia, July 2007.

Brown, M. ; Hubbard, T. & Kujath, M.. (2007). Development of a long-range untethered frictional microcrawler. *Journal of Micromechanics and Microengineering,* Vol. 17, pp. 1025-1033.

CMC Microsystems, http://www.cmc.ca/index.htm, site accessed Aug. 17, 2009.

Chang, S. T. ; Paunov, V. N. ; Petsey, D. N. & Velev, O. D. (2007). Remotely powered self-propelling particles and micropumps based on miniature diodes. *Nature Materials*, Vol. 6, March 2007, pp. 235-240.

Cugat, O. ; Delamare, J. & Reyne, G. (2003). Magnetic micro-actuators and systems (MAGMAS). *IEEE Transactions on Magnetics*, Vol. 39, No. 5, pp. 3607-3612.

Dario, P. ; Velleggi, R. ; Carrozza, M. C. ; Montesi, M. C. & Cocco, M. (1992). Microactuators for microrobots : a critical survey. *Journal of Micromechanics and Microengineering*, Vol. 2, pp. 141-157.

Donald, B. ; Levey, C. ; McGray, C. ; Rus, D. & Sinclair, M. (2003). Power delivery and locomotion of untethered microactuators. *Journal of Microelectromechanical Systems*, Vol. 12, No. 6, pp. 947-959.

Donald, B. ; Levey, C. ; McGray, C. ; Paprotny, I. & Rus D. (2006). An untethered, electrostatic, globally controllable MEMS micro-robot. *Journal of Microelectromechanical Systems*, Vol. 15, No. 1, pp. 1-15.

Donald, B. ; Levey, C. & Paprotny, I. (2008). Planar microassembly by parallel actuation of MEMS microrobots, *Journal of Microelectromechanical Systems*, Vol. 17, No. 4, pp. 789-808.

Dreyfus, R. ; Baudry, J. ; Roper, M. L. ; Fermigier, M. ; Stone, H. A. ; and Bibette, J. (2005). Microscopic artificial swimmers. *Nature*, Vol. 437, pp. 862-865.

Dubin, L. E. (1957). On curves of minimal length with a constraint on average curvature and with prescribed initial an dterminal positions and tangents. *American Journal of Mathematics*, Vol. 79, No. 3, pp. 497-516.

Ebefors, T. ; Mattsson, J. U. ; Kalvesten, E. & Stemme, G. (2000). A robust micro conveyer realized by arrayed polyimide joint actuators, *Journal of Micromechanics and Microengineering*, Vol. 10, pp. 337-349.

Edqvist, E. ; Snis, N. ; Casanova Mohr, R. ; Scholz, O. ; Corradi, P. ; Gao, J. ; Dieguez, A. ; Wyrsch, N. & Johannson, S. (2009). Evaluation of building technology for mass producible millimeter-sized robots using flexible printed circuit boards. *Journal of Microelectromechanics and Microengineering*, Vol. 19, 075011.

Fan, L. ; Tai, Y. & Muller, R. (1988). Integrated movable micromechanical structures for sensors and actuators. *IEEE Transactions on Electron Devices*, Vol. 35, No. 6, pp. 724-730.

Firebaugh, S. L. & Piepmeier, J. A. (2008). The RoboCup Nanogram League : an opportunity for problem-based undergraduate education in microsystems. *IEEE Transactions on Education*, Vol. 51, No. 3, pp. 394-399.

Floyd, S. ; Pawashe, C. & Sitti, M. (2008). An untethered magnetically actuated micro-robot capable of motion on arbitrary surfaces. *Proceedings of the IEEE International Conference on Robotics and Automation*, pp. 419-424, Pasadena, CA, May 19-23, 2008.

[a] Frutiger, D. R. ; Vollmers, K. ; Kratochvil, B. E. & Nelson, B. J. (2008). Small, fast and under control : wireless resonant magnetic micro-agents. Proceedings of the International Symposium on Experimental Robotics, July 2008.

[b] Frutiger, D. R. ; Kratochvil, B. E. ; Vollmers, K. & Nelson, B. J. (2008). Magmites – wireless resonant magnetic microrobots. Proceedings of the IEEE International Conference on Robotics and Automation, pp. 1770-1771, Pasadena, CA, May 19-23, 2008.

Frutiger, D. R. ; Vollmers, K. ; Kratochvil, B. E. & Nelson, B. J. (2009). Small, fast and under control: wireless resonant magnetic micro-agents. *International Journal of Robotics Research*, November, 2009.

Gates, B. (2007). A robot in every home. *Scientific American*, Jan., pp. 58-65.

Gauthier, M. & Piat, E. (2004). An electromagnetic micromanipulation system for single cell manipulation. *Journal of Micromechatronics*, Vol. 2, No. 2, pp. 87-119.

Hayt, Jr., W. H. & Buck, J. A. (2006). *Engineering Electromagnetics*, 7th Ed. McGraw-Hill Higher Education, ISBN 0-07-252495-2, New York, NY.

Hermes, H. (1974). On local and global controllability. *SIAM J. Contr.*, vol. 12, pp. 252–261.

Huikai, X. ; Erdmann, L. ; Zhu, X. ; Gabriel, K. & Fedder, G. (2002). Post-CMOS processing for high-aspect-ratio integrated silicon microstructures. *Journal of Microelectromechanical Systems*, Vol. 11, No. 2, pp. 93-101.

Kladitis, P. E. & Bright, V. M. (2000). Prototype microrobots for micro-positioning and micro-unmanned vehicles. Sensors and Actuators, Vol. 80, pp. 132-137.

Kovacs, G. (1998). *Micromachined Transducers Sourcebook*, pp. 277-289, WCB McGraw-Hill, ISBN 0-07-290722-3, Boston.

Kratochvil, B. E. ; Dong, L. ; Zhang, L. ; Abbott, J. J. and Nelson, B. J. (2008). Nanohelices as motion converters. Proceedings of the 2008 IEEE/RSJ International Conference on Intelligent Robots and Systems, Nice, France, Sept. 22-26, 2008.

Kratochvil, B. E. ; Frutiger, D. ; Vollmers, K. & Nelson, B. J. (2009). Visual servoing and characterization of resonant magnetic actuators for decoupled locomotion of multiple untethered mobile microrobots, *Proceedings of the IEEE International Conference on Robotics and Automation*, pp. 1010-1015, Kobe, Japan, May 12-17, 2009.

Lal, A. ; Duggirala, R. & Li, H. (2005). Pervasive power : a radioisotope-powered piezoelectric generator. *IEEE Pervasive Computing*, Vol. 4, No. 1, pp. 53-61.

Lionis, G & Kyriakopoulos, K. J. (2007). PWM control for a micro-robot moving on a discrete curvature trajectory set. Proceedings of the 2007 IEEE International Conference on Robotics and Automation, pp. 2324-2329, Rome, Italy, April 10-14, 2007.

Liu, C. (2006). Foundations of MEMS, Pearson Education, Inc., ISBN 0-13-147286-0, Upper Saddle River, New Jersey.

Lozano-Perez, T. ; Mason, M. T. & Taylor, R. H. (1984). Automatic synthesis of fine-motion strategies for robots. *International Journal of Robotics Research*, Vol. 3, No. 1, pp. 3-24.

Madou, M. (1997). *Fundamentals of Microfabrication*, CRC Press, ISBN 0-8493-0826-7, Boca Raton, FL.

Manohara M. ; Morikawa, E.; Choi, J. & Sprunger, P.T. (1999). Transfer by direct photo etching of poly(vinylidene fluoride) using x-rays. *Journal of Microelectromechanical Systems*, Vol. 8, pp. 417-422.

Markus, K. ; Koester, D. ; Cowen, A. ; Mahadevan, R. ; Dhuler, V. ; Roberson, D. & Smith, L. (1995). MEMS infrastructure : the multi-user MEMS processes (MUMPS). *Proceedings of the SPIE*, Vol. 2639, pp. 54-63.

[a]Martel, S. (2005). Special surface for power delivery to wireless micro-electro-mechanical systems. *Journal of Micromechanics and Microengineering*, Vol. 15, pp. S251-S258.

[b]Martel, S. (2005). Fundamental principles and issues of high-speed piezoactuated three-legged motion for miniature robots designed for nanometer-scale operations. *The International Journal of Robotics Research*, Vol. 24, No. 7, pp. 575-588.

Martel, S. (2007). Magnetic resonance propulsion, control and tracking at 24 Hz of an untethered device in the carotid artery of a living animal : an important step in the development of medical micro- and nanorobots. *Proceedings of the 295h Annual International Conference of the IEEE EMBS Cite Internationale*, pp. 1475-1478, Lyon France, August 23-26, 2007.

Martel, S. ; Felfoul, O. & Mohammadi, M. (2008). Flagellated bacterial nanorobots for medical interventions in the human body. *Proceedings of the 2nd Biennial IEEE/RAS-EMBS International Conference on Biomedical Robotics and Biomechatronics*, pp. 264-269, Scottsdale, Arizona, Oct. 19-22.

Mathieu, J.-B. ; Beaudoin, G. & Martel, S. (2006). Method of propulsion of a ferromagnetic core in the cardiovascular system through magnetic gradients generated by an MRI system. *IEEE Transactions on Biomedical Engineering*, Vol. 53, pp. 292-299.

MEMSCAP, Inc., http://www.memscap.com/index.php, site accessed Aug. 17, 2009.

Nathanson, H. ; Newell, W., Wickstrom, R. & Davis, J. (1978). The resonant gate transistor. *IEEE Transactions on Electron Devices*, Vol. 14, pp. 117-133.

Nguyen, A. T. & Martel, S. (2006). Miniaturization of a piezo-actuation system embedded in an instrumented autonomous robot. *Proceedings of the 4th International IEEE-NEWCAS Conference*, pp. 261-264, Gatineau, Quebec.

Ohmichi, O. ; Yamagata, Y. & Higuchi, T. (2002). Micro-impact drive mechanisms using optically excited thermal expansion. *Journal of Microelectromechanical Systems*, Vol. 6, No. 3, pp. 200-207.

Patil, A. ; Patil, V. ; Shin, D. W. ; Choi, J. ; Paik, D. & Yoon, S. (2008). Issues and challenges facing rechargeable thin film lithium batteries. *Materials Research Bulletin*, Vol. 43, pp. 1913-1942,

Pawashe, C. ; Floyd, S. & Sitti, M. (2008). Dynamic modelling of stick slip motion in an untethered magnetic micro-robot. Proceedings of Robotics : Science and Systems IV, Zurich, Switzerland, June 2008.

[a]Pawashe, C. ; Floyd, S. & Sitti, M. (2009). Modeling and experimental characterization of an untethered magnetic micro-robot. *The International Journal of Robotics Research*, Vol. 28, pp. 1077-1094.

[b]Pawashe, C. ; Floyd, S. & Sitti, M. (2009). Multiple magnetic microrobot control using electrostatic clamping. *Applied Physics Letters*, Vol. 94, 164108.

Petersen, K. (1978). Dynamic micromechanics on silicon : techniques and devices. *IEEE Transactions on Electron Devices*, Vol. 25, No. 10, pp. 1241-1250.

Piepmeier, J. A. & Firebaugh, S. L. (2008). Vision-based control of a scratch drive microrobot. *Proceedings of the Southeastern Symposium on System Theory*, pp. 352-355, New Orleans, Louisiana.

Piepmeier, J. A. & Firebaugh, S. L. (2010). PWM control accuracy for scratch drive actuators, submitted to the *IEEE International Conference on Robotics and Automation*, Anchorage, Alaska, May 2010.

Ross, R. (2007). Robotic fly takes off. Technology Review, http://www.technology review.com/Infotech/19068/, created July 19, 2007, accessed Aug. 12, 2009.

Saitou, K. ; Wang, D.-A. & Wou, S. J. (2000). Externally resonated linear microvibromotor for microassembly. *Journal of Microelectromechanical Systems*, Vol. 9, No. XX, pp. 336-346.

Sievers, T . & Fatikow, S. (2005). Visual servoing of a mobile microrobot inside a scanning electron microscope. *Proceedings of the IEEE/RSJ International Conference on Intelligent Robots and Systems*, pp. 1350-1354, Edmonton, Alberta, March 2008.

Sitti, M. (2007). Microscale and nanoscale robotics systems. *IEEE Robotics & Automation Magazine*, Vol. 14, No. 1, pp. 53-60.

Smits, J. G. (1989). Is micromechanics becoming a new subject for academic courses or the design of a piezoelectric on silicon microrobot, *Proceedings of the 8th University/Government/Industry Microelectronics Symposium*, pp. 141-145, Westborough, MA. 1989.

Sniegowski, J. & Garcia, E. (1996). Surface-micromachined gear trains driven by an on-chip electrostatic microengine. *IEEE Electron Device Letters*, Vol. 17, No. 7, pp. 366-368.

Spong, M. ; Hutchinson, S. & Vidyasagar, M. (2006) *Robot Modeling and Control*. John Wiley & Sons, Inc. Hoboken, NJ.

Sul, O. ; Falvo, M. ; Taylor, R. ; Washburn, S. & Superfine, R. (2006). Thermally-actuated untethered impact-driven locomotive microdevices. *Applied Physics Letters*, Vol. 89, 203512.

Tamaz, S. ; Gourdeau, R. ; Chanu, A. ; Mathieu, J.-B. & Martel, S. (2008). Real-time MRI-based control of a ferromagnetic core for endovascular navigation. *IEEE Transactions on Biomedical Engineering*, Vol. 55, No. 7, pp. 1854-1863.

Tea, N. ; Milanovic, V. ; Zincke, C. ; Suehle, J. ; Gaitan, M. ; Zaghloul, M. & Geist, J. (1997). Hybrid postprocessing etching for CMOS-compatible MEMS. *Journal of Microelectromechanical Systems*, Vol. 6, No. 4, pp. 363-372.

Tang, W. ; Nguyen, T. ; Judy, M. & Howe, R. (1990). Electrostatic-comb drive of lateral polysilicon resonators. *Sensors and Actuators A*, Vol. 21, No. 1-3, pp. 328-331.

Vollmers, K. ; Frutiger, D. ; Kratochvil, B. & Nelson, B. (2008). Wireless resonant magnetic microactuator for untethered mobile microrobots. *Applied Physics Letters*, Vol. 92, 144103.

Wang, L.-P. (2003). Design, fabrication and measurement of high-sensitivity piezoelectric microelectromechanical systems accelerometers. *Journal of Microelectromechanical Systems*, Vol. 12, pp. 433-439.

Wood, R. J. ; Avadhanula, S. ; Menon, M. & Fearing, R. S. (2003). Microrobotics using composite materials : the micromechanical flying insect thorax. *Proceedings of the 2003 IEEE International Conference on Robotics and Automation*, pp. 1842-1849, Taipei, Taiwan, Sept. 14-19, 2003.

Yamahata, C. ; Collard, D. ; Domenget, A. ; Hosogi, M. ; Kumemura, M. ; Hashiguchi, G. & Fujita, H. (2008). Silicon nanotweezers with subnanometer resolution for the micromanipulation of biomolecules. *Journal of Microelectromechanical Systems*, Vol. 17, No. 3, pp. 623-631.

Yamazaki, A. ; Sendoh, M. ; Ishiyama, K. ; Aria, K. I. ; Kato, R. ; Nakano, M. & Fukunaga, H. (2004). Wireless micro swimming machine with magnetic thin film. *Journal of Magnetism and Magnetic Materials*, Vol. 272-276, pp. e1741-e1742.

Yasuda, T ; Shimoyama, I. & Miura, H. (1994). Microrobot actuated by a vibration energy field. *Sensors and Actuators A*, Vol. 43, pp. 366-370.

Yasuda, T ; Shimoyama, I. & Miura, H. (1995). Microrobot locomotion in a mechanical vibration field. *Advanced Robotics*, Vol. 9, pp. 165-176.

Yesin, K. B.; Vollmers, K. & Nelson, B. J. (2006). Modeling and control of untethered biomicrorobots in a fluidic environment using electromagnetic fields. International Journal of Robot Research, Vol. 25, pp. 527-534.

14

Automated camera calibration for robot soccer

Donald G Bailey and Gourab Sen Gupta
School of Engineering and Advanced Technology, Massey University
Palmerston North, New Zealand

1. Introduction

Robot soccer has become popular over the last decade not only as a platform for education and entertainment but as a test bed for adaptive control of dynamic systems in a multi-agent collaborative environment (Messom, 1998). It is a powerful vehicle for exploration and dissemination of scientific knowledge in a fun and exciting manner. The robot soccer environment encompasses several technologies—embedded micro-controller based hardware, wireless radio-frequency data transmission, dynamics and kinematics of motion, motion control algorithms, real-time image capture and processing and multi-agent collaboration.

The vision system is an integral component of modern autonomous mobile robots. With robot soccer, the physical size of the robots in the micro-robot and small robot leagues limits the power and space available, precluding the use of local cameras on the robots themselves. This is overcome by using a global vision system, with one (or more for larger size fields) cameras mounted over the playing field. The camera or cameras are connected to a vision processor that determines the location and orientation of each robot and the location of the ball relative to the playing field. This data is then passed to the strategy controller, which determines how the team should respond to the current game situation, plans the trajectories or paths of the robots under its control, and transmits the appropriate low-level motor control commands to the robots, which enable them to execute the plan.

To manage complexity in collaborative robot systems, a hierarchical state transition based supervisory control (STBS) system can be used (Sen Gupta et al., 2002; Sen Gupta et al., 2004). However, the performance of such a system, or indeed any alternative higher-level control system, deteriorates substantially if the objects are not located accurately because the generic control functions to position and orient the robots are then no longer reliable.

The high speed and manoeuvrability of the robots make the game very dynamic. Accurate control of high-speed micro-robots is essential for success within robot soccer. This makes accurate, real-time detection of the position and orientation of objects of particular importance as these greatly affect path-planning, prediction of moving targets and obstacle avoidance. Each robot is identified in the global image by a "jacket" which consists of a pattern of coloured patches. The location of these coloured patches within the image is used to estimate the position and orientation of the robot within the playing area. The cameras must therefore be calibrated to provide an accurate mapping between image coordinates and world coordinates in terms of positions on the playing field (Bailey & Sen Gupta, 2004).

The portable nature of robot soccer platforms means that every time the system is set up, there are differences in the camera position and angle relative to the field. Each team has their own camera, and both cannot be mounted exactly over the centre of the playing area. It is also difficult to arrange the camera position so that it is perfectly perpendicular to the playing surface. Consequently, each camera is looking down on the playing area at a slight angle, which introduces a mild perspective distortion into the image. The size of the playing area, combined with constraints on how high the camera may be mounted, require that a wide angle lens be used. This can introduce significant barrel distortion within the image obtained. Both of these effects are readily apparent in Fig. 1. The limited height of the camera combined with the height of the robots means that each detected robot position is also subject to parallax error. These factors must all be considered, and compensated for, to obtain accurate estimates of the location and orientation of each robot.

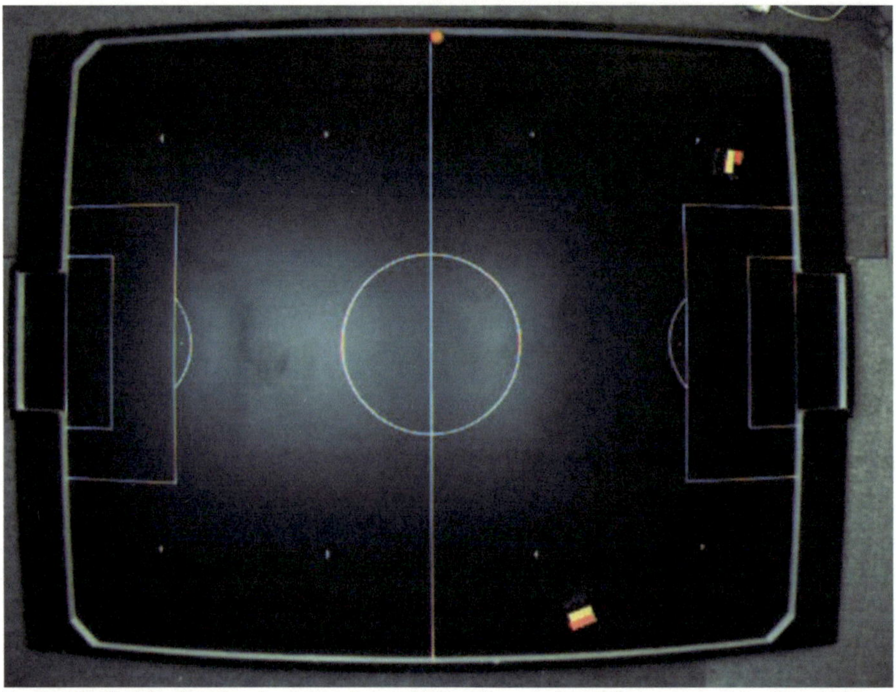

Fig. 1. Example field, clearly showing lens distortion and mild perspective distortion.

2. Effects of distortion

At the minimum, any calibration must determine the location of the playing field within the image. The simplest calibration (for single camera fields) is to assume that the camera is aligned with the field and that there is no distortion. The column positions of the goal mouths, C_{Left} and C_{Right}, and the row positions of the field edges at, or near, the centreline, R_{Top} and R_{Bottom}, need to be located within the image. Then, given the known length, L, and width, W, of the field, the estimated height of the camera, H, and the known height of the robots, h, an object positioned at row R and column C within the image may be determined relative to the centre of the field as

$$x = \left(C - \frac{C_{Right} + C_{Left}}{2} \right) \left(\frac{L}{C_{Right} - C_{Left}} \right) \left(\frac{H-h}{H} \right)$$
$$y = \left(R - \frac{R_{Bottom} + R_{Top}}{2} \right) \left(\frac{W}{R_{Bottom} - R_{Top}} \right) \left(\frac{H-h}{H} \right) \quad (1)$$

The first term sets the centre of the field as the origin, the second scales from pixel units to physical units, and the third term accounts for parallax distortion resulting from the height of the robot (assuming that the camera is positioned over the centre of the field).
While this calibration is simple to perform, it will only be accurate for an ideal camera positioned precisely in the centre of the field and aligned perpendicularly to the playing field. Each deviation from ideal will introduce distortions in the image and calibration errors.

2.1 Lens distortion

The most prevalent form of lens distortion is barrel distortion. It results from the lens having a slightly higher magnification in the centre of the image than at the periphery. Barrel distortion is particularly noticeable with wide-angle lenses such as those used with robot soccer. While there are several different physical models of the lens distortion based on the known characteristics of the lens (Basu & Licardie, 1995; Pers & Kovacic, 2002), the most commonly used model is a generic radial Taylor series relating the ideal, undistorted coordinates (x_u,y_u) to the distorted coordinates in the image (x_d,y_d):

$$x_d = x_u\left(1+\kappa_1 r_u^2 + \kappa_2 r_u^4 + ...\right) + \left(\varepsilon_1\left(r_u^2 + 2x_u^2\right) + 2\varepsilon_2 x_u y_u\right)\left(1+\varepsilon_3 r_u^2 + ...\right)$$
$$y_d = y_u\left(1+\kappa_1 r_u^2 + \kappa_2 r_u^4 + ...\right) + \left(2\varepsilon_1 x_u y_u + \varepsilon_2\left(r_u^2 + 2y_u^2\right)\right)\left(1+\varepsilon_3 r_u^2 + ...\right) \quad (2)$$

Both sets of coordinates have the centre of distortion as the origin, and $r_u^2 = x_u^2 + y_u^2$. The set of parameters κ and ε characterise a particular lens. Note that the centre of distortion is not necessarily the centre of the image (Willson & Shafer, 1994). For most lenses, two radial and two tangential terms are sufficient (Li & Lavest, 1996), and most calibration methods limit themselves to estimating these terms. A simple, one parameter, radial distortion model is usually sufficient to account for most of the distortion (Li & Lavest, 1996):

$$r_d = r_u(1+\kappa r_u^2) \quad (3)$$

This forward transform is most commonly used because in modelling the imaging process, the image progresses from undistorted coordinates to distorted image coordinates. Sometimes, however, the reverse transform is used. This swaps the roles of the two sets of coordinates. Since the model is an arbitrary Taylor series expansion, either approach is valid (although the coefficients will be different for the forward and reverse transforms). The first order reverse transform is given by

$$r_u = r_d(1+\kappa r_d^2) \quad (4)$$

Effect on position

If the simple calibration was based on distances from the centre of the image, the position error would increase with radius according to the radially dependent magnification. However, since the calibration of eq. (1) sets the positions at the edges and ends of the field, the position error there will be minimal. The absolute position errors should also be zero near the centre of the image, increase with radius to a local maximum between the centre and edges of the playing area, and decrease to zero again on an ellipse through the table edge and goal points. Outside this ellipse, the errors will increase rapidly with distance, having the greatest effect in the corners of the playing field.

Effect on orientation

Determining the effect of radial distortion on the angle is more complicated. Consider a point in the undistorted image using radial coordinates (r_u,φ). At this point, the lens distortion results in a magnification

$$M = \frac{r_d}{r_u} \tag{5}$$

Next, consider a test point offset from this by a small distance, r, at an angle, θ_u, with a magnification, M_2. If the magnification is constant (M_2=M) then everything is scaled equally, and by similar triangles, the angle to the test point will remain the same. This implies that if the test point is in the tangential direction ($\theta_u - \varphi$=90°), there will be no angle error. Similarly, since the distortion is radial, if the test point is aligned ($\theta_u - \varphi$=0) the test point will be stretched radially, but the angle will not change. These two considerations imply that it is best to consider offsets in the tangential and radial direction. This geometry is shown in Fig. 2.

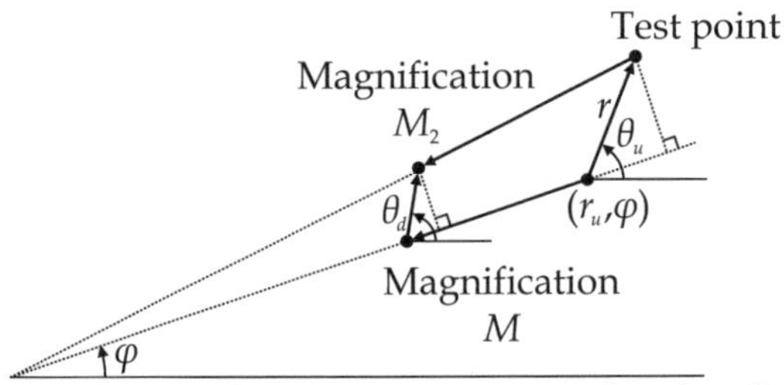

Fig. 2. Geometry for calculating the change as a result of radial lens distortion. The distortion and scales have been exaggerated to make the effects visible.

After distortion, the angle to the test point becomes

$$\begin{aligned}\tan(\theta_d - \varphi) &= \frac{M_2 r \sin(\theta_u - \varphi)}{M_2\left(r_u + r\cos(\theta_u - \varphi)\right) - M r_u} \\ &= \frac{M_2 r \sin(\theta_u - \varphi)}{M_2 r \cos(\theta_u - \varphi) + (M_2 - M) r_u}\end{aligned} \tag{6}$$

The test distance, r, is small, therefore

$$\begin{aligned} M_2 - M &\approx \Delta r_u^2 \frac{dM}{dr_u^2} \\ &= \left((r_u - r\cos(\theta_u - \varphi))^2 - r_u^2 \right) \frac{dM}{dr_u^2} \\ &\approx 2rr_u \cos(\theta_u - \varphi) \frac{dM}{dr_u^2} \end{aligned} \tag{7}$$

so

$$\tan(\theta_d - \varphi) \approx \frac{M}{M + 2r_u^2 \frac{dM}{dr_u^2}} \tan(\theta_u - \varphi) \tag{8}$$

When the forward map of eq. (3) is used, eq. (8) becomes

$$\tan(\theta_d - \varphi) \approx \frac{1 + \kappa r_u^2}{1 + 3\kappa r_u^2} \tan(\theta_u - \varphi) \tag{9}$$

Since the magnification changes faster with increasing radius, the angle error ($\theta_d - \theta_u$) will also be larger further from the centre of distortion, and increase more rapidly in the periphery.

2.2 Perspective distortion

Perspective distortion results when the line of sight of the camera is not perpendicular to the plane of the playing area. This will occur when the camera is not directly over the centre of the playing area, and it must be tilted to fit the complete playing area within the field of view. A perspective transformation is given by

$$\begin{aligned} x_d &= \frac{h_1 x_u + h_2 y_u + h_3}{h_7 x_u + h_8 y_u + h_9} \\ y_d &= \frac{h_4 x_u + h_5 y_u + h_6}{h_7 x_u + h_8 y_u + h_9} \end{aligned} \tag{10}$$

where (x_u, y_u) and (x_d, y_d) are the coordinates of an undistorted and distorted point respectively. This is often represented in matrix form using a homogenous coordinate system (Hartley & Zisserman, 2000):

$$\begin{bmatrix} kx_d \\ ky_d \\ k \end{bmatrix} = \begin{bmatrix} h_1 & h_2 & h_3 \\ h_4 & h_5 & h_6 \\ h_7 & h_8 & h_9 \end{bmatrix} \begin{bmatrix} x_u \\ y_u \\ 1 \end{bmatrix} \quad \text{or} \quad \mathbf{P}_d = \mathbf{H}\mathbf{P}_u \tag{11}$$

The 3x3 transformation matrix, **H**, incorporates rotation, translation, scaling, skew, and stretch as well as perspective distortion. Just considering perspective distortion and for small tilt angles **H** simplifies to

$$\mathbf{H}_{perspective} = \begin{bmatrix} 1 & 0 & 0 \\ 0 & 1 & 0 \\ p_x & p_y & 1 \end{bmatrix} \quad (12)$$

The effect of this is to change the scale factor, *k*, in eq. (11), giving a position dependent magnification. As a consequence, parallel lines converge, meeting somewhere on the vanishing line given by

$$p_x x_u + p_y y_u + 1 = 0 \quad (13)$$

Effect on position

Since the simple calibration sets four points on the edges of the playing area, these points will have no error. In direction that the camera is tilted, *k* will be greater than 1, shrinking the scene. The radial error will be positive, and the tangential errors will be towards the direction line. In the opposite direction, the magnification is greater than 1. The radial error will be negative, and the tangential errors will be away from the direction line. There will be a line approximately across the middle of the playing area where

$$p_x x_u + p_y y_u + 1 = 1 \quad (14)$$

which will have no error. The angle of the line, and the severity of the errors will depend on the angle and extent of the camera tilt respectively.

Effect on orientation

Again the distortion will depend on the change of magnification with position. As the content becomes more compressed (closer to the vanishing line) angle distortion will increase, because the slope of the magnification becomes steeper and the angle errors will be greater. Without loss of generality, consider the camera tilted in the x direction (p_y=0). Consider a test point offset from (x_u,y_u) by distance r at an angle θ_u. After distortion, the offset becomes:

$$\begin{aligned} \Delta y &= \frac{y_u + r\sin\theta_u}{1 + p_x(x_u + r\cos\theta_u)} - \frac{y_u}{1 + p_x x_u} \\ \Delta x &= \frac{x_u + r\cos\theta_u}{1 + p_x(x_u + r\cos\theta_u)} - \frac{x_u}{1 + p_x x_u} \end{aligned} \quad (15)$$

Hence, the angle of the test point after distortion is given by

$$\begin{aligned} \tan\theta_d &= \frac{\Delta y}{\Delta x} = \frac{(y_u + r\sin\theta_u)(1 + p_x x_u) - y_u(1 + p_x(x_u + r\cos\theta_u))}{(x_u + r\cos\theta_u)(1 + p_x x_u) - x_u(1 + p_x(x_u + r\cos\theta_u))} \\ &= \frac{r\sin\theta_u(1 + p_x x_u) - y_u p_x r\cos\theta_u}{r\cos\theta_u} \\ &= (1 + p_x x_u)\tan\theta_u - y_u p_x \end{aligned} \quad (16)$$

Along the line of sight (y_u=0) there is no distortion radially or tangentially. However, other orientations become compressed as the magnification causes foreshortening, especially as it approaches the vanishing line. At other positions, angles that satisfy

$$\tan\theta_u = \frac{y_u}{x_u} \tag{17}$$

are not distorted. This is expected, since the perspective transformation will map straight lines onto straight lines. Orientations perpendicular to the line of sight (θ_u=90°) are not distorted. Other orientations are compressed by the perspective foreshortening.
The mild perspective distortion encountered with the robot soccer system will introduce mild distortion. However, angle errors will be largest on the side of the field where the image appears compressed.

2.3 Parallax distortion

In the absence of any other information, the camera is assumed to be at a known height, H, directly above the centre of the robot soccer playing area. This allows the change in scale associated with the known heights of the robots to be taken into account. Since the robot jackets are always a fixed height above the playing surface, parallax correction simply involves scaling the detected position relative to the position of the camera. Errors in estimating the camera position will only introduce position errors; they will not affect the angle.

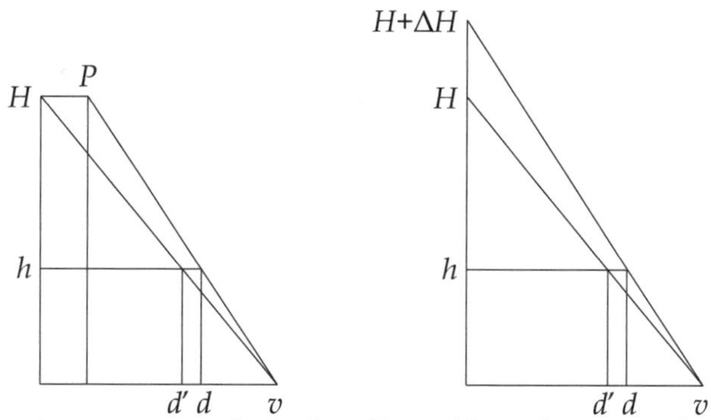

Fig. 3. Parallax correction geometry. Left: the effect of lateral error; right: the effect of height error. Note, the robot height, h, has been exaggerated for clarity.

Effect on position

Consider the geometry shown in Fig. 3. When the camera is offset laterally by P, an object at location d in the playing area will appear at location v by projecting the height

$$v = \frac{(d-P)H}{H-h} + P = \frac{dH - Ph}{H-h} \tag{18}$$

However, if it is assumed that the camera is not offset, the parallax correction will estimate the object position as d'. The error is given by

$$d' - d = \frac{v(H-h)}{H} - d = \frac{-Ph}{H} \tag{19}$$

The lateral error is scaled by the relative heights of the robot and camera. This ratio is typically 40 or 50, so a 5 cm camera offset will result in a 1 mm error in position. Note that the error applies to everywhere in the playing area, independent of the object location.
An error in estimating the height of the camera by ΔH will also result in an error in location of objects. In this case, the projection of the object position will be

$$v = \frac{d(H+\Delta H)}{H+\Delta H-h} \tag{20}$$

Again, given the assumptions in camera position, correcting this position for parallax will result in an error in estimating the robot position of

$$d' - d = \frac{v(H-h)}{H} - d = \frac{-dh\Delta H}{(H+\Delta H-h)H} \tag{21}$$

Since changing the height of the camera changes the parallax correction scale factor, the error will be proportional to the distance from the camera location. There will be no error directly below the camera, and the greatest errors will be seen in the corners of the playing area.

2.4 Effects on game play

When considering the effects of location and orientation errors on game play, two situations need to be considered. The first is local effects, for example when a robot is close to the ball and manoeuvring to shoot the ball. The second is when the robot is far from play, but must be brought quickly into play.
In the first situation, when the objects are relatively close to one another, what is most important is the relative location of the objects. Since both objects will be subject to similar distortions, they will have similar position errors. However, the difference in position errors will result in an error in estimating the angle between the objects (indeed this was how angle errors were estimated earlier in this section). While orientation errors may be considered of greater importance, these will correlate with the angle errors from estimating the relative position, making orientation errors less important for close work.
In contrast with this, at a distance the orientation errors are of greater importance, because shooting a ball or instructing the robot to move rapidly will result in moving in the wrong direction when the angle error is large. For slow play, this is less significant, because errors can be corrected over a series of successive images as the object is moving. However at high speed (speeds of over two metres per second are frequently encountered in robot soccer), estimating the angles at the start of a manoeuvre is more critical.
Consequently, good calibration is critical for successful game play.

3. Standard calibration techniques

In computer vision, the approach of Tsai (Tsai, 1987) or some derivation is commonly used to calibrate the relationship between pixels and real-world coordinates. These approaches estimate the position and orientation of the camera relative to a target, as well as estimating the lens distortion parameters, and the intrinsic imaging parameters. Calibration requires a dense set of calibration data points scattered throughout the image. These are usually provided by a 'target' consisting of an array of spots, a grid, or a checkerboard pattern. From the construction of the target, the relative positions of the target points are well known. Within the captured image of the target, the known points are located and their correspondence with the object established. A model of the imaging process is then adjusted to make the target points match their measured image points.

The known location of the model enables target points to be measured in 3D world coordinates. This coordinate system is used as the frame of reference. A rigid body transformation (rotation and translation) is applied to the target points. This uses an estimate of the camera pose (position and orientation in world coordinates) to transform the points into a camera centred coordinate system. Then a projective transformation is performed, based on the estimated lens focal length, giving 2D coordinates on the image plane. Next, these are adjusted using the distortion model to account for distortions introduced by the lens. Finally, the sensing element size and aspect ratio are used to determine where the control points should appear in pixel coordinates. The coordinates obtained from the model are compared with the coordinates measured from the image, giving an error. The imaging parameters are then adjusted to minimise the error, resulting in a full characterisation of the imaging model.

The camera and lens model is sufficiently non-linear to preclude a simple, direct calculation of all of the parameters of the imaging model. Correcting imaging systems for distortion therefore requires an iterative approach, for example using the Levenberg-Marquardt method of minimising the mean squared error (Press et al., 1993). One complication of this approach is that for convergence, the initial estimates of the model parameters must be reasonably close to the final values. This is particularly so with the 3D rotation and perspective transformation parameters.

Planar objects are simpler to construct accurately than full 3D objects. Unfortunately, only knowing the location of points on a single plane is insufficient to determine a full imaging model (Sturm & Maybank, 1999). Therefore, if a planar target is used, several images must be taken of the target in a variety of poses to obtain full 3D information (Heikkila & Silven, 1996). Alternatively, a reduced model with one or two free parameters may be obtained from a single image. For robot soccer, this is generally not too much of a problem since the game is essentially planar.

A number of methods for performing the calibration for robot soccer are described in the literature. Without providing a custom target, there are only a few data points available from the robot soccer platform. The methods range from the minimum calibration described in the previous section through to characterisation of full models of the imaging system.

The basic approach described in section 2 does not account for any distortions. A simple approach was developed in (Weiss & Hildebrand, 2004) to account for the gross characteristics of the distortion. The playing area was divided into four quadrants, based on the centreline, and dividing the field in half longitudinally between the centres of the goals. Each quadrant was corrected using bilinear interpolation. While this corrects the worst of the position errors resulting from both lens and perspective distortion, it will only partially correct orientation errors. The use of a bilinear transformation will also result in a small jump in the orientation at the boundaries between adjacent quadrants.

A direct approach of Tsai's calibration is to have a chequered cloth (as the calibration pattern) that is rolled out over the playing area (Baltes, 2000). The corners of the squares on the cloth provide a 2D grid of target points for calibration. The cloth must cover as much as possible of the field of view of the camera. A limitation of this approach is that the calibration is with respect to the cloth, rather than the field. Unless the cloth is positioned carefully with respect to the field, this can introduce other errors.

This limitation may be overcome by directly using landmarks on the playing field as the target locations. This approach is probably the most commonly used and is exemplified in (Ball et al., 2004) where a sequence of predefined landmarks is manually clicked on within the image of the field. Tsai's calibration method is then used to determine the imaging model by matching the known locations with their image counterparts. Such approaches based on manually selecting the target points within the image are subject to the accuracy and judgement of the person locating the landmarks within the image. Target selection is usually limited to the nearest pixel. While selecting more points will generally result in a more accurate calibration by averaging the errors from the over-determined system, the error minimisation cannot remove systematic errors. Manual landmark selection is also very time-consuming.

The need to locate target points subjectively may be overcome by automating the calibration procedure. Egorova (Egorova et al., 2005) uses the bounding box to find the largest object in the image, and this is used to initialise the transform. A model of the field is transformed using iterative global optimisation to make the image of the field match the transformed model. While automatic, this procedure takes five to six seconds using a high end desktop computer for the model parameters to converge.

A slightly different approach is taken by Klancar (Klancar et al., 2004). The distortion correction is split into two stages: first the lens distortion is removed, and then the perspective distortion parameters are estimated. This approach to lens distortion correction is based on the observation that straight lines are invariant under a perspective (or projective) transformation. Therefore, any deviation from straightness must be due to lens distortion (Brown, 1971; Fryer et al., 1994; Park & Hong, 2001). This is the so-called 'plumb-line' approach, so named because when it was first used by (Brown, 1971), the straight lines were literally plumb-lines hung within the image. (Klancar et al., 2004) uses a Hough transform to find the major edges of the field. Three points are found along each line: one on the centre and one at each end. A hyperbolic sine radial distortion model is used (Pers & Kovacic, 2002), with the focal length optimised to make the three target points for each line as close to collinear as possible. One limitation of Klancar's approach is the assumption that the centre of the image corresponds with the centre of distortion. However, errors within the location of the distortion centre results in tangential distortion terms (Stein, 1997) which are not considered with the model. The second stage of Klancar's algorithm is to use the convergence of parallel lines (at the vanishing points) to estimate the perspective transformation component.

None of the approaches explicitly determines the camera location. Since they are all based on 2D targets, they can only gain limited information on the camera height, resulting in a limited ability to correct for parallax distortion. The limitations of the existing techniques led us to develop an automatic method that overcomes these problems by basing the calibration on a 3D model.

4. Automatic calibration procedure

The calibration procedure is based on the principles first described in (Bailey, 2002). A three stage solution is developed based on the 'plumb-line' principle. In the first stage, a parabola is fitted to each of the lines on the edge of the field. Without distortion, these should be straight lines, so the quadratic component provides data for estimating the lens distortion. A single parameter radial distortion model is used, with a closed form solution given for determining the lens distortion parameter. In the second stage, homogenous coordinates are used to model the perspective transformation. This is based on transforming the lines on the edge of the field to their known locations. The final stage uses the 3D information inherent in the field to obtain an estimate of the camera location (Bailey & Sen Gupta, 2008).

4.1 Edge detection

The first step is to find the edge of the playing field. The approach taken will depend on the form of the field. Our initial work was based on micro-robots, where the playing field is bounded by a short wall. The white edges apparent in Fig. 1 actually represent the inside edge of the wall around the playing area, as shown in Fig. 4. In this case, the edge of the playing area corresponds to the edge between the white of the wall and the black of the playing surface. While detecting the edge between the black and white sounds straightforward, it is not always as simple as that. Specular reflections off the black regions can severely reduce the contrast in some situations, as can be seen in Fig. 5, particularly in the bottom right corner of the image.

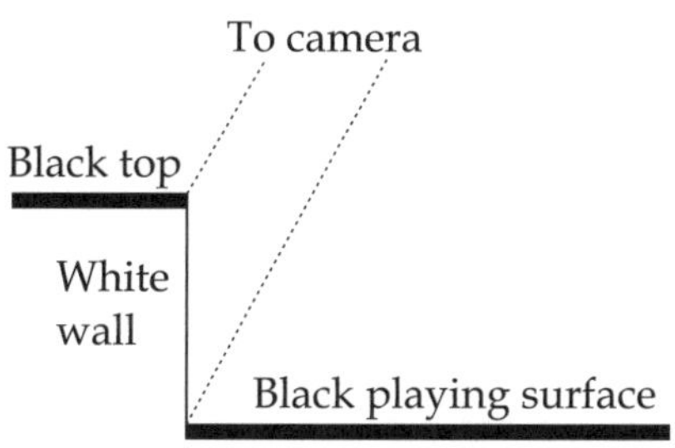

Fig. 4. The edge of the playing area.

Two 3x3 directional Prewitt edge detection filters are used to detect both the top and bottom edges of the walls on all four sides of the playing area. To obtain an accurate estimate of the calibration parameters, it is necessary to detect the edges to sub-pixel accuracy. Consider first the bottom edge of the wall along the side of the playing area in the top edge of the image. Let the response of the filtered image be $f[x,y]$. Within the top 15% of the image, the maximum filtered response is found in each column. Let the maximum in column x be located on row $y_{max,x}$. A parabola is fitted to the filter responses above and below this maximum (perpendicular to the edge), and the edge pixel determined to sub-pixel location as (Bailey, 2003):

$$edge[x] = y_{max,x} + \frac{f[x,y_{max,x}+1]-f[x,y_{max,x}-1]}{4f[x,y_{max,x}]-2\left(f[x,y_{max,x}+1]+f[x,y_{max,x}-1]\right)} \tag{22}$$

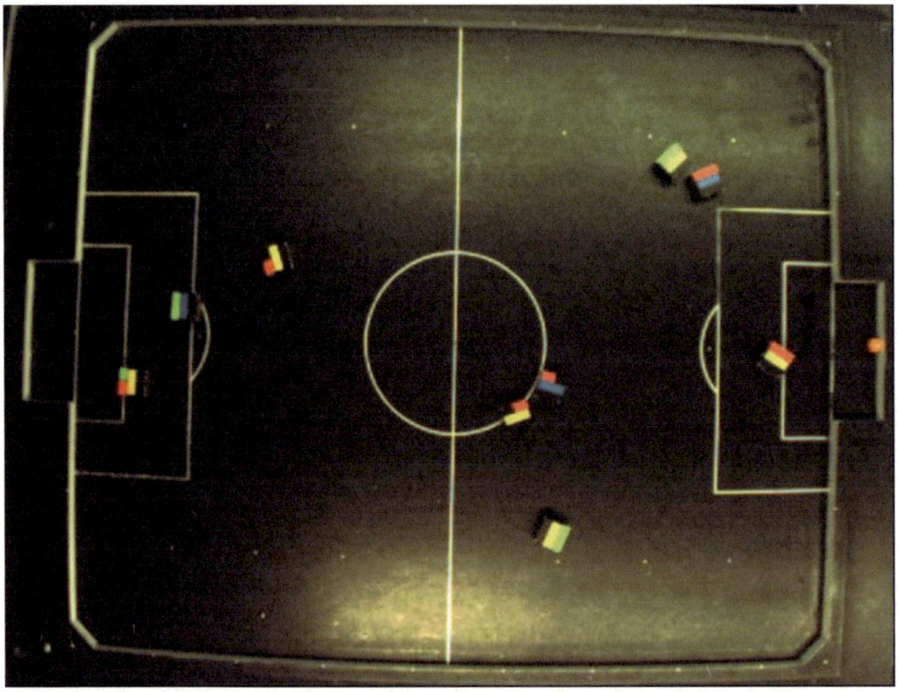

Fig. 5. As a result of lighting and specular reflection, the edge of the playing area may be harder to detect.

A parabola is then fitted to all the detected edge points ($x, edge[x]$) along the length of the edge. Let the parabola be $y(x) = ax^2 + bx + c$. The parabola coefficients are determined by minimising the squared error

$$E = \sum_x \left(ax^2 + bx + c - edge[x] \right)^2 \tag{23}$$

The error is minimised by taking partial derivatives of eq. (23) with respect to each of the parameters a, b, and c, and solving for when these are equal to zero. This results in the following set of simultaneous equations, which are then solved for the parabola coefficients.

$$\begin{bmatrix} \sum x^4 & \sum x^3 & \sum x^2 \\ \sum x^3 & \sum x^2 & \sum x \\ \sum x^2 & \sum x & \sum 1 \end{bmatrix} \begin{bmatrix} a \\ b \\ c \end{bmatrix} = \begin{bmatrix} \sum x^2 edge[x] \\ \sum x.edge[x] \\ \sum edge[x] \end{bmatrix} \tag{24}$$

The resulting parabola may be subject to errors from noisy or misdetected points. The accuracy may be improved considerably using robust fitting techniques. After initially estimating the parabola, any outliers are removed from the data set, and the parabola refitted to the remaining points. Two iterations are used, removing points more than 1 pixel from the parabola in the first iteration, and removing those more that 0.5 pixel from the parabola in the second iteration.

A similar process is used with the local minimum of the Prewitt filter to detect the top edge of the wall. The process is repeated for the other walls in the bottom, left and right edges of the image. The robust fitting procedure automatically removes the pixels in the goal mouth from the fit. The results of detecting the edges for the image in Fig. 1 are shown in Fig. 6.

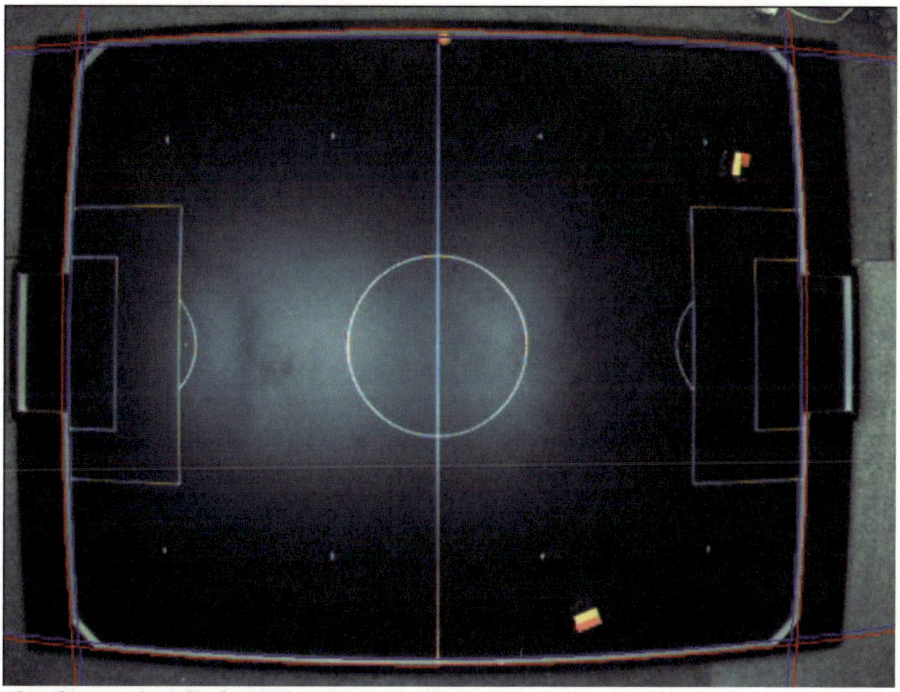

Fig. 6. The detected walls from the image in Fig. 1.

4.2 Estimating the distortion centre

Before correcting for the lens distortion, it is necessary to estimate the centre of distortion. With purely radial distortion, lines through the centre will remain straight. Therefore, considering the parabola components, a line through the centre of distortion will have no curvature (a=0). In general, the curvature of a line will increase the further it is from the centre. It has been found that the curvature, a, is approximately proportional to the axis intercept, c, when the origin is at the centre of curvature (Bailey, 2002).

The x centre, x_0, maybe determined by considering the vertical lines within the image (the left and right ends of the field) and the y centre, y_0, from the horizontal lines (the top and bottom sides of the field). Consider the horizontal centre first. With just two lines, one at each end of the field, the centre of distortion is given by

$$x_0 = \frac{a_2 c_1 - a_1 c_2}{a_2 - a_1} \tag{25}$$

With more than two lines available, this may be generalised by performing a least squares fit between the intercept and the curvature:

$$x_0 = \frac{\sum c_i a_i \sum c_i - \sum a_i \sum c_i^2}{\sum c_i a_i \sum 1 - \sum a_i \sum c_i} \tag{26}$$

The same equations may be used to estimate the y position of the centre, y_0.
Once the centre has been estimated, it is necessary to offset the parabolas to make this the origin. This involves substituting

$$\begin{aligned}\hat{x} &= x - x_0 \\ \hat{y} &= y - y_0\end{aligned} \tag{27}$$

into the equations for each parabola, $y = ax^2 + bx + c$ to give

$$\begin{aligned}\hat{y} &= a(\hat{x} + x_0)^2 + b(\hat{x} + x_0) + c - y_0 \\ &= a\hat{x}^2 + (2ax_0 + b)\hat{x} + (ax_0^2 + bx_0 + c - y_0)\end{aligned} \tag{28}$$

and similarly for $x = ay^2 + by + c$ with the x and y reversed.
Shifting the origin changes the parabola coefficients. In particular, the intercept changes, as a result of the curvature and slope of the parabolas. Therefore, this step is usually repeated two or three times to progressively refine the centre of distortion. The centre relative to the original image is then given by the sum of successive offsets.

4.3 Estimating the aspect ratio

For pure radial distortion, the slopes of the a vs c curve should be the same horizontally and vertically. This is because the strength of the distortion depends only on the radius, and not on the particular direction. When using an analogue camera and frame grabber, the pixel clock of the frame grabber is not synchronised with the pixel clock of the sensor. Any difference in these clock frequencies will result in aspect ratio distortion with the image stretched or compressed horizontally by the ratio of the clock frequencies. This distortion is not usually a problem with digital cameras, where the output pixels directly correspond to sensing elements. However, aspect ratio distortion can also occur if the pixel pitch is different horizontally and vertically.
To correct for aspect ratio distortion if necessary, the x axis can be scaled as $\hat{x} = x / R$. The horizontal and vertical parabolas are affected by this transformation in different ways:

$$\begin{aligned}y &= ax^2 + bx + c \\ &= aR^2\hat{x}^2 + bR\hat{x} + c\end{aligned} \tag{29}$$

and

$$\hat{x} = \frac{x}{R} = \frac{a}{R}y^2 + \frac{b}{R}y + \frac{c}{R} \tag{30}$$

respectively. The scale factor, R, is chosen to make the slopes of a vs c to be the same horizontally and vertically. Let s_x be the slope of a vs c for the horizontal parabolas and s_y be the slope for the vertical parabolas. The scale factor is then given by

$$R = \sqrt{s_x / s_y} \tag{31}$$

4.4 Estimating the lens distortion parameter

Since the aim is to transform from distorted image coordinates to undistorted coordinates, the reverse transform of eq. (4) is used in this work. Consider first a distorted horizontal line. It is represented by the parabola $y_d = ax_d^2 + bx_d + c$. The goal is to select the distortion parameter, κ, that converts this to a straight line. Substituting this into eq. (4) gives

$$\begin{aligned} y_u &= y_d\left(1+\kappa\left(x_d^2+y_d^2\right)\right) \\ &= \left(ax_d^2+bx_d+c\right)\left(1+\kappa\left(x_d^2+\left(ax_d^2+bx_d+c\right)^2\right)\right) \\ &= c(1+\kappa c^2)+b(1+3\kappa c^2)x_d+\left(a+c\kappa(3ac+3b^2+1)\right)x_d^2+\ldots \end{aligned} \tag{32}$$

where the ... represents higher order terms. Unfortunately, this is in terms of x_d rather than x_u. If we consider points near the centre of the image (small x) then the higher order terms are negligible so

$$\begin{aligned} x_u &= x_d(1+\kappa r_d^2) \\ &\approx x_d(1+\kappa y_d^2) \\ &\approx x_d(1+\kappa c^2) \end{aligned} \tag{33}$$

or

$$x_d \approx \frac{x_u}{1+\kappa c^2} \tag{34}$$

Substituting this into eq. (32) gives

$$y_u = c(1+\kappa c^2)+\frac{b(1+3\kappa c^2)}{1+\kappa c^2}x_u+\frac{a+c\kappa(3ac+3b^2+1)}{\left(1+\kappa c^2\right)^2}x_u^2+\ldots \tag{35}$$

Again, assuming points near the centre of the image, and neglecting the higher order terms, eq. (35) will be a straight line if the coefficient of the quadratic term is set to zero. Solving this for κ gives

$$\kappa = \frac{-a}{c(3ac+3b^2+1)} \tag{36}$$

Each parabola (in both horizontal and vertical directions) will give separate estimates of κ. These are simply averaged to get a value of κ that works reasonably well for all lines. (Note that if there are any lines that pass close to the origin, a weighted average should be used because the estimate of κ from such lines is subject to numerical error (Bailey, 2002).)

Setting the quadratic term to zero, and ignoring the higher order terms, each parabola becomes a line

$$\begin{aligned} y_u &= \frac{b(1+3\kappa c^2)}{1+\kappa c^2}x_u+c(1+\kappa c^2) \\ &= m_y x_u + d_y \end{aligned} \tag{37}$$

and similarly for the vertical lines. The change in slope of the line at the intercept reflects the angle distortion and is of a similar form to eq. (9). Although the result of eq. (37) is based on the assumption of points close to the origin, in practise, the results are valid even for quite severe distortions (Bailey, 2002).

4.5 Estimating the perspective transformation

After correcting for lens distortion, the edges of the playing area are straight. However, as a result of perspective distortion, opposite edges may not necessarily be parallel. The origin is also at the centre of distortion, rather than in more convenient field-centric coordinates. This change of coordinates may involve translation and rotation in addition to just a perspective map. Therefore the full homogenous transformation of eq. (11) will be used. The forward transformation matrix, **H**, will transform from undistorted to distorted coordinates. To correct the distortion, the reverse transformation is required:

$$\mathbf{P}_u = \mathbf{H}^{-1}\mathbf{P}_d \tag{38}$$

The transformation matrix, **H**, and its inverse $\mathbf{H}^{-1}$, have only 8 degrees of freedom since scaling **H** by a constant will only change the scale factor k, but will leave the transformed point unchanged. Each line has two parameters, so will therefore provide two constraints on **H**. Therefore, four lines, one from each side of the playing field, are sufficient to determine the perspective transformation.
The transformation of eq. (38) will transform points rather than lines. The line (from eq. (37)) may be represented using homogenous coordinates as

$$\begin{bmatrix} m_y & -1 & d_y \end{bmatrix} \begin{bmatrix} x \\ y \\ 1 \end{bmatrix} = 0 \text{ or } \mathbf{LP} = 0 \tag{39}$$

where **P** is a point on the line. The perspective transform maps lines onto lines, therefore a point on the distorted line ($\mathbf{L}_d\mathbf{P}_d$=0) will lie on the transformed line ($\mathbf{L}_u\mathbf{P}_u$=0) after correction. Substituting into eq. (11) gives

$$\mathbf{L}_u = \mathbf{L}_d\mathbf{H} \tag{40}$$

The horizontal lines, $y = m_y x + d_y$, need to be mapped to their known location on the sides of the playing area, at y=Y. Substituting into eq. (40) gives three equations in the coefficients of **H**:

$$\begin{aligned} 0 &= m_y h_1 - h_2 + d_y h_3 \\ -1 &= m_y h_4 - h_5 + d_y h_6 \\ Y &= m_y h_7 - h_8 + d_y h_9 \end{aligned} \tag{41}$$

Although there are 3 equations, there are only two independent equations. The first equation constrains the transformed line to be horizontal. The last two, taken together, specify the vertical position of the line. The two constraint equations are therefore

$$\begin{aligned} 0 &= m_y h_1 - h_2 + d_y h_3 \\ 0 &= Y m_y h_4 - Y h_5 + Y d_y h_6 + m_y h_7 - h_8 + d_y h_9 \end{aligned} \tag{42}$$

Similarly, the vertical lines, $x = m_x y + d_x$, need to be mapped to their known locations at the ends of the field, at x=X.

$$\begin{aligned} 0 &= -h_4 + m_x h_5 + d_x h_6 \\ 0 &= -X h_1 + X m_x h_2 + X d_x h_3 - h_7 + m_x h_8 + d_x h_9 \end{aligned} \tag{43}$$

For the robot soccer platform, each wall has two edges. The bottom edge of the wall maps to the known position on the field. The bottom edge of each wall will therefore contribute two equations. The top edge of the wall, however, is subject to parallax, so its absolution position in the 2D reference is currently unknown. However, it should be still be horizontal or vertical, as represented by the first constraint of eq. (42) or (43) respectively. These 12 constraints on the coefficients of **H** can be arranged in matrix form (showing only one set of equations for each horizontal and vertical edge):

$$\begin{bmatrix} m_y & -1 & d_y & 0 & 0 & 0 & 0 & 0 & 0 \\ 0 & 0 & 0 & m_y Y & -Y & d_y Y & m_y & -1 & d_y \\ \hline 0 & 0 & 0 & -1 & m_x & d_x & 0 & 0 & 0 \\ -X & m_x X & d_x X & 0 & 0 & 0 & -1 & m_x & d_x \end{bmatrix} \begin{bmatrix} h_1 \\ h_2 \\ \vdots \\ h_9 \end{bmatrix} = \mathbf{0} \text{ or } \mathbf{D}\hat{\mathbf{H}} = 0 \tag{44}$$

Finding a nontrivial solution to this requires determining the null-space of the 12x9 matrix, **D**. This can be found through singular value decomposition, and selecting the vector corresponding to the smallest singular value (Press et al., 1993). The alternative is to solve directly using least squares. First, the square error is defined as

$$E = \mathbf{D}\hat{\mathbf{H}}(\mathbf{D}\hat{\mathbf{H}})^T = \mathbf{D}\hat{\mathbf{H}}\hat{\mathbf{H}}^T\mathbf{D}^T \tag{45}$$

Then the partial derivative is taken with respect to the coefficients of $\hat{\mathbf{H}}$:

$$\frac{\partial E}{\partial \hat{\mathbf{H}}} = \mathbf{D}^T\mathbf{D}\hat{\mathbf{H}} = 0 \tag{46}$$

$\mathbf{D}^T\mathbf{D}$ is now a square 9x9 matrix, and $\hat{\mathbf{H}}$ has eight independent unknowns. The simplest solution is to fix one of the coefficients, and solve for the rest. Since the camera is approximately perpendicular to the playing area, h_9 can safely be set to 1. The redundant bottom line of $\mathbf{D}^T\mathbf{D}$ can be dropped, and the right hand column of $\mathbf{D}^T\mathbf{D}$ gets transferred to the right hand side. The remaining 8x8 system may be solved for h_1 to h_8. Once solved, the elements are rearranged back into a 3x3 matrix for **H**, and each of the lines is transformed to give two sets of parallel lines for the horizontal and vertical edges.

The result of applying the distortion correction to the input image is shown in Fig. 7.

4.6 Estimating the camera position

The remaining step is to determine the camera position relative to the field. While in principle, this can be obtained from the perspective transform matrix if the focal length and sensor size are known, here they will be estimated directly from measurements on the field. The basic principle is to back project the apparent positions of the top edges of the walls on two sides. These will intersect at the camera location, giving both the height and lateral position, as shown in Fig. 8.

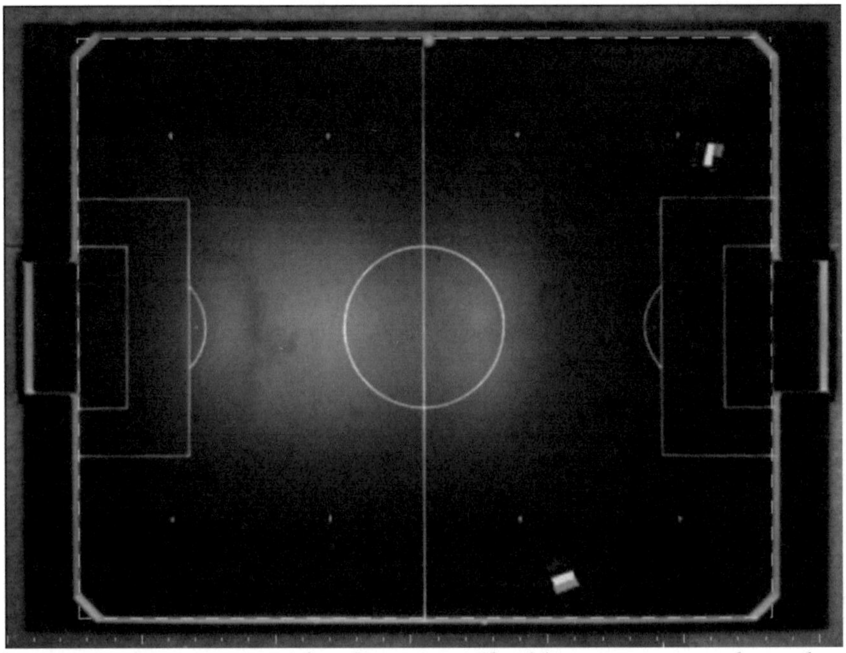

Fig. 7. The image after correcting for distortion. The blue + corresponds to the centre of distortion, and the red + corresponds to the detected camera position. The camera height is indicated in the scale on the bottom (10 cm per division).

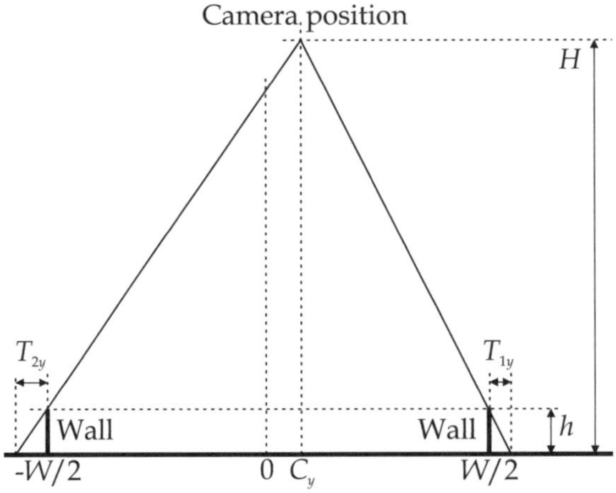

Fig. 8. Geometry for estimating the camera position.

The image from the camera can be considered as a projection of every object onto the playing field. Having corrected for distortion, the bottom edges of the walls will appear in their true locations, and the top edges of the walls are offset by parallax.
Let the width of the playing area be W and wall height be h. Also let the width of the projected side wall faces be T_{1y} and T_{2y}. The height, H, and lateral offset of the camera from the centre of the field, C_y, may be determined from similar triangles:

$$\frac{H}{\frac{W}{2} - C_y + T_{1y}} = \frac{h}{T_{1y}} \tag{47}$$

Rearranging gives:

$$T_{1y} = \frac{h\left(\frac{W}{2} - C_y\right)}{H - h} \tag{48}$$

and similarly for the other wall

$$T_{2y} = \frac{h\left(\frac{W}{2} + C_y\right)}{H - h} \tag{49}$$

Equations (48) and (49) can be solved to give the camera location

$$C_y = \left(\frac{T_{2y} - T_{1y}}{T_{2y} + T_{1y}}\right)\frac{W}{2} \tag{50}$$

$$H = \frac{hW}{T_{1y} + T_{2y}} + h \tag{51}$$

Similar geometrical considerations may be applied along the length of the field to give

$$C_x = \left(\frac{T_{2x} - T_{1x}}{T_{2x} + T_{1x}}\right)\frac{L}{2} \tag{52}$$

$$H = \frac{hL}{T_{1x} + T_{2x}} + h \tag{53}$$

where L is the length of the playing field and T_{1x} and T_{2x} are the width of the projected end walls.
Equations (50) to (53) give four independent equations for three unknowns. Measurement limitations and noise usually result in equations (51) and (53) giving different estimates of the camera height. In such situations, it is usual to determine the output values (C_x, C_y, and H) that are most consistent with the input data (T_{1x}, T_{2x}, T_{1y}, and T_{2y}). For a given camera location, the error between the corresponding input and measurement can be obtained from eq. (48) as

$$E_{1y} = \frac{h\left(\frac{W}{2} - C_y\right)}{H - h} - T_{1y} \tag{54}$$

and similarly for each of the other inputs. The camera location can then be chosen that minimises the total squared error

$$E^2 = E_{1y}^2 + E_{2y}^2 + E_{1x}^2 + E_{2x}^2 \tag{55}$$

This can be found by taking partial derivatives of eq. (55) with respect to each of the camera location variables and solving for the result to 0:

$$\frac{\partial E^2}{\partial C_y} = \frac{4h^2 C_y}{(H-h)^2} - \frac{2h\left(T_{2y} - T_{1y}\right)}{H-h} = 0 \tag{56}$$

or

$$\frac{2hC_y}{H-h} = T_{2y} - T_{1y} \tag{57}$$

Similarly

$$\frac{\partial E^2}{\partial C_x} \Rightarrow \frac{2hC_x}{H-h} = T_{2x} - T_{1x} \tag{58}$$

The partial derivative with respect to the camera height is a little more complex because H appears in the denominator of each of the terms. The partial derivative of the errors across the width of the field is

$$\frac{\partial E_y^2}{\partial H} = \frac{\partial E_{1y}^2}{\partial H} + \frac{\partial E_{2y}^2}{\partial H} = \frac{-h}{(H-h)^2}\left(\frac{4h\left(\frac{W^2}{4} + C_y^2\right)}{H-h} - W\left(T_{1y} + T_{2y}\right) - 2C_y\left(T_{2y} - T_{1y}\right)\right) \tag{59}$$

This can be simplified by eliminating C_y through substituting eq. (57)

$$\frac{\partial E_y^2}{\partial H} = \frac{-h}{(H-h)^2}\left(\frac{hW^2}{H-h} - W\left(T_{1y} + T_{2y}\right)\right) \tag{60}$$

Finally, combining the partial derivatives along the length of the field with those across the width of the field gives:

$$\frac{\partial E^2}{\partial H} = \frac{\partial E_y^2}{\partial H} + \frac{\partial E_x^2}{\partial H} = \frac{-h}{(H-h)^2}\left(\frac{hW^2}{H-h} - W\left(T_{1y} + T_{2y}\right) + \frac{hL^2}{H-h} - L\left(T_{1x} + T_{2x}\right)\right) = 0 \tag{61}$$

Solving for H gives

$$H = \frac{\left(W^2 + L^2\right)h}{W\left(T_{1y} + T_{2y}\right) + L\left(T_{1x} + T_{2x}\right)} + h \tag{62}$$

Finally, the result from eq. (62) can be substituted into equations (57) and (58) to give the lateral position of the camera:

$$\left(C_x, C_y\right) = \frac{\frac{1}{2}\left(W^2 + L^2\right)}{W\left(T_{1y} + T_{2y}\right) + L\left(T_{1x} + T_{2x}\right)}\left(T_{2x} - T_{1x}, T_{2y} - T_{1y}\right) \tag{63}$$

The detected position of the camera is overlaid on the undistorted image in Fig. 7.

5. Applying the corrections

While it is possible to apply the distortion correction to the image prior to detecting the objects, in practise this is computationally inefficient. Only a relatively small number of objects need to be detected, and the distortion is not so severe as to preclude reliable detection directly within the distorted image. Therefore, the robots and ball positions are detected within the distorted image, with the position (and orientation in the case of the robot players) returned in distorted image coordinates. The procedure for correcting the image coordinates follows the calibration procedure described in the previous section.

5.1 Correcting object position

First, the detected feature location (x_f, y_f) is offset relative to the centre of distortion from eq. (27) and corrected for aspect ratio distortion if necessary:

$$(x_d, y_d) = \left((x_f - x_0)/R, y_f - y_0\right) \tag{64}$$

The lens distortion is then corrected by applying the radially dependent magnification from eq. (4)

$$(x_u, y_u) = \left(1 + \kappa r_d^2\right)(x_d, y_d) \tag{65}$$

This point is then transformed into field-centric coordinates and corrected for perspective distortion by applying eq. (38)

$$\begin{bmatrix} kx_p \\ ky_p \\ k \end{bmatrix} = \mathbf{H}^{-1} \begin{bmatrix} x_u \\ y_u \\ 1 \end{bmatrix} \tag{66}$$

where $\mathbf{H}^{-1}$ is the inverse of the matrix obtained from solving eq. (46) in fitting the field edges. The resulting point is normalised by dividing through the left hand side of eq. (66) by k.
Finally, the feature point is corrected for parallax error. From similar triangles

$$\frac{\hat{x}_f - C_x}{H - h_f} = \frac{x_p - C_x}{H} \tag{67}$$

where h_f is the known height of the object and $(\hat{x}_f, \hat{y}_f)$ is the corrected location of the object feature point. Equation (67), and its equivalent in the y direction, may be rearranged to give the corrected feature location:

$$\left(\hat{x}_f, \hat{y}_f\right) = \frac{H - h_f}{H}\left(x_p, y_p\right) + \frac{h_f}{H}\left(C_x, C_y\right) \tag{68}$$

The first term in eq. (68) is the scale factor that corrects for the height of the object, and the second term compensates for the lateral position of the camera as in eq. (19).

5.2 Correcting object orientation

As outlined in section 2, the distortion will affect the detected orientation of objects within the image. The simplest approach to correct the object orientation, θ, is to also transform a test point (x_t, y_t) that is offset a small distance, r, from the object location in the direction specified by the orientation:

$$(x_t, y_t) = (x_f, y_f) + r(\cos\theta, \sin\theta) \quad (69)$$

The offset should be of similar order to the offset used to measure the orientation in the distorted image (for example half the width of the robot). The corrected orientation may then be determined from the angle between the corrected test point and the corrected object location:

$$\hat{\theta} = \tan^{-1}\left(\frac{\hat{y}_t - \hat{y}_f}{\hat{x}_t - \hat{x}_f}\right) \quad (70)$$

6. Results and discussion

As the image in Fig. 7 shows, the calibration method is effective at correcting distortion around the edge of the field. However, to have confidence that the model is actually correcting points anywhere in the playing area, it is necessary to check the transformation at a number of points scattered throughout the image. The calibration procedure was tested on three fields. The first was a small (150 cm x 130 cm) 3-aside micro-robot playing field, captured using a 320x240 analogue camera (Bailey & Sen Gupta, 2004). The second two were larger (220 cm x 180 cm) 5-aside fields, captured using a 656x492 digital Firewire camera (Bailey & Sen Gupta, 2008).

On the small field, a set of 76 points was extracted from throughout the playing area using the field lines and free kick markers as input, as indicated in Fig. 9. The RMS residual error after correcting the validation points was 1.75 mm, which corresponds to about 30% of the width of a pixel. The lateral position of the camera was in error by 1.2 cm, which results in a negligible parallax error (from eq. (19)). The height of the camera was over-estimated by approximately 9 cm. This will give a maximum parallax error in the corners of the playing field (from eq. (21)) of approximately 1.1 mm, which is again a fraction of a pixel.

For the larger fields (shown in Fig. 1 and Fig. 5) both resulted in a significantly improved image that appeared to be free from major distortions (see Fig. 7). In both cases, the lateral position of the camera was also measured with good accuracy, with a total lateral error of 0.9 cm and 1.0 cm respectively for the two fields. Again, the consequent parallax error (from eq. (19)) is negligible. The error in the height, however, was significantly larger, with an under-estimate of 8.6 cm for the image in Fig. 1 and an under-estimate of 33 cm for the image in Fig. 5. Even this large height error results in an error of less than 3 mm in the corners of the field (from eq. (21)). This is still less than one pixel at the resolution of the image, and the error is significantly smaller over the rest of the field.

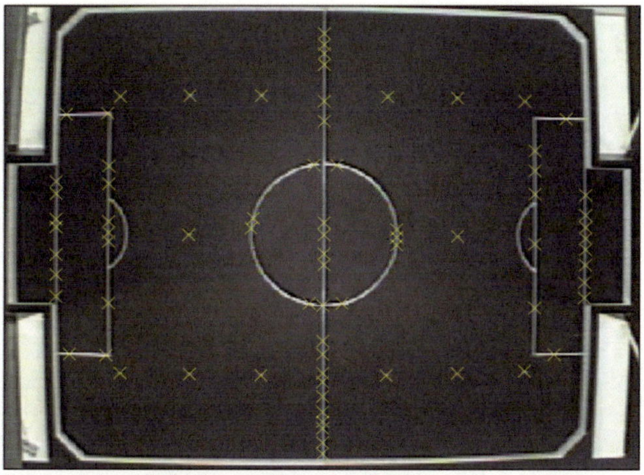

Fig. 9. The 3-aside field with validation points marked.

The larger height errors are not completely unexpected, because the height is estimated by back projecting the relatively small parallax resulting from a low wall. The wall only occupied 3 pixels in the image of the small field and approximately 5 pixels in the larger field. Measurement of this parallax requires sub-pixel accuracy to gain any meaningful results. Any small errors in measuring the wall parallax are amplified to give a large error in the estimate of the camera height. The lateral position is not affected by measurement errors to the same extent, because it is based on the relative difference in parallax between the two sides.

The cause of the large height error was examined in some detail in (Bailey & Sen Gupta, 2008). The under-estimate of the height was caused by the measured parallax of the walls being larger than expected (although the error was still sub-pixel). Since the parallax appears in the denominator of eq. (62), any over-estimate of the parallax will result in an under-estimate of the height of the camera. Two factors contributed to this error. First, a slight rounding of the profile at the top of the wall combined with specular reflection resulted in the boundary between the white and black extending over the top of the wall, increasing the apparent width. A second factor, which exacerbates this in Fig. 5, is that the position of the lights gave a stronger specular component in the vicinity of the walls.

The other fields were less affected for the following reasons. Firstly, the lights were positioned over the field rather than outside it. The different light angle means that these fields were less prone to the specular reflection effects on the top corner of the wall. Secondly, the other fields were in a better condition, having been repainted more recently, and with less rounding of the top edge of the walls. Consequently, the wall parallax was able to be measured more accurately and the resultant errors were less significant.

6.1 Future work

The next step is to extend the work presented to the larger 11-aside field. These fields have two cameras, one over each half of the field. The calibration principles will be the same. The biggest difference is that the centreline will form one of the edges of the calibration, and this does not have a height associated with it. This will limit the accuracy of the parallax

correction data, although in principle the height of three walls should provide sufficient data for estimating the camera location.
A further extension is to the Robocup league, which has no walls. Again two cameras are required, one to capture each half of the field, as shown in Fig. 10. The image processing algorithms will need to be modified for line detection rather than edge detection, and another mechanism found to estimate the camera position. One approach currently being experimented with is to place poles of known height in each corner and at each end of the centre-line as can be seen in Fig. 10. The parabola based lens and perspective distortion correction will be based on the edges of the field and centreline, and the markers on the poles detected and back projected to locate the camera.

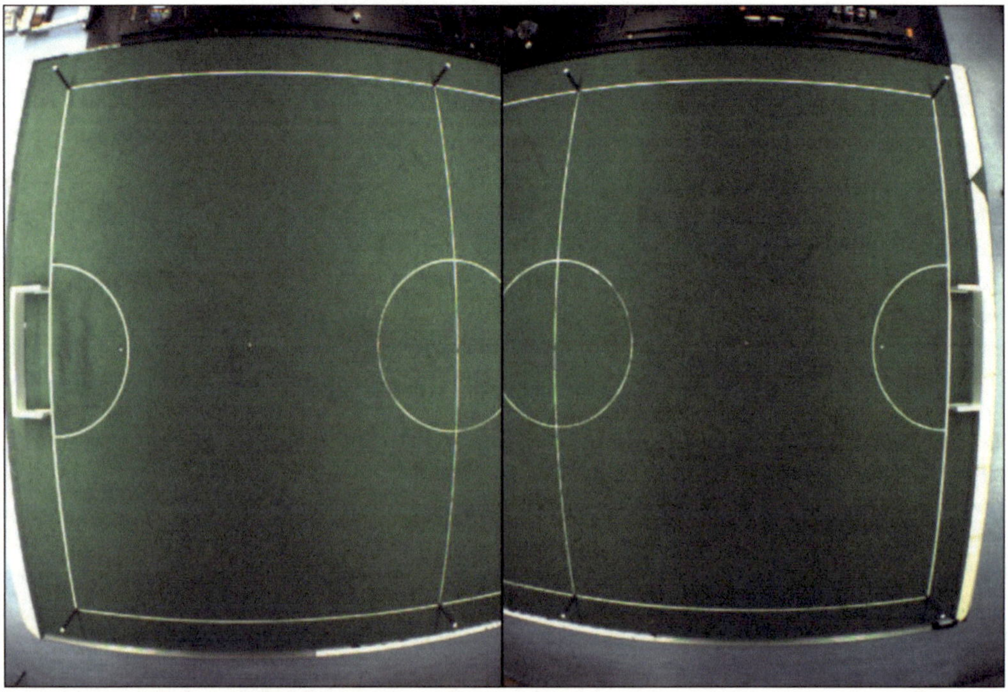

Fig. 10. Larger Robocup field without walls captured from two separate cameras. The poles placed in each corner of the field allow calibration of camera position.

7. Conclusion

The new calibration method requires negligible time to execute. Apart from the command to perform the calibration, it requires no user intervention, and is able to determine the model parameters in a fraction of a second. The model parameters are then used to automatically correct both the positions and orientations of the robots as determined from the distorted images. It is demonstrated that just capturing data from around the field is sufficient for correcting the whole playing area. The residual errors are significantly less than one pixel, and are limited by the resolution of the captured images.
The lateral position of the camera was able to be estimated to within 1 cm accuracy. The scaling effect of the parallax correction makes this error negligible. The height of the camera

is harder to measure accurately, because it is back-projecting the short height of the playing field walls. On two fields, it was within 8 cm, but on a third field the height was under-estimated by 33 cm. However, even with this large error, the parallax error introduced in estimating the robot position less than one pixel anywhere on the playing field. Accurate height estimation requires good lighting, devoid of specular reflections near the walls, and for the walls to be in good condition.
The significant advantage of this calibration procedure over others described in the literature is that it is fully automated, and requires no additional setup or user intervention. While not quite fast enough to process every image, the procedure is sufficiently fast to perform recalibration even during set play (for example while preparing for a free kick) or a short timeout. It is also sufficiently accurate to support sub-pixel localisation and orientation.

8. Acknowledgements

This research was performed within the Advanced Robotics and Intelligent Control Centre (ARICC). The authors would like to acknowledge the financial support of ARICC and the School of Electrical and Electronic Engineering at Singapore Polytechnic.

9. References

Bailey, D. & Sen Gupta, G. (2004). Error assessment of robot soccer imaging system, *Proceedings of Image and Vision Computing New Zealand (IVCNZ'04)*, pp 119-124, Akaroa, New Zealand, 21-23 November, 2004.

Bailey, D. & Sen Gupta, G. (2008). Automatic estimation of camera position in robot soccer, *Proceedings of 2008 International Machine Vision and Image Processing Conference (IMVIP2008)*, pp 91-96, Portrush, Northern Ireland, 3-5 September, 2008.

Bailey, D.G. (2002). A new approach to lens distortion correction, *Proceedings of Image and Vision Computing New Zealand 2002*, pp 59-64, Auckland, New Zealand, 26-28 November, 2002.

Bailey, D.G. (2003). Sub-pixel estimation of local extrema, *Proceedings of Image and Vision Computing New Zealand 2003*, pp 414-419, Palmerston North, New Zealand, 26-28 November, 2003.

Ball, D.M.; Wyeth, G.F. & Nuske, S. (2004). A global vision system for a robot soccer team, *Proceedings of 2004 Australasian Conference on Robotics and Automation*, pp 1-7, Canberra, 6-8 December, 2004.

Baltes, J. (2000). Practical camera and colour calibration for large rooms, *Proceedings of RoboCup-99: Robot Soccer World Cup III*, pp 148-161, New York, USA, 2000.

Basu, A. & Licardie, A. (1995). Alternative models for fish-eye lenses. *Pattern Recognition Letters*, vol. 16, no. 4, pp. 433-441.

Brown, D.C. (1971). Close range camera calibration. *Photogrammetric Engineering*, vol. 37, no. 8, pp. 855-866.

Egorova, A.; Simon, M.; Wiesel, F.; Gloye, A. & Rojas, R. (2005). Plug and play: fast automatic geometry and color calibration for cameras tracking robots, *Proceedings of RoboCup 2004: Robot Soccer World Cup VIII*, pp 394-401, Lisbon, Portugal, 2005.

Fryer, J.G.; Clarke, T.A. & Chen, J. (1994). Lens distortion for simple C-mount lenses. *International Archives of Photogrammetry and Remote Sensing*, vol. 30, no. 5, pp. 97-101.

Hartley, R. & Zisserman, A. (2000). *Multiple view geometry in computer vision*. Cambridge University Press.

Heikkila, J. & Silven, O. (1996). Calibration procedure for short focal length off-the-shelf CCD cameras, *Proceedings of International Conference on Pattern Recognition*, pp 166-170, Vienna, Austria, 25-29 October, 1996.

Klancar, G.; Kristan, M. & Karba, R. (2004). Wide-angle camera distortions and non-uniform illumination in mobile robot tracking. *Robotics and Autonomous Systems*, vol. 46, no. 2, pp. 125-133.

Li, M. & Lavest, J.M. (1996). Some aspects of zoom lens camera calibration. *IEEE Transactions on Pattern Analysis and Machine Intelligence*, vol. 18, no. 11, pp. 1105-1110.

Messom, C.H. (1998). Robot soccer - sensing, planning, strategy and control, a distributed real time intelligent system approach, *Proceedings of Third International Symposium on Artificial Life and Robotics (AROB)*, pp 422-426, Oita, Japan, 19-21 January, 1998.

Park, S.W. & Hong, K.S. (2001). Practical ways to calculate camera lens distortion for real-time camera calibration. *Pattern Recognition*, vol. 34, no. 6, pp. 1199-1206.

Pers, J. & Kovacic, S. (2002). Nonparametric, model-based radial lens distortion correction using tilted camera assumption, *Proceedings of Computer Vision Winter Workshop 2002*, pp 286-295, Bad Aussee, Austria, 4-7 February, 2002.

Press, W.H.; Flannery, B.P.; Teukolsky, S.A. & Vetterling, W.T. (1993). *Numerical recipes in C: the art of scientific computing*. Cambridge University Press.

Sen Gupta, G.; Messom, C.H. & Sng, H.L. (2002). State transition based supervisory control for robot soccer system, *Proceedings of International Workshop on Electronic Design, Test and Applications (DELTA)*, pp 338-342, Christchurch, New Zealand, 29-31 January, 2002.

Sen Gupta, G.; Messom, C.H. & Demidenko, S. (2004). State transition based (STB) role assignment and behaviour programming in collaborative robotics, *Proceedings of The Second International Conference on Autonomous Robots and Agents (ICARA 2004)*, pp 385-390, Palmerston North, New Zealand, 13-15 December, 2004.

Stein, G.P. (1997). Lens distortion calibration using point correspondences, *Proceedings of IEEE Computer Society Conference on Computer Vision and Pattern Recognition*, pp 602-608, San Juan, Puerto Rico, 17-19 June, 1997.

Sturm, P.F. & Maybank, S.J. (1999). On plane-based camera calibration: a general algorithm, singularities, applications, *Proceedings of IEEE Computer Society Conference on Computer Vision and Pattern Recognition*, pp 432-437, Fort Collins, Colorado, USA, 23-25 June, 1999.

Tsai, R.Y. (1987). A versatile camera calibration technique for high-accuracy 3D machine vision metrology using off-the-shelf TV cameras and lenses. *IEEE Journal of Robotics and Automation*, vol. 3, no. 4, pp. 323-344.

Weiss, N. & Hildebrand, L. (2004). An exemplary robot soccer vision system, *Proceedings of CLAWAR/EURON Workshop on Robots in Entertainment, Leisure and Hobby*, pp, Vienna Austria, 2-4 December, 2004.

Willson, R.G. & Shafer, S.A. (1994). What is the center of the image? *Journal of the Optical Society of America A*, vol. 11, no. 11, pp. 2946-2955.

15

Optimal Offensive Player Positioning in the Simulated Robotic Soccer

Vadim Kyrylov
Rogers State University
USA
Serguei Razykov
Simon Fraser University
Canada

1. Introduction

1.1 Background

This work deals with modeling rational player behavior in the simulated robotic soccer game. Although we are using the setting for the RoboCup 2D simulated soccer, we want to propose a method that is not based on the particular features of this implementation. This will result in a more general method that could be equally applicable to 3D soccer and also to the soccer video games other than RoboCup.

The ability by the soccer player making rational decisions about where to move on the field without the ball is the critical factor of success both in a real-life and simulated game. With the total of 22 soccer players in two teams, normally only one player on each side is controlling the ball or trying to get the ball possession. The rest 20 players are doing something else; indeed each must be deliberately moving to some place on the field. For an average player, this is happening more than 90 per cent of the time. Thus, unlike other behaviors dealing with rather rare events such as passing the ball, one should be expecting about 90 per cent impact on the whole game from any improvement in player positioning.

Here we propose the algorithm for determining by the artificial player a reasonably good position on the soccer field for himself when the ball is beyond his control. We limit the consideration to the *offensive* positioning, i.e. when the ball is possessed by own team. In the offensive situation, the major purpose of co-coordinated moving to some individual position on the field by each player without the ball is creating opportunities for receiving passes and eventually scoring the opponent goal. Therefore, player positioning is not a standalone task; rather, it is part of a coordination mechanism and is closely related to decision making about passing the ball and the communication between players.

In this study, however, we deliberately isolate positioning from the communication and ball passing assuming that all these features are developed in a concert. In the early publication, one of the co-authors proposed a solution to the ball passing problem with respect to multiple optimality criteria, i.e. a Pareto-optimal solution (Kyrylov, 2008). For the purpose of optimal offensive positioning, here we are using similar multi-criteria approach.

1.2 The Two-Layer Player Positioning Model

During the last 10 years, RoboCup scholars have addressed the positioning problem in different ways. Before overviewing them, first we propose a *generic two-level model* that includes all existing methods for player positioning. Besides providing a common base for the comparison of the existing methods, it helps to demonstrate the unique features of our approach.

The first, *upper level*, determines so-called *reference* position for the player where he should be moving to unless this player is involved in intercepting or kicking the ball. Player coordination on this level is addressed only implicitly. The second, *lower level*, is responsible for fine tuning this reference position by transforming it into the *target* one, i.e. the place on the field where the player should be. Good methods thus facilitate coordination with teammates. On both levels, either heuristic decision making rules are applied or some optimality criteria are used. Such rules and criteria are normally reflecting the algorithm designer's vision of the soccer game. In many cases these rules are acquired from the real soccer played by humans. This is different from the decision making rules derived by learning algorithms; they do not always have direct counterparts in the body of knowledge accumulated by humans about the soccer game.

1.3 Previous Work

The first ever comprehensive study of the player positioning problem in the simulated soccer was presumably made in (Stone, Veloso, & Riley, 1999). In this method called *strategic positioning by attraction and repulsion* (SPAR), the higher-level reference position can be regarded as a fixed point in the field assigned to each player with respect to its role in the team formation. For each player the lower control level is dealing with the target position which is the point on the field where attraction and repulsion factors reach some balance. This allowed the player to deviate from its default position within some area in order to move to the target position calculated with respect to current situation. This method proved to be a good start, as after it was implemented in *CMUnited*, this simulated soccer team became the World RoboCup winner in 1998. Yet later on it was criticized for the limited flexibility on the upper control level. More sophisticated method named *Situation-Based Strategic Positioning* and first proposed in 1999 (Reis, Lau, & Oliveira, 2008) has addressed these higher-level shortcomings; unsurprisingly, *FCPortugal* in which it was implemented, has beaten *CMUnited*. This method was based on a set of logical rules that reflected the subjective views of the algorithm designers on player positioning.

The development that followed, did not demonstrate much improvement on the lower control level, though. The next very successful team, *UvA Trilearn* (De Boer & Kok, 2002) that has outplayed *FCPortugal* in several competitions, implemented somewhat simpler player positioning method. In our classification, the early version of this method dated 2001-2002 is completely located on the higher control layer. Indeed, at any time, the reference position was determined as a weighted sum of the player home position in the formation and current location of the ball. So this position is changing over time and maintains relative locations of the players in the formation thus implementing the team strategy. However, this method does not take into account fine details such as the opportunity to receive a pass. Presumably for this reason it was later improved using so-called *coordination graphs* (Kok, Spaan, & Vlassis, 2003). This lower-level model combines decision making about passing the ball and receiving the pass in an elegant way; implemented in *UvA Trilearn*, it helped to become the World RoboCup winner in 2003. However, from the today standpoint, this model could be further improved, as it was using heuristics rather than rigor multi-criteria optimization.

One more group of the improvements to player positioning is a set of methods based on learning algorithms (Andou, 1998; Nakashima, Udo, & Ishibuchi, 2003; Kalyanakrishnan, Liu, & Stone, 2007). Like the coordination graph, some of these methods do not treat positioning as a standalone player skill, which makes theoretical comparisons difficult. More difficulties arise while trying to elicit meaningful decision making rules and especially address the convergence issue of learned rules to optimal decision making algorithms based on explicitly formulated decision criteria. Thus we are leaving methods based on learning beyond the scope of this study.

The *main objective* of this work is to provide an improved solution for low-level decision making about player positioning based on the known tactical principles of the real-life soccer played by humans (Beim, 1977; Johnes & Tranter, 1999). Because the recent studies addressed the lower control level rather superficially, it looks like our closest prototype is still SPAR (Stone, Veloso, & Riley, 1999), the historically first method that was using both control levels. On the lower level, this method maximizes the following utility function:

$$U(P) = \sum_{i=1}^{n} w_{Oi} dist(P, O_i) + \sum_{j=1}^{n-1} w_{Ti} dist(P, T_j) - w_B dist(P, B) - w_G dist(P, G), \tag{1}$$

where

P – the desired position for the player in anticipation of a pass;
w_{Oi}, w_{Ti}, w_B, w_G – some non-negative weights;
n – the number of agents on each team;
O_i– the current position of each opponent, $i = 1, \ldots, n$;
T_j – the current position of each teammate, $j = 1, \ldots, (n-1)$;
B – the current position of the teammate with ball;
G – the position of the opponent goal;
$dist(A,C)$ – distance between positions A and C.

One advantage of this utility function is robustness; the two sums suppress the sensor noise. As the authors put it, this method repulses the player without ball from other players and encourages advancing to the opponent goal while seeking its best position. This statement indeed is reflecting the soccer tactics to some extent.

There are also apparent shortcomings. From the tactical standpoint, the aggregated criterion (1) does not encourage players to deliberately get open for receiving a pass; this can only happen by chance. Also the contribution of the remotely located players is exaggerated; increasing distance from the closest opponent by say 5 meters has same effect as increasing distance by 5 meters for the opponent located on the other end of the field. From the mathematical standpoint, the authors clearly indicate that this is a vector optimization problem; indeed, each addendum in (1) could stand for a criterion. However, the reduction to single-criteria optimization is questionable. Aggregating several criteria in a linear combination is indeed theoretically acceptable if all criteria functions and constraints are convex (Miettinen, 1998). However, the first two addendums in (1) are multi-modal functions of P; hence they are non-convex. So this single-criterion optimization does not guarantee that all potentially acceptable player positions will be considered. In other words, some options that are making the *Pareto set* might be left out. This issue could be resolved by using the Pareto optimality principle.

Dynamics is one more general difficulty with optimizing player position on low level. Reaching an optimal position takes some time, and the player must be persistent in doing so. In our experiments we have found that some random factors in the simulated soccer tend to make the computation of the optimal target very unstable from cycle to cycle. A closely related problem is choosing the right time horizon for predicting the situation. As for now, we have found in the literature neither a solution to this problem nor even a discussion of it with respect to player positioning. Thus we would speculate that the reluctance of RoboCup scholars to systematically improve player positioning on the lower level could be explained by these difficulties.

1.4 The Unique Contribution of This Study

The unique contribution of this work is in that we (1) investigate and resolve the time horizon issue; (2) propose a new set of decision making criteria based on real-life soccer tactics; and (3) implement a Pareto-optimal solution to the optimization problem with these criteria.

Section 2 addresses the time horizon issue. Section 3 explains how feasible alternatives for decision making cold be generated. Section 4 introduces five optimality criteria and the algorithm for finding Pareto-optimal solution. Section 5 provides experimental results and presents the conclusions.

2. Predicting the Situation for Player Positioning

Because the essence of player positioning is planning his actions for some future time interval, identifying the time constraints for such planning is critically important. We see two such constraints: the prediction time horizon τ_1 and the planning time horizon τ_2.

2.1 Prediction Time Horizon

A straightforward approach is just setting some fixed prediction time horizon τ_1, say 1 to 3 seconds, and trying to extrapolate the movements of the players and the ball for this time interval. The simplest way is just using linear extrapolations of player positions. However, our experiments have shown that this method does not work well. The major problem is in that, once the ball is kicked by some player in new direction, the rest players revise their previous intentions and change their directions of movement. Neglecting these abrupt changes if the next ball kick occurs before the prediction time expires would result in poor decisions. On the other hand, forecasting new movements by the players when the ball gets under close control by the teammate, although theoretically possible, is impractical because of too high computational costs.

While trying to resolve this issue, it makes sense to see how human players are acting in this situation (Beim, 1977; Johnes & Tranter, 1999).

Because we consider the offensive positioning, the ball must be under the control by some teammate. This teammate may be in two different states: (a) chasing the freely rolling ball while staying the fastest player to it and thus maintaining the ball possession or (b) keeping the ball close to him while being able to pass it at any time.

Figure 1 illustrates case (a). The yellow player #7 (**C**) is about to receive the ball (**B**) passed by his teammate #8 (**A**). The decision maker in question is player #9 (**E**). He must figure it out where to better move to render himself for receiving the ball passed by player #7 if the latter decides to do so.

In this case, an experienced human player without the ball can easily predict the situation and estimate time remaining until the ball will be intercepted. So nothing abrupt will likely occur while the ball is rolling freely. Thus a human player predicts the situation and plans his actions accordingly by determining the target position and moving to it.

This gives the idea of the time horizon used by the player without ball #9 for predicting the situation on the field. Based on this prediction, he determines the good position **F** and concentrates on reaching it. No communication is really necessary, and it does not make much sense substantially changing the originally optimized target position while the ball is rolling. Only minor corrections to it may be necessary as the world model is updated.

Thus the prediction time horizon for the positioning problem is the time τ_1 needed for the ball to cover distance **BD**.

2.2 Planning Time Horizon

The planning time horizon is the time interval τ_2 available for the player to execute the intended action to go to some *target* position. This execution time is needed to determine the spatial constraints for the aspired player movements. It does not make sense for the player to go to the areas on the field that require time longer than τ_2 for getting there.

During the execution time τ_2 the player must concentrate on reaching good position **F** expecting that the teammate would make a pass once he gets close control of the moving ball. Assuming that the ball is passed without any delay, the estimated minimal available time for reaching this position is the time needed for the ball to roll along the segment **BD** and then to cover segment **DF**. This determines the planning time horizon in our model.

2.3 Predicting Player Movements

For determining the best location of the future ball interception point **F** it is critical to know the anticipated positions of the opponent players and the teammates at time τ_1. It is reasonable to assume that during this time the behavior of all players is staying persistent. While the ball is rolling freely along segment **AD** in Figure 1, its movement follows the laws of physics and is predictable. All players (if they are rational) are acting in the ways that are also rather easy to predict if their decision making model is known.

Therefore, for predicting player movements we are using the prediction horizon τ_1 that is equal to the time remaining until the ball will be intercepted by a teammate. (If the fastest to the ball is the opponent player, the situation changes to defensive; this lies outside the scope of this study.) With this approach, once the ball has been kicked, the player without the ball estimates the interception point and time τ_1. This is the minimal time remaining until the ball could be likely kicked in the new direction substantially changing the situation. Because the sensor information is imperfect, the time horizon and the target position are updated as new information arrives. Still our experiments have shown that these updates are only slightly changing the decision thus not affecting the persistent behavior. This can be expected, as the player is watching the vicinity of the ball location more frequently.

The model for predicting the situation comprises three components: the ball, the friendly and the opponent player movements. The ball movement can be predicted with high precision, as the underlying physics is simple; the typical model assumes straight movement with the known speed decrement. The movements by teammates can be also predicted with precision, as their control algorithms and perceived states of the world to some extent are available to the player in question. The fastest teammate to the ball will be intercepting it by moving with the maximal speed; so its position can be predicted even more precisely. The rest teammates will be moving towards the best positions determined with yet another precisely known algorithm which could be used for predicting their positions. However, in our method we do not use such a sophisticated approach; in regards of each teammate without the ball, we assume that it is just moving with a constant velocity. This velocity is estimated by the decision making player using his own world model. Of the opponent players, the fastest one will be presumably also chasing the ball by following the trajectory that can be predicted fairly precisely. For the rest opponents possible options include assuming same positioning algorithm as for the teammates or using the opponent model estimated and communicated by the online coach.

In the case when the ball is kickable by the teammate (i.e. when it is either holding the ball, or is closely dribbling, or prepares to pass), we would suggest that τ_1 should be set to a small constant value τ_{min} which should be determined experimentally.

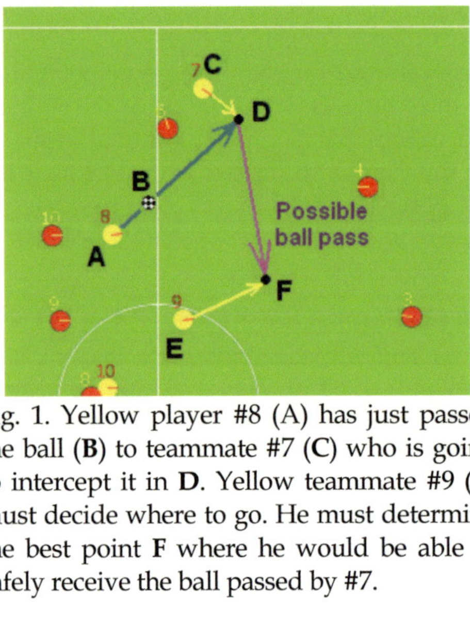

Fig. 1. Yellow player #8 (A) has just passed the ball (**B**) to teammate #7 (**C**) who is going to intercept it in **D**. Yellow teammate #9 (**E**) must decide where to go. He must determine the best point **F** where he would be able to safely receive the ball passed by #7.

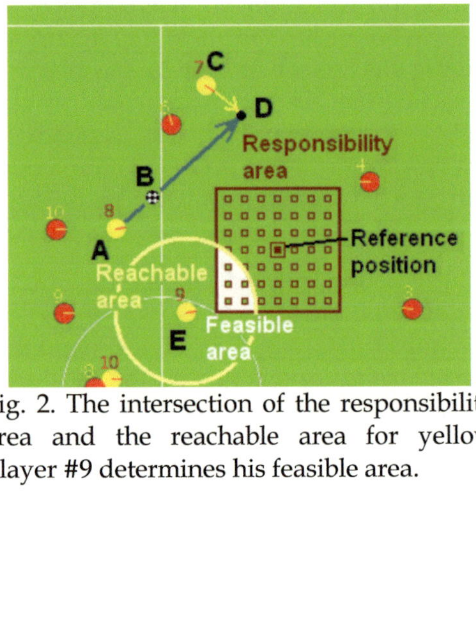

Fig. 2. The intersection of the responsibility area and the reachable area for yellow player #9 determines his feasible area.

3. Identifying the Feasible Options

Decision making is always a choice from a set of alternatives. In the discussed problem, the player first generates a set of feasible options (i.e. positions on the soccer field) and evaluates those using different criteria. Then the multi-criteria algorithm is applied to find the single best option by balancing the anticipated risks and rewards.

The feasible options are drawn from a set of alternative positions in the vicinity of the reference position. Because the decision space (xy-plane) is continuous, it contains infinite number of such positions. With highly nonlinear and non-convex decision criteria, searching such space systematically would be hardly possible. Therefore, we use a discrete approximation, with the alternative positions forming on the xy-plane a grid about the reference position. To reduce computations, we would like to keep the number of points in this grid minimal. The grid step determines the total number of the alternatives to be searched. The rule of thumb is setting the step equal to the radius of the player reach for kicking the ball. Increasing it might result in lost opportunities. Using a smaller step makes sense only if we have enough computational time to further fine tune the balance of different optimality criteria (see more about it in the next section).

In Figure 2 the three areas determining the set of feasible options for the yellow player #9 are shown. The square with the reference position as its center, defines the *responsibility area*. By staying in it, this player is maintaining the team formation. The circle around the player approximates the *reachable area*, i.e. the set of points on the field that he can reach in time τ_2 or less. The *feasible area* is intersection of these two areas, i.e. the set of all feasible positions for this player.

Thus the player is only interested in those positions in his responsibility area that could be reached in time τ_2. This allows eliminating part of the grid that is lying beyond the player reach. One more constraint that helps eliminating poor alternatives is the maximal distance from the reference position. The alternatives lying outside the field or creating risk of offside are also eliminated.

Figure 3 shows the example situation on the field with the predicted positions of all objects at the moment when the ball is intercepted. The head of the black arrow is the anticipated interception point. Figure 4 shows the alternative positions for the red player #8. The player area of responsibility is filled with small gray points; the reference position being the center of this area. The bigger blue points show the reachable positions of which this player must choose the best.

4. Criteria for Decision Making and the Optimization Algorithm

Each position in the feasible set has its own pros and cons that must be evaluated by the intelligent player and the best option must be selected. Thus we arrive to a multi-criteria optimization problem. To correctly formalize it, we need to specify the optimality criteria and choose the algorithm for finding a balanced solution with respect to these criteria.

4.1 Optimality Criteria

The decision criteria for choosing the best option should be reflecting the soccer tactics; in particular they should be measuring anticipated rewards and risks. We propose slightly different criteria sets for attackers, midfielders, and defenders because their tactical roles differ (Beim, 1977; Johnes & Tranter, 1999).

For the attackers the criteria set is, as follows.

1. All players must maintain the formation thus implementing the team strategy. So the distance between the point in the feasible set and the reference position should be minimized.

2. All attackers must be open for a direct pass. Thus the angle between the direction to the ball interception point and the direction to the opponent located between the evaluated position and the interception point must be maximized.
3. All players must maintain open space. This means that the distance from the evaluated point to the closest opponent should be maximized.
4. The attackers must keep an open path to the opponent goal to create the scoring opportunity. So the distance from the line connecting the evaluated point and the goal center to the closest opponent (except the goalie) should be maximized. This criterion is only used in the vicinity of the opponent goal.
5. The player must keep as close as possible to the opponent offside line to be able to penetrate the defense. So, the player should minimize the x-coordinate distance between the point in the feasible set and the offside line (yet not crossing this line).

Fig. 3. Red player #5 has passed the ball to red #7. Arrows show the predicted positions of objects when the ball will be intercepted.

Fig. 4. The area of responsibility for red player #8 (gray dots) and the reachable positions (dark dots).

Note that each criterion appears to have equal tactical importance; this observation will be used while discussing the optimization procedure below.
Criteria for midfielders and defenders differ in that they do not contain criteria 4 and 5 that encourage the opponent defense penetration. Instead, these players should be creating opportunities for launching the attack. This is achieved by minimizing the opponent player presence between the evaluated position and the direction to the opponent side of the field.

4.2 Optimization Algorithm

All proposed criteria are conflicting, as it is hardly possible to optimize all them simultaneously; a reasonable balance must be sought instead. This situation is well known in the literature on systems analysis and economics; a special paradigm called the *Pareto optimality principle* allows to eliminate wittingly inappropriate so-called dominated alternatives (Miettinen, 1998). These are the points in the feasible set that could be outperformed by at least some other point by at least one criterion. So only the non-dominated alternatives making so-called *Pareto set* should be searched for the 'best' balance of all criteria. Balancing requires additional information about the relative importance of these criteria, or their weights. If the criteria functions and the feasible set are all convex, then the optimal point could be found by minimizing the weighed sum of the criteria (assuming that they all must be minimized) (Miettinen, 1998). However, on the xy-plane, which is the soccer field, several local maxima for criteria 2, 3, and 4 exist; they all are around the predicted locations of opponent players. Therefore, in our case there is no hope for such a simple solution as using the weighed sum.

The way out has been proposed in our recent work (Kyrylov, 2008), where a method for searching the balanced optimal point in the finite Pareto set was presented. This method is based on the sequential elimination of the poorest alternatives using just one criterion at a time. With N alternatives in the Pareto set, it requires N-1 steps. The criterion for the elimination on each step is selected randomly with the probability proportional to the weight of this criterion. Hence more important criteria are being applied more frequently. The sole remaining option after N-1 steps is the result of this optimization. This method works for any non-convex and even disconnected Pareto set. Its computational complexity is $O(N^2)$.

In this application, we have further simplified the decision making procedure by assuming that all criteria have equal importance. Thus instead of randomly selecting the criteria on each step of elimination, our procedure is looping through the criteria in the deterministic order.

If the total number of the alternatives is too small, this would result in only near-optimal decision. Better balancing of the conflicting criteria is possible with increased N. So we propose to estimate the available computational time in current simulation cycle and select larger N if time permits. This optimization algorithm is scalable indeed. It is also robust, because even with small N the decisions returned by it are still fairly good.

If this optimization ends in still rather poor option, the player elects just to move towards the reference position; making decision to pass the ball or not is left up to the teammate, anyway. This teammate may elect to dribble or to pass the ball to some other player whose position on the field is better.

Although we proposed five optimality criteria, for the purpose of illustration we have aggregated them all in just two: the *Risk*, which is combination of criteria 2 and 3, and *Gain* which aggregates criteria 1, 4, and 5. The signs of the individual criteria in these aggregates were chosen so that both *Risk* and *-Gain* must be minimized. As a result, it is easy to visually explore the criteria space because it has only two dimensions.

Figures 5 and 6 illustrate the configuration of the Pareto set in the decision and criteria space, respectively. Of the total of 21 points in the Pareto set, 20 are eliminated one by one in the order shown on the labels near each point in Figure 6; the remaining point is the sought solution. Note that the Pareto frontier is non-convex.

The optimal point is reachable and is located at less than the maximal distance of the reference position. It is lying on the way towards the opponent goal and far away from the predicted positions of the two threatening opponents, yellow #10 and #6. This point is open for receiving the pass by red player #8 from the anticipated interception point where red #7 is about to arrive faster that his opponent yellow #11. This is indeed a well-balanced solution to the positioning problem for red player #8. With non-aggregated five criteria we can only expect even better decisions.

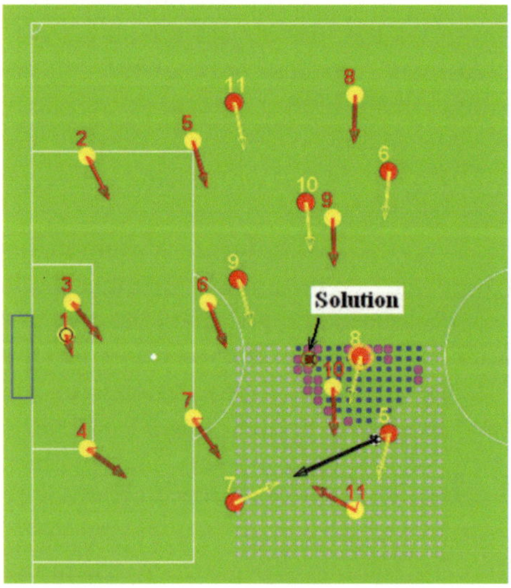

Fig. 5. The Pareto set for red player #8 (bigger dots) and the optimal solution.

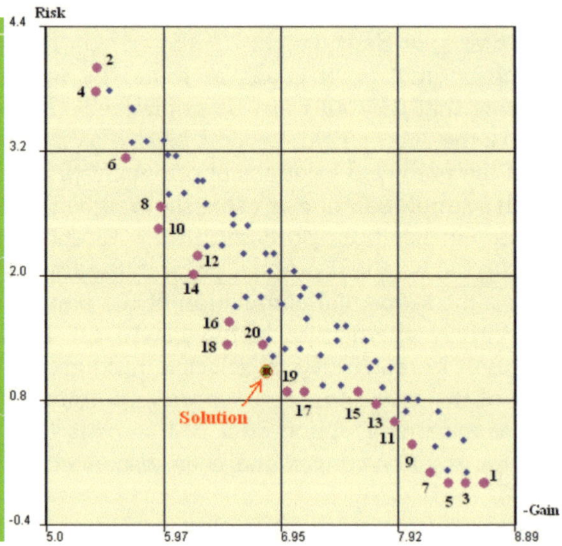

Fig. 6. The criteria space. Numbers at the points in the Pareto set show the elimination order. Note that this set is not convex.

5. Experimental Results and Conclusion

We have conducted experiments with the purpose to estimate the sole contribution of the proposed method for the lower-level optimized player positioning compared with only strategic, higher-level positioning.

Measuring the player performance using existing RoboCup teams is difficult because new features always require careful fine tuning with the existing ones. For this reason, we decided to compare two very basic simulated soccer teams. The only difference was in that the experimental team had player positioning on two levels and the control team just on one level. Players in both teams had rather good direct ball passing and goal scoring skills and no dribbling or holding the ball at all. Thus any player, once gaining the control of the ball, was forced to immediately pass it to some teammate. In this setting, the ball was rolling freely more than 95 per cent of the time, thus providing ideal conditions for evaluating the proposed method.

To further isolate the effects of imperfect sensors, we decided to use *Tao of Soccer*, the simplified soccer simulator with complete information about the world; it is available as the open source project (Zhan, 2009). Using the RoboCup simulator would require prohibitively long running time to sort out the effects of improved player positioning among many ambiguous factors.

The higher-level player positioning was implemented similar to used in *UvA Trilearn* (De Boer, & Kok, 2002); this method proved to be reasonably good indeed. Assuming that both goals are lying on x-coordinate axis, the coordinates of the reference position for *i*-th player are calculated as follows:

$$x_i = w^*xhome_i + (1\text{-}w)^*xball + \Delta x_i,$$
$$y_i = w^*xhome_i + (1\text{-}w)^*yball, \quad (2)$$

where w is the weight ($0<w<1$), ($xhome_i$, $yhome_i$) and ($xball$, $yball$) are the fixed home and the current ball positions respectively, Δx_i is the fixed individual adjustment of x-coordinate whose sign differs for the offensive and defensive situations and the player role in the team formation. Our improving was in introducing the shift Δx_i in this method.
Because players in the control team were moving to the reference positions without any fine tuning, ball passing opportunities were occurring as a matter of chance. In the experimental team, rather, players were creating these opportunities on purpose.
The team performance was measured by the game score difference. Figure 7 shows the histogram based on 100 games each 10 minutes long.

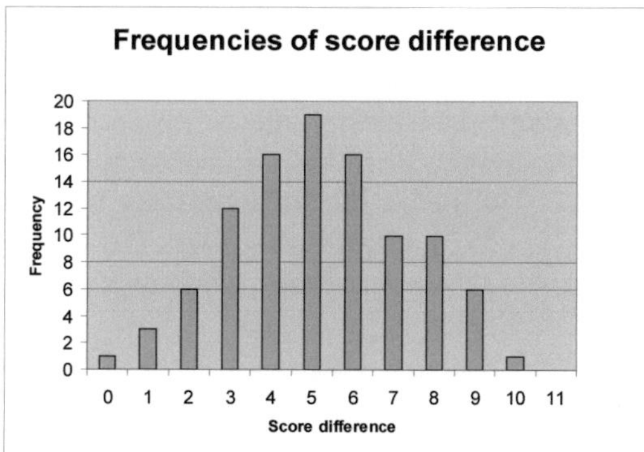

Fig. 7. A histogram of the score difference in 100 games.

In this experiment only one game has ended in a tie; in all the rest 99 games the experimental team won. The mean and the standard deviation of the score difference are 5.20 and 2.14, respectively. By approximating with Gaussian distribution, we get 0.9925 probability of not losing the game. The probability to have the score difference greater than 1 is 0.975 and greater than 2 is 0.933. This gives the idea of the potential contribution of the low-level position optimization. With the smaller proportion of the time when the ball is rolling freely, this contribution will decrease. So teams favoring ball passing would likely benefit from our method more than teams that prefer dribbling.
The experimental results demonstrate that, by locally adjusting their positions using the proposed method, players substantially contribute to the simulated soccer team performance by scoring on the average about five extra goals than the opponent team that does not have this feature. This confirms that optimized player positioning in the simulated soccer is the critical factor of success.
Although this method has been developed for simulated soccer, we did not rely much on the specifics of the simulation league. Nor have we built our method upon the specifics of the two dimensions. Therefore, we believe that the main ideas presented in this work could

be reused with minor modifications in the 3D simulated soccer and in other RoboCup leagues. These ideas could be also reused by video games developers. Besides soccer, our general approach is applicable to different sports games.

6. References

Andou, T. (1998). Refinement of Soccer Agents' Positions Using Reinforcement Learning. In: *RoboCup 1997: Robot Soccer World Cup I*, H. Kitano (Ed.), 373-388, Springer-Verlag, ISBN 3540644733, Berlin Heidelberg New York

Beim, G. (1977). *Principles of Modern Soccer*. Houghton Mifflin Harcourt, ISBN 0395244153, Boston, MA:.

De Boer, R. & Kok, J. (2002). *The Incremental Development of a Synthetic Multi-Agent System: The UvA Trilearn 2001 Robotic Soccer Simulation Team*. Master's Thesis. University of Amsterdam, Amsterdam

Fisher, R.A. (1990). *Statistical Methods, Experimental Design, and Scientific Inference,* Oxford University Press, ISBN 0198522290, Oxford, NY

Johnes, H. & Tranter, T. (1999). *The Challenge of Soccer Strategies: Defensive and Attacking Tactics,* Reedswain, ISBN 189094632X, Spring City, PA

Kalyanakrishnan, S.; Liu, Ya. & Stone, P. (2007). Half Field Offense in RoboCup Soccer: A Multiagent Reinforcement Learning Case Study. In *RoboCup-2006 Robot Soccer World Cup X*, G. Lakemeyer, E. Sklar, D. Sorrenti, T. Takahashi (Eds.), 72-85, Springer-Verlag, ISBN 3540740236, Berlin Heidelberg New York

Kok, J.; Spaan, M. & Vlassis, N. (2003) Multi-Robot Decision Making Using Coordination Graphs, *Proceedings of the 11th International Conference on Advanced Robotics (ICAR)*, pp. 1124–1129, Coimbra, Portugal, June 2003.

Kyrylov V. (2008). A Robust and Scalable Pareto Optimal Ball Passing Algorithm for the Robotic Soccer. In *Soccer Robotics*, P. Lima Ed.), 153-166, Advanced Robotics Institute, ISBN 9783902613219, Vienna

Miettinen, K. (1998). *Nonlinear Multiobjective Optimization,* Kluwer Acaemic Publishers, ISBN 0792382781, Berlin

Nakashima, T.; Udo, M. & Ishibuchi, H. (2003). Acquiring the positioning skill in a soccer game using a fuzzy Q-learning, *Proceedings of IEEE International Symposium on Computational Intelligence in Robotics and Automation, 16-20 July 2003*, v.3, 1488- 1491

Reis, L. P.; Lau, N. & Oliveira, E. C. (2008). Situation Based Strategic Positioning for Coordinating a Team of Homogeneous Agents. In: *Balancing Reactivity and Social Deliberation in Multi-Agent Systems: From RoboCup to Real-World Applications*, M. Hannenbauer, J. Wendler, E. Pagello (Eds.), 175-197, Springer-Verlag, ISBN 3540423273, Berlin Heidelberg New York

Stone, P.; Veloso, M. & Riley, P. (1999). The CMUnited-98 Champion Simulator Team. In: *RoboCup 1998: Robot Soccer World Cup II*, Springer-Verlag, M. Asada and H. Kitano (Eds.), 61-76, Springer-Verlag, ISBN 3540663207, Berlin Heidelberg New York

Zhan, Yu. (2006). *Tao of Soccer*: An open source project, https://sourceforge.net/projects/soccer/